Cases on the Societal Effects of Persuasive Games

Dana Ruggiero
Bath Spa University, UK

A volume in the Advances in Multimedia and Interactive Technologies (AMIT) Book Series

Managing Director: Lindsay Johnston
Production Editor: Jennifer Yoder
Development Editor: Erin O'Dea
Acquisitions Editor: Kayla Wolfe
Typesetter: Lisandro Gonzalez
Cover Design: Jason Mull

Published in the United States of America by
Information Science Reference (an imprint of IGI Global)
701 E. Chocolate Avenue
Hershey PA 17033
Tel: 717-533-8845
Fax: 717-533-8661
E-mail: cust@igi-global.com
Web site: http://www.igi-global.com

Copyright © 2014 by IGI Global. All rights reserved. No part of this publication may be reproduced, stored or distributed in any form or by any means, electronic or mechanical, including photocopying, without written permission from the publisher.
Product or company names used in this set are for identification purposes only. Inclusion of the names of the products or companies does not indicate a claim of ownership by IGI Global of the trademark or registered trademark.

Library of Congress Cataloging-in-Publication Data

CIP Data Pending
ISBN 978-1-4666-6206-3 (hardcover)
ISBN 978-1-4666-6207-0 (ebook)
ISBN 978-1-4666-6209-4 (print & perpetual access)

This book is published in the IGI Global book series Advances in Multimedia and Interactive Technologies (AMIT) (ISSN: 2327-929X; eISSN: 2327-9303)

British Cataloguing in Publication Data
A Cataloguing in Publication record for this book is available from the British Library.

All work contributed to this book is new, previously-unpublished material. The views expressed in this book are those of the authors, but not necessarily of the publisher.

Advances in Multimedia and Interactive Technologies (AMIT) Book Series

ISSN: 2327-929X
EISSN: 2327-9303

Mission

Traditional forms of media communications are continuously being challenged. The emergence of user-friendly web-based applications such as social media and Web 2.0 has expanded into everyday society, providing an interactive structure to media content such as images, audio, video, and text.

The **Advances in Multimedia and Interactive Technologies (AMIT) Book Series** investigates the relationship between multimedia technology and the usability of web applications. This series aims to highlight evolving research on interactive communication systems, tools, applications, and techniques to provide researchers, practitioners, and students of information technology, communication science, media studies, and many more with a comprehensive examination of these multimedia technology trends.

Coverage

- Multimedia Services
- Social Networking
- Multimedia Technology
- Digital Technology
- Digital Images
- Multimedia Streaming
- Digital Communications
- Audio Signals
- Digital Games
- Mobile Learning

IGI Global is currently accepting manuscripts for publication within this series. To submit a proposal for a volume in this series, please contact our Acquisition Editors at Acquisitions@igi-global.com or visit: http://www.igi-global.com/publish/.

The Advances in Multimedia and Interactive Technologies (AMIT) Book Series (ISSN 2327-929X) is published by IGI Global, 701 E. Chocolate Avenue, Hershey, PA 17033-1240, USA, www.igi-global.com. This series is composed of titles available for purchase individually; each title is edited to be contextually exclusive from any other title within the series. For pricing and ordering information please visit http://www.igi-global.com/book-series/advances-multimedia-interactive-technologies/73683. Postmaster: Send all address changes to above address. Copyright © 2014 IGI Global. All rights, including translation in other languages reserved by the publisher. No part of this series may be reproduced or used in any form or by any means – graphics, electronic, or mechanical, including photocopying, recording, taping, or information and retrieval systems – without written permission from the publisher, except for non commercial, educational use, including classroom teaching purposes. The views expressed in this series are those of the authors, but not necessarily of IGI Global.

Titles in this Series

For a list of additional titles in this series, please visit: www.igi-global.com

Video Surveillance Techniques and Technologies
Vesna Zeljkovic (New York Institute of Technology, Nanjing Campus, China)
Information Science Reference • copyright 2014 • 369pp • H/C (ISBN: 9781466648968) • US $215.00 (our price)

Techniques and Principles in Three-Dimensional Imaging An Introductory Approach
Martin Richardson (De Montfort University, UK)
Information Science Reference • copyright 2014 • 324pp • H/C (ISBN: 9781466649323) • US $200.00 (our price)

Computational Solutions for Knowledge, Art, and Entertainment Information Exchange Beyond Text
Anna Ursyn (University of Northern Colorado, USA)
Information Science Reference • copyright 2014 • 511pp • H/C (ISBN: 9781466646278) • US $180.00 (our price)

Perceptions of Knowledge Visualization Explaining Concepts through Meaningful Images
Anna Ursyn (University of Northern Colorado, USA)
Information Science Reference • copyright 2014 • 418pp • H/C (ISBN: 9781466647039) • US $180.00 (our price)

Exploring Multimodal Composition and Digital Writing
Richard E. Ferdig (Research Center for Educational Technology - Kent State University, USA) and Kristine E. Pytash (Kent State University, USA)
Information Science Reference • copyright 2014 • 352pp • H/C (ISBN: 9781466643451) • US $175.00 (our price)

Multimedia Information Hiding Technologies and Methodologies for Controlling Data
Kazuhiro Kondo (Yamagata University, Japan)
Information Science Reference • copyright 2013 • 497pp • H/C (ISBN: 9781466622173) • US $190.00 (our price)

www.igi-global.com

701 E. Chocolate Ave., Hershey, PA 17033
Order online at www.igi-global.com or call 717-533-8845 x100
To place a standing order for titles released in this series,
contact: cust@igi-global.com
Mon-Fri 8:00 am - 5:00 pm (est) or fax 24 hours a day 717-533-8661

Editorial Advisory Board

Daniel Ashton, *Bath Spa University, UK*
Sara de Frietas, *Curtin University, Australia*
James Newman, *Bath Spa University, UK*
Kate Pullinger, *Bath Spa University, UK*
Michael Wagner, *Drexel University, USA*

Table of Contents

Detailed Table of Contents

This case illustrates the way in which the football management simulation game, Football Manager (Sports Interactive), enhances the processes through which players formulate their social identities, which extend beyond the boundaries of gameplay itself. The case discusses the findings of my interviews with Football Manager players, which provides an in-depth examination of experiences associated with the game, both during gameplay and the way in which it functions within the wider social contexts of their lives. I discuss these findings in relation to social identity theory (Tajfel, 1978, 1979; Tajfel & Turner, 1979), through the way in which the game promotes players' sense of in-group affiliation, as well as promoting positive shared experiences between players. In this way, the current case presents an interesting insight into the social functions of the game and its role within the social narratives and identities of its players. From this, I conclude the utility of Football Manager as a persuasive game for formulating players' social identities, which may lead to further positive social impacts.

In this chapter, I present a post-mortem covering three consecutive offerings of a course on persuasive games at the university level over a three-year period from 2010 – 2013. The course, "Designing Persuasive Games," is part of a larger, multi-disciplinary program on digital media and game design. In this course, students are invited to engage both with theory and praxis, the process of "practicing" theory (Shaffer, 2004), by not only reading and writing about persuasive games but also through the design and development of one. Here, I present the overall design of the course across the three offerings and describe the most significant aspects of the course, from a pedagogical perspective, that I believe to be of value to others designing similar courses. These aspects include choosing a game engine, scaling projects to retain rhetoric, modding as praxis, and player experience testing. A sample grading rubric for persuasive games is also included at the conclusion of this chapter.

In designing De BurgemeesterGame—The Mayor Game—we aimed to develop a game that would be used and appreciated by a target population that was hardly used to being trained and had little affinity with applied gaming: mayors. To make sure that the (learning) goals, the context, the characteristics of the target population, and the creative design were all integrated into the game, we chose to work in a consortium with a focus group. We included engaging elements like simple gameplay based on actual processes, authentic scenarios presented in the way of dilemmas, time pressure, and collaboration. This resulted in a game that was accepted by the target population and has been played by more than half of all mayors in The Netherlands. Mayors feel the game challenges them to explore their decision making during crisis management and stimulates them to discuss this with other mayors.

In this chapter, I define six factors that determine the conceptualization of persuasive strategies for advergames. Advergames are understood here as "digital games specifically designed for a brand with the aim of conveying an advertising message" (De la Hera Conde-Pumpido, In Press). These six factors have been used for the analysis of the advergame Tem de Tank (DDB Amsterdam & Flavour, 2010), which was launched in 2010 by Volkswagen to introduce the Volkswagen Polo BlueMotion. The reason for selecting this game as a case study for this chapter is that, although the advergame's goals were properly defined, the game contains, in my opinion, a series of problems in terms of persuasion. Therefore, this game is a perfect case study to exemplify how the factors presented here can be useful to identify problems in the persuasive strategy of an advergame.

Maintaining the healthcare of young people living with long-term medical conditions is dependent upon them acquiring a range of self-care skills. Encouraging them to attain these as well as assessing their competency in them beyond the healthcare setting is challenging. The development of educational computer games like Health Heroes, Re-Mission, and Sparx have been shown to successfully improve self-care, communication, and adherence to medicines in young patients. Therefore, this medium might be an alternative means for delivery of healthcare information. In this chapter, we propose that by encapsulating healthcare processes in Game-Based Learning (GBL) either by computer games or by applying the principles of gamification, a more fun, structured, and objective process would be created, one to which young people can relate. The framework we suggest will provide doctors with an insight into how GBL could be used positively in a healthcare setting as well as provide a basis for application to other disciplines where knowledge and skill acquisition can be challenging.

The Tavistock method, commonly known as group relations, was originated from the work of British psychoanalyst Wilfred Ruprecht Bion. The Tavistock method's basic premise is that an aggregate cluster of persons becomes a group when interaction between members occurs. Within a group, there is organizational politics, and there are two features of organizational politic that should be considered when investigating its relationships with employee attitudes and behaviors. First, perceptions are more important than reality. Second, organizational politics may be interpreted as either beneficial or detrimental to an individual's well-being. Thus, organizational politics perceptions may result in differing responses to organizational policies and practices depending on whether politics are viewed as an opportunity or as a threat. How well one survives within an organization is correlated with how well one navigates these organizational politics. The Tavistock method is utilized as a game to assess and train individuals on organizational politics.

This case identifies and exemplifies a potential subset of persuasive games, called emergently persuasive games. These are games that focus more on unspecified, player-led persuasion as opposed to persuasion based on specific, designer-led outcomes. The game described in this case is *SF0*, and its players have been observed to have become more outgoing, creative, and wise, despite only an incidental, general level of pre-designed persuasion being advertised. It is demonstrated that the ambiguous rules of the game allow the players to customise the gameplay based on their everyday needs, and therefore decide for themselves whether and how they want to be persuaded. These creative interpretations of the rules are actively encouraged, rather than being discouraged as they would be in other games. The ongoing player-discussion of conflicting interpretations facilitates a very effective constructivist environment for self-improvement and understanding. Data was analysed from 24 players of *SF0*, and a Grounded Theory was generated both to explain the general observations of the player data and to identify the diverse ways in which real-world benefit has arisen.

Foreword

Persuasion is a ubiquitous feature of life. We encounter it when we consider the menu at a fast-food restaurant, when we shop for new running shoes, and when we log onto Facebook. We're offered suggestions on appropriate conduct when we use the furniture in public spaces, when we visit public restrooms, and when we make comments online. We're given guidance on what to believe and how to make our value judgments when we go to art galleries, when we peruse TripAdvisor, and perhaps especially when we act in game worlds.

When games such as chess and its German Kriegsspiel descendants first appeared, they were used as training vehicles to educate their players on the complexities and tactics of warfare. To this day, they remain persuasive in their design, connecting to clear real-world referents. What is perhaps less obvious about games is that persuasion is always a core part of their design. All game systems, rule sets, and mechanics invite us to act in particular ways – whether this concerns how we line up goals in foosball, how we lumber around in *Noby Noby Boy*, or how we approach our drawings in *Draw Something*. Games and persuasion share a special relationship: they present persuasive appeals in their design, and we bring them to life through the act of playing.

With regards to theories on persuasive games, two perspectives have dominated research to date, those of B.J. Fogg and Ian Bogost. Fogg (2003) defines a persuasive technology tool as "an interactive product designed to change attitudes or behaviours or both by making desired outcomes easier to achieve" and adopts the position that persuasion must be intended by the designer. Bogost (2007), in contrast, defines persuasive games as "videogames that mount procedural rhetorics effectively," not requiring intentionality on behalf of the designer, but only positing that for a game to be persuasive, its system should embody a rhetorical position.

What the literature has been lacking until this point are practical examinations and reflections on what happens when persuasive games are deployed. *Case Studies on the Effects of Persuasive Games* does just this. It draws together a diverse collection of cases of persuasive games tackling social change of different types, including civil servant training, healthcare, and education. It sheds light on game designs that successfully trigger persuasion, including pervasive games that foster cross-cultural empathy, quest-based games that invite their players to be the people they want to be, and mobile games that convince even those not playing them to reduce their power consumption. It also gives us insight into games that ultimately fail to support persuasion as well as why – for example, why RPGs set in environments such as Second Life and based on existing fiction are apparently a poor medium for conveying persuasive messages. Cases like these enable us to bring the aforementioned theoretical frames to life, and give us data with which to critique them.

Importantly, many of the cases to be found in this book foreground the experiences of players – how they played, what they played, why they played, and what happened afterwards. For a design practice to mature, it is crucial to reflect on the experiences of those we have designed for in light of our expectations. Other disciplines have been exploring this facet of design for some time – including architecture by way of design patterns communicating collective wisdom about building traditions, and experience design by way of its embrace of "third wave" perspectives that focus on phenomenological experience. Design practices that neglect to share and reflect on how their products are experienced by audiences will always be inherently limited.

One message that stands out in this volume is that games can and are successfully being used to persuasive effect even if they were not intentionally designed to do so. In "Emergently-Persuasive Games," Neil Dansey discusses how in *SF0*, ostensibly a game about interpreting and completing ambiguous and metaphorical quests both online and offline, players use the quests as an impetus to take on life challenges and partake in social interactions they would have otherwise lacked the personal motivation for. In "*Football Manager* as a Persuasive Game for Social Identity Formation," Linda Kaye explores how through *Football Manager* players develop a sense of group affiliation with other *Football Manager* players, laying the grounds for new friendships, as well as strengthening existing ones. This is unlikely to have been a design objective of *Football Manager*'s designers, and yet turns out to be one of the game's effects.

Another striking message that comes through is that players are at the heart of persuasion and games. Some players are specifically drawn by the allure of persuasion and life change, as in the case of *SF0*. Other players, if the conditions are right, will serve as persuasive proxies for communicating persuasive messages. In "Communicating the Obvious," Matthias Svahn and Annika Waern analyse how their game *Agents Against Power Waste* was able to convince entire families to reduce their power consumption, even though it was only the children of these families who were signed up as players of the game.

Maybe best of all, in drawing together and juxtaposing diverse examples of persuasive games in use, *Case Studies on the Effects of Persuasive Games* presents implications and open questions for the design of persuasive games. One of these is that player communities in both single and multiplayer games can serve an important yet often overlooked role in persuasion, forming support networks that can and do support change. Another is that some players are clearly seeking out persuasion in games, regardless of a designer's intent. Perhaps this means we can sometimes be subtler in how we design persuasion for these players as we know they are already receptive, in the way that a game like *Unmanned* quietly promotes its anti-drone message through its presentation of drone attacks as banal events, unworthy, even, of undivided attention. Alternatively, we could also be more extreme in our approaches, perhaps taking a leaf from the world of Nordic LARP, which foregrounds emotional involvement, failing over winning, and immersion to the point of dreaming in character – again, because we know the audience is receptive.

This book exemplifies that there isn't a right way to design a persuasive game; rather, there are many possible right ways, depending on players and context. In fact, reflecting on both Fogg's and Bogost's perspectives on persuasive technology and games in light of these cases, neither perspective singularly manages to pin down what is going on in the complex relationship between player, game, and persuasion. If persuasive game design is to mature as a design tradition, we must fold the real examples and experiences of players back into how we understand and conceptualise it. *Case Studies on the Societal Effects of Persuasive Games* moves us towards this.

Rilla Khaled
University of Malta, Malta

Rilla Khaled *is an Associate Professor at the Institute of Digital Games at the University of Malta and has a PhD in Computer Science from Victoria University of Wellington in New Zealand. Her research focuses on how to design more effective and meaningful serious and persuasive games, the interactions between games, gamification, and culture, how to adapt game design methods to foster creativity and design diversity, and game design and AI. She is currently the lead game designer for the EU FP7-funded ILearnRW project. Her current personal research project, REFLECT, introduces reflective game design, an alternative approach to the design, deconstruction, and interpretation of play experiences. With a background in software engineering and cross-cultural psychology, Rilla's interests extend from the software-level design and development of games up to their higher-level effects on and as cultural practices.*

REFERENCES

Bogost, I. (2007). *Persuasive games*. Cambridge, MA: MIT Press.

Fogg, B. J. (2003). *Persuasive technology*. San Francisco, CA: Morgan Kauffman.

Preface

Persuasive games are a natural progression from the rise of video games in popular culture and a technology push that is sweeping the world (Ruggiero, 2013, 2014). Video games create opportunities where people can learn from their experiences in a virtual world and apply that knowledge base to real world problems. Persuasive games are games that use content, affect, and persuasive techniques to change the values, wants, beliefs, and actions of players. The study of persuasive games is an interdisciplinary field of research necessitating a range of knowledge in an array of academic subjects; persuasive games are a form of video games (Salen & Zimmerman, 2005), learning (Gee, 2003; Prensky, 2006), and affect (Bogost, 2003; Evans, 2011).

Persuasive games have been studied in various contexts with differing definitions over the past 40 years. Predating the invention of the computer, humans have used play and games for teaching necessary skills and socialization for millennia (Huizinga, 1955; Abt, 1970). Games explicitly created to change attitudes and behavior date back to 1790, when British publishers of the *New Game of Human Life* advised parents to play the board game with their children and "request their attention to a few moral and judicious observations explanatory of each character as they proceed and contrast the happiness of a virtuous and well spent life with the fatal consequences arriving from vicious and immoral pursuits" (Lepore, 2007, para. 3).

In 1843, a board game released in the US called *Mansion of Happiness* gave instructions that instructed players to make good and moral decisions to gain the seat of happiness. Moreover, Milton Bradley created the *Checkered Game of Life* in 1860 with the intention "to forcibly impress upon the minds of youth the great moral principles of virtue and vice." While a commercial success that helped launch Bradley's board game business, there is no evidence that it had any moral affect on the minds of children (Lepore, 2007 para. 3).

The 1960s and 1970s witnessed a surge of multiplayer simulations. Given credibility by the Rand Foundation, which developed a number of persuasive games for use in the Cold and Vietnam wars; most of these were intended for education, training, and exploring alternative courses of action (Abt, 1970) with some persuasive purposes. For example, sociologists at Johns Hopkins developed *The Life Career Game, The Family Game, The Representative Democracy Game, The Community Response Game*, and *The Consumer Game* with game aims at the player learning the necessity to defer gratification through persuasion techniques (Avedon & Sutton-Smith, 1971).

Persuasive games today are an established part of video game landscapes. They have attracted the attention of the media (Armstrong, 2005; Ochalla, 2007), academics (Bogost, 2003, 2007, 2008; Frasca, 2001; Gee, 2003, 2007), and funding agencies such as the MacArthur Foundation (MacArthur Foundation, 2012). Websites, both academic and commercial, feature persuasive games such as Ben Sawyer's Serious Games Initiatives Website, which has a repository of information about serious games (including persuasive games) and has served as a news source since 2002. Ian Bogost has established a commercial site, www.persuasivegames.com and a non-commercial site www.watercoolergames.org to serve as a forum for the uses of video games in advertising, politics, education, and other everyday activities. In 2008, serious game creators and scholars, Mary Flanagan and Helen Nessenbaum, launched www.valuesatplay.org to harness the power of video games in the service of humanistic principles. To achieve this goal, Values@Play sponsors an annual "Better Game Contest" and encourages developers to consider the effects of their games and to include positive principles like equity, creativity, diversity, and negotiation (Flanagan & Nessenbaum, 2012).

THE CHALLENGE

How can video games "modify or change values, wants, and beliefs of others"? Attempts at analyzing persuasion date back to ancient Greece; according to Aristotle, persuasion is achieved through rhetoric, and three parts that include ethos, pathos, and logos (Cooper, 1932). Ethos uses claims about the persuader's moral character and his or her trustworthiness, an important aspect of the persuasion process if it is to be effective (Evans, 2011). Pathos is an emotional appeal to secure the goodwill of the listener, while logos is the reasoned argument that appeals to the listener's rational mind. Aristotle's categorization has been elaborated over time but is still useful for analysis of persuasion (Bogost, 2007). An important addition to the definition of persuasion is Burke, who in 1969 defined it as "the use of words by human agents to form attitudes or induce actions in other human agents." While the term "words" is limiting, Burke does also include non-verbal means of communication.

However, decades of research on advertising and marketing have confirmed that persuasion is a complex phenomenon dependent on many interrelated factors that make the cross-effects of these factors difficult to separate (O'Keefe, 2002). Factors such as the interest of receivers of the message, their level of education, their knowledge of the issue, their cultural background, their feelings about the originator of the message, the medium used for the persuasive message, and competing factors all influence the success or failure of an attempt at persuasion (Petty & Cacciopo, 1981).

Persuasive games use several mechanisms of persuasion that have been posited by game researchers, including immersion, flow, engagement, persuasive rhetoric, and persuasive ethos.

Immersion is the experience of being transported to an elaborately simulated place that takes over all attention and becomes enveloping (Murray, 1997). Technology has increased the power of immersion through video games. According to Kahled (2007), "It seems that games as persuasive technology hold much promise for changing people's attitudes: games are by nature interactive, and people tend to retain more impressions" (p. 17). Related to immersion is the concept of agency, which Murray (1997) describes as the satisfying power to take meaningful action and see the results of our decisions and choices. Agency helps immersion build when our actions in the video game are appropriate to the game narrative, strengthening the belief in the consistency of the game world.

Flow has been a theory posited by some game theorists (Amory, 2006) that games are compelling because players are in a highly energized state of concentration and focus (Csikszentmihalyi, 1990). Flow is achieved when the level of the challenge and the level of the player's skill are in dynamic tension, creating a highly focused state of mind. Amory (2006) posits that the player can assimilate tacit knowledge through the process that is then assimilated and constructed after emerging from the state of flow.

Engagement is closely related to flow where the player finds the game so engrossing that they assimilate facts and values without realizing they are doing so (Quinn, 2005). Research (Padgett, Strickland, & Coles, 2006; Thomas & Cahill, 1997; Tuzun, 2007) supports the claims made above that games increase engagement through flow, immersion, and agency. Accordingly, when players are more engaged with the game, they are more likely to see the situation from the perspective presented in the game. Yee and Bailenson (2006) placed college-age students in a virtual environment where they used avatars that resembled elderly people. The researchers tested the attitudes of the subjects towards the elderly and found that the computer simulation increased empathy toward people with similar traits to that of the avatar and decreased players' stereotypes of the elderly. Another study on engagement by Goldsworthy, Barab, and Goldsworthy (2000) found that adolescents with ADHD who played a simulation game performed significantly better than the control group on measures of engagement. Both of these studies demonstrate that games with a persuasive message can affect engagement through manipulation of the immersion elements within the game context.

Both procedural rhetoric and ethos are Aristotelian theories of persuasion updated by Bogost (2007) and Evans (2011). While Bogost defines procedural rhetoric as "the practice of using processes persuasively" (2007, p. 2), Evans argues that rhetoric is not enough and includes ethos or "persuasion by empathy, fact, and integrity" (2011, p. 71). By demonstrating that every action in the game has consequences, which are built into the game structure by the designers, the rhetoric and ethos of these procedures not only allow the player to learn through game play but also are a more effective and longer-lasting way of assimilating information (Bogost, 2007). The mechanisms that make a persuasive game work as a persuasive tool can be seen through current examples of persuasive games.

Having established how games can persuade, is there a way to determine how effective they are? How many students who play *Food Force,* a game that puts them in the shoes of a World Health Organization food worker and that has been downloaded 4 million times, retain an interest in the politics of food distribution after game play concludes? *America's Army,* a game developed by the U.S. Army, has had players dedicate more than 160 million hours of game time (Clarren, 2006), but how many of those players actually enlist in the Army? Games that are intended to lead to actions are easier to evaluate because you can measure the effect. For example, the effectiveness of a game that is aimed at persuading people to visit a Website can be calculated by the number of players who clicked through from the game to the Website.

When measuring a game without such concrete goals, such as the intention to influence the players' attitude, this influence is more difficult to measure. While O'Keefe (2002) states that persuasion can be measured by comparing attitudes, other persuasion theorists such as Miller (2002) flatly state that no means exist for directly observing or measuring an attitude where only a minimal relationship is often observed between indicators and attitudinal behaviors. According to Miller (2002), "Persuasion is seldom, if ever, a one-message proposition; people are constantly in the process of being persuaded" (p. 14).

ORGANIZATION OF THE BOOK

This book is organized into 13 cases. A brief description of each follows.

Chapter 1 illustrates the way in which a football manager simulation game enhances the processes through which players formulate their social identities both within and beyond the game itself.

Chapter 2 presents a post-mortem covering three consecutive offerings of a course on persuasive games at university level. Design and pedagogical perspectives are described and best practices suggested for future courses.

Chapter 3 describes the design and implementation of a game that would be used and appreciated by a target population that are not used to being trained and had little affinity with applied gaming: mayors. Persuasive design was used to create an engaging and informative experience that explored decision making during crisis management.

Chapter 4 defines six factors that determine the conceptualization of persuasive strategies for advergames. These factors are used to analyse an advergame launched by Volkswagen.

Chapter 5 proposes a framework for encapsulating healthcare processes in game-based learning to provide doctors with an insight into how game-based learning could be used positively for young people living with long-term medical conditions.

Chapter 6 describes a persuasive game aimed at enhancing the walking activities of elderly residents in neighbourhoods characterized by a low socioeconomic status. First, the design process is described, followed by the user studies, concept testing, and iterative prototyping. Second, the final prototype is presented and framed using the theoretical Persuasive Game Design model.

Chapter 7 analyses the manner in which values are conveyed to the player in a persuasive game for reducing household energy consumption.

Chapter 8 looks at the Tavistock method as a game to assess and train individuals on organizational politics. The basic premise of the method is considered, adapted using game mechanics, and implemented to test the responses to organizational politics.

Chapter 9 identifies and exemplifies a potential subset of persuasive games, called emergently persuasive games, which focus more on unspecified, player-led persuasion as opposed to specific, designer-led outcomes.

Chapter 10 describes the game design and study of a large-scale field experiment where a persuasive pervasive game was put to use to influence households' attitudes towards electricity consumption. The chapter focuses on how the game impacted the psychological process of persuasion in responding families and individuals.

Chapter 11 describes a science outreach program that was designed with the goal of teaching basic physics and math to middle school students and encouraging them to take an early interest in science. The main tool in achieving this goal is a series of online games described in this chapter.

Chapter 12 describes a persuasive game designed for an urban context with a high level of multi-ethnic presence. The chapter looks at a quest-based game where players are ordinary citizens who are plunged into an alternative reality where they deal with gestures and tasks of everyday life in a foreign context.

Chapter 13 examines the strategies and tactics of persuasion used by the players in a role-playing game organized online. Six different categories of closure are described, and each is analysed in terms of conveying a persuasive message to a target player.

Dana Ruggiero
Bath Spa University, UK

REFERENCES

Abt, C. (1970). *Serious games*. New York: The Viking Press.

Amory, A. (2006). Game object model version II: A theoretical framework for educational game development. *Educational Technology Research and Development, 54*, 1–27.

Armstrong, R. (2005, April 21). Bang, bang! You're fed. *The Independent.*

Avedon, E. M., & Sutton-Smith, B. (1971). *The study of games*. New York: J. Wiley and Sons.

Bogost, I. (2003). *Persuasive games.* Retrieved September 11, 2011 from www.persuasivegames.com

Bogost, I. (2007). *Persuasive games: The expressive power of video games*. Cambridge, MA: The MIT Press.

Bogost, I. (2008). *How wrong I was about political games in 2008.* Retrieved September 11, 2013 from www.watercoolergames.org/archives/000888.shtml

Clarren, R. (2006, September 16). Virtually dead in Iraq. *Salon*. Retrieved October 12, 2013 from www.salon.com/ent/feature/2006/09/16/americasarmy/

Cooper, L. (1932). *The rhetoric of Aristotle*. Englewood Cliffs, NJ: Prentice-Hall.

Csikszentmihalyi, M. (1990). *Flow: The psychology of optimal experience*. New York: Harper & Row.

Evans, M. A. (2011). Procedural ethos: Confirming the persuasive in serious games. *International Journal of Gaming and Computer-Mediated Simulations, 3*(4), 70–80. doi:10.4018/jgcms.2011100105

Flanagan, M., & Nissenbaum, H. (2008). *Values @ play.* Retrieved December 12, 2013 from www.valuesatplay.org

Frasca, G. (2001). *Videogames of the oppressed: Videogames as a means for critical thinking and debate*. Atlanta, GA: Georgia Institute of Technology.

Gee, J. P. (2003). *What video games have to teach us about learning and literacy*. New York, NY: Palgrave Macmillan.

Gee, J. P. (2007). *Good video games + good learning: Collected essays on video games, learning and literacy*. New York: Peter Lang.

Goldsworthy, R. C., Barab, S. A., & Goldsworthy, E. L. (2000). The STAR project: Enhancing adolescents' social understanding through video-based, multimedia scenarios. *Journal of Special Education Technology*, *15*(2), 13–26.

Huizinga, J. (1955). Nature and significance of paly as a cultural phenomenon. In K. Salen, & E. Zimmerman (Eds.), *The game design reader: A rules of play anthology* (pp. 96–120). Cambridge, MA: MIT Press.

Khaled, R. (2007). *Culturally-relevant persuasive technology*. Wellington, New Zealand: Victoria University of Wellington.

Lepore, J. (2007). The meaning of life. *New Yorker (New York, N.Y.)*, *83*(13), 38–43.

MacArthur Foundation. (2011, September 15). *Digital learning competition*. Retrieved on February 1, 2014 from http://www.dmlcompetition.net/

Miller, G. R. (2002). On being persuaded. In J. P. Dillard, & M. Pfau (Eds.), *The persuasion handbook: Developments in theory and practice* (pp. 3–16). Thousand Oaks, CA: Sage Publications.

Murray, J. (1997). *Hamlet on the holodeck: The future of narrative in cyberspace*. Cambridge, MA: MIT Press.

O'Keefe, D. J. (2002). *Persuasion: Theory and research*. Thousand Oaks, CA: Sage Publications, Inc.

Ochalla, B. (2007). *Who says video games have to be fun? The rise of serious games*. Retrieved November 3, 2013 from www.gamasutra.com/view/feature/1465/who_says_video_games_have_to_be_.php

Padgett, L. S., Strickland, D., & Coles, C. D. (2006). Case study: Using a virtual reality computer game to teach fire safety skills to children diagnosed with fetal alcohol syndrome. *Journal of Pediatric Psychology*, *31*(1), 65–70. doi:10.1093/jpepsy/jsj030 PMID:15829610

Petty, R. E., & Cacioppo, J. T. (1981). *Attitudes and persuasion: Classic and contemporary approaches*. Dubuque, IA: Wm. C. Brown Company.

Prensky, M. (2006). *Don't bother me mom – I'm learning*. St. Paul, MN: Paragon House.

Quinn, C. (2005). *Engaging learning: Designing e-learning simulation games*. Hoboken, NJ: Pfeiffer Essential Resources.

Ruggiero, D. (2013). Persuasive games as social action agents: Challenges and implications in learning and society. *International Journal of Gaming and Computer-Mediated Simulations*, *5*(4), 75–85. doi:10.4018/ijgcms.2013100104

Ruggiero, D. (2014). Spent: The effect of a persuasive game on affective learning. In *Proceedings from Computer Human Interaction 2014*. Toronto, Canada: ACM.

Salen, K., & Zimmerman, C. (Eds.). (2005). *The game design reader: A rules of play anthology*. Cambridge, MA: MIT Press.

Thomas, R., & Cahill, J. (1997). Using an interactive computer game to increase skill and self-efficacy regarding safer gender. *Health Education & Behavior*, *24*(1), 71–86. doi:10.1177/109019819702400108 PMID:9112099

Tuzun, H. (2007). Blending video games with learning: Issues and challenges with classroom implementations in the Turkish context. *British Journal of Educational Technology*, *38*(3), 465–477. doi:10.1111/j.1467-8535.2007.00710.x

Yee, N., & Bailenson, J. N. (2006). *Walk a mile in digital shoes: The impact of embodied perspective-taking on the reduction of negative stereotyping in immersive virtual environments*. Paper presented at PRESENCE 2006. Palo Alto, CA.

Acknowledgment

First, I would like to say thank you to all of the authors who contributed to this edited volume. Your hard work and dedication to this book has been tireless and has made editing it a successful adventure. To the editorial team, Michael Wagner, Kate Pullinger, Daniel Ashton, Sara de Frietas, and James Newman, thank you for your contributions to the process and your guidance. I would also like to thank Erin O'Dea for her support in the book writing process; it's been a learning experience.

Dana Ruggiero
Bath Spa University, UK

Chapter 1
Football Manager as a Persuasive Game for Social Identity Formation

Linda K. Kaye
Edge Hill University, UK

EXECUTIVE SUMMARY

This case illustrates the way in which the football management simulation game, Football Manager (Sports Interactive), enhances the processes through which players formulate their social identities, which extend beyond the boundaries of gameplay itself. The case discusses the findings of my interviews with Football Manager players, which provides an in-depth examination of experiences associated with the game, both during gameplay and the way in which it functions within the wider social contexts of their lives. I discuss these findings in relation to social identity theory (Tajfel, 1978, 1979; Tajfel & Turner, 1979), through the way in which the game promotes players' sense of in-group affiliation, as well as promoting positive shared experiences between players. In this way, the current case presents an interesting insight into the social functions of the game and its role within the social narratives and identities of its players. From this, I conclude the utility of Football Manager as a persuasive game for formulating players' social identities, which may lead to further positive social impacts.

BACKGROUND

I am a Senior Lecturer in the Department of Psychology at Edge Hill University. My research focuses on the psychology of digital gaming, with a particular focus on the social contexts of gameplay. Within this, I am particularly interested in the positive social and psychological impacts which digital games can provide when

DOI: 10.4018/978-1-4666-6206-3.ch001

Copyright ©2014, IGI Global. Copying or distributing in print or electronic forms without written permission of IGI Global is prohibited.

considering gaming as a functional part of our everyday lives. The background to the case relates to the fact that there is currently very little understanding of the way in which specific games may be related to our everyday social experiences. In particular, issues surrounding how they may provide enjoyable social experiences, promote development of friendships, and facilitate the formation of social groups remain relatively unclear. Based on this, I was keen to investigate these important issues as part of my own research.

Having existing connections with Football Manager players made the processes of this research relatively straightforward. The individuals who kindly took part in my interviews were very keen to share their experiences, and were willing to recommend their own friends to be suitable participants for further interviews. In fact, I had first met one of these participants; "Peter", when he had contacted me to ask whether I could assist in taking part in a short documentary which comprised a group media project he and his group were completing for their second year of their university course. This project was aptly named "Football Manager stole my life", in which "Peter" and friends reflected on their experiences of playing the game Football Manager, and deconstructed the issue of addiction, and their time spent playing it. Within this, I was interviewed as an expert in the area of digital games, in which insights into the social experiences of digital games comprised the main focus of the interview. It was my engagement in assisting with this project which first opened my eyes to the possibility of pursuing some research specifically in relation to this game. I therefore got in contact with "Peter" and his friends, and suggested the possibility of them "returning the favour", in being interviewed for my own research. They happily obliged.

SETTING THE STAGE

This case describes the way in which the football management simulation game Football Manager (Sports Interactive) enhances the processes through which players formulate their social identities, which extend beyond the boundaries of gameplay. The game not only provides opportunities for players to make decisions and consider tactics, but provides social experiences in which players compete in online leagues, or can engage in network play with other players. This presents an interesting case for examining the different social experiences associated with playing this particular digital game. That is, the direct social experiences of play (e.g., social interactions, competition) may be distinct from other those of other games, due the nature of the game features. Specifically, whereas other "e-sports" may present greater opportunities for direct competitive gameplay and interactions, Football Manager may present alternative social opportunities, such as "indirect social experiences" which occur outside of gameplay (e.g., social cohesion, conversations relating to the game).

Prior to the case described in this chapter, little research has been dedicated to specifically examining the digital game Football Manager, and how it functions within players' everyday social experiences. In fact, to date, little research has been dedicated to examining these issues for any form of digital game. The need for further research, particularly in relation to sports games and the activity of sport more generally, is reflected in the recommendations by Leonard (2004), particularly in calling for further qualitative research to investigate such issues. Based on this assertion, I felt it was important to explore both the in-game and "outside-of-game" social experiences associated with playing Football Manager. This was intended to provide an in-depth examination of gamers' experiences associated with the game, both during gameplay and the way in which it functioned within the wider social contexts of their lives. To achieve the aims of my research, I interviewed four Football Manager players who had a range of experiences with this specific game.

All interviews were conducted on a one-to-one basis with the exception of one, which consisted of a paired interview format. The rationale for this particular choice of format related to the fact that these two participants were good friends who regularly played Football Manager together. I felt this was a unique recruitment opportunity which was too valuable to overlook, particularly as my research aims were to explore shared social experiences of the game. Previous studies have utilised this approach has an effective method for creating a supportive social context through which to explore a range of issues (Highet, 2003; Mauthner, 1997; Mayall, 2000; Michell, 1997). All my other interviews followed a traditional one-to-one format.

I presented all the interview questions in an open-ended format, to allow participants to explore issues in sufficient depth. These were presented in a semi-structured way, to allow flexibility in the narratives which emerged through the process. The first theme of the interviews explored direct social experiences associated with playing

Table 1. Demographics of the sample

Pseudonym	Gender	Age	FM Experience (In Years)
Susan	Female	19	2
Peter	Male	25	>5
Kirk	Male	20	4
James	Male	26	>5

Football Manager (e.g., "How does playing with others affect how you feel when playing Football Manager?") This aimed to develop an understanding of the way in which social gameplay was related to players' positive and/or negative experiences. Similarly, the second theme aimed to examine the role of "indirect" social experiences of playing Football Manager (e.g., How important is it that you have friends who also play Football Manager?) This was particularly relevant given the fact that no research, to date, has examined these issues. I also presented my participants with the opportunity to discuss other issues they felt were important or relevant, by allowing time at the end of the interview for an exploration of these issues.

CASE DESCRIPTION

The Game

Football Manager, developed by Sports Interactive, is a football management simulation game, in which players take control of a football club of their choice, and compete in online leagues. Additionally, players are required to make decisions surrounding tactics and player transfers. However, no control can be asserted over the players on the pitch which makes this game somewhat distinct from other sports games, in which this is the key feature of gameplay. The most recent version of the game (Football Manager 13) has extended from its predecessors through the development of "Network play" in which games can be more easily played competitively with friends, in somewhat equivalent ways to other multiplayer games. However, the classic version of the game primarily consists individual gameplay with the aim to exceed other players' league standings. Given that direct competition is not necessarily a primary feature of the game, alternative forms of competition are often utilised between players. For example, players' league successes can be compared between players, to promote a sense of competitiveness in achieving higher league standings than others. These experiences are therefore underpinned by social processes which can motivate gameplay, and promote a sense of enjoyment in players. These experiences are largely relevant to everyday "offline" interactions, rather than those in online contexts, which are often attributed as being key social gaming experiences.

In attempting to understand the psychological underpinnings of the way in which Football Manager "brings people together", to facilitate shared experiences, I turned to the established theories within Social Psychology. Here it occurred to me that one theory in particular was very useful in understanding and explaining these issues. This particular theory is social identity theory, which I will discuss in detail in the subsequent section.

Theoretical Framework

A theoretical framework I felt underpinned my research particularly effectively was social identity theory (Tajfel, 1978, 1979; Tajfel & Turner, 1979). This approach describes the extent to which an individual identifies with a social group, as a result of how they perceive themselves (the self) in terms of their group membership. Identity with the group therefore merges the distinction between the individual's personal sense of self, with that of their sense of self as a group member (i.e. the "collective self"). An individual's self-concept is therefore largely defined by the belonging to certain social groups (Tajfel & Turner, 1979). Here, a distinction between "in-groups" and "out-groups" is made, in which solidarity within in-groups is said to promote a sense of self-esteem through such affiliation (Abrams & Hogg, 1988). This is relevant both to individual and collective self-esteem (Hogg & Abrams, 1990). The mechanism through which self-esteem is said to be promoted is through the perception of being valued as a part of the in-group which contributes to positive identity (Ellemers, Haslam, Platow & Knippenberg, 2003). Additionally, these positive appraisals translate into wider positive outcomes, for feelings of life satisfaction and well-being (Kong, Zhao & You, 2013; Isiklar, 2012).

The identification with the "in-group" can be established through associated experiences, as well as through the power of symbols (Jenkins, 2008) and choice of language and communication (Postmes, 2003). A unique characteristic of the in-group relates to the fact that group membership is meaningful to all individuals which promotes a sense of shared representation and associated behaviour. Three inter-related processes underpin social identity theory (Tajfel, 1974; 1978):

1. **Social Categorisation:** In which individuals see themselves and others as categories rather than as individuals
2. **Social Comparison:** In which individuals assess the worth of groups through comparing their relative features
3. **Social Identification:** In which individuals' identity is formulated by their experiences within a social group or situation

These processes are proposed to interact with one another, in which individuals define themselves at the group level, which determines inter-group behaviour (Tajfel, 1974, 1978).

A large amount of theory has been established which explains the mechanisms behind identity to particular groups or communities, both in everyday life and in the virtual world. Indeed, Jones (1998) provides a detailed analysis of how membership to communities and establishment of connections in virtual spaces may be distinct from those in the "real world". For instance, he states that "real world" community memberships cannot be formulated in the same way of virtual ones, given that interactions are much more discreet in virtual worlds (e.g., signing up to Internet forums). He further proposes that it is the discourses associated with virtual communities which form connections and a sense of community between individuals (Jones, 1998). Conversely, other scholars have argued that "community" is simply defined by the interactions with other individuals with a "common interest" (Licklider & Taylor, 1968), simply through social networking (Bender, 1978), or which encompasses "virtual togetherness" through interpersonal collective interaction in a technological culture (Bakardjieva, 2007). Other scholars have provided a more clear definition of what characterises "community" in suggesting it to be a collection of individuals who occupy a structural location in an institution or society (Agre, 2003). Within this, individuals engage in "collective cognition" which is learnt through their collective experiences (Agre, 2003).

Regardless of the definition used to characterise "community membership", it appears that these processes may operate in distinct ways between the "real world" and the virtual world. Some scholars would contend that "real world" interactions are better suited to the strengthening of social identity compared to virtual-based ones, such as those in computer-mediated communication (CMC), given the increased proximity, frequent contact, and personal acquaintanceship with others (Lea, Spears & Rogers, 2003). However, one advantage of CMC for development of social identity relates to the fact that it is more suited to developing wider, more diverse connections with a greater number of individuals, than those more traditional interactions. However, it remains unclear as to the extent to which these processes may be relevant for digital game-players. That is, although social identity for being a part of the in-group as Football Manager player is highly conceivable based on this theoretical premise, it is unclear whether this formulates through the processes within the virtual space of Football Manager, or through processes such as social discourses associated with the game which occur in the "real world". Examining social identity theory and its relevance to the formation of in-group membership for digital games presents an interesting area of research, which formed the basis for my line of enquiry in the case described in this chapter.

Social identity theory has previously been applied to understanding the way in which media entertainment influences social identity (Reid, Giles & Abrams, 2004). Specifically, it has been suggested that individuals engage in specific types of media and this determines the way in which they form their identities and group affiliations (Reid et al., 2004). Conversely, other research has demonstrated how established social identities determine the extent to which individuals may seek out specific types of media (Harwood, 1999). Given this, it is conceivable that these links holds true for individuals who engage in digital gaming, in strengthening a sense of social identity with other individuals who play games. Specifically for Football Manager players, experiences of the game are anticipated to be associated with indicators of social identity with other players of the game (in-group members). Indeed, these notions are largely reflected in previous evidence suggesting the way in which digital games can function as a socio-cultural artefact of players' daily lives, specifically through the way sports-themed games function as part of players' identity formation (Crawford & Gosling, 2009). In this way, these authors assert that it is impossible to isolate game-playing from the social influences in daily life. Here they refer to an "audience research approach" in which researchers attempt to understand gaming within everyday life, and functions as a resource for social interactions, identities and performances (Crawford & Gosling, 2009).

These assertions were reflected in the narratives of my case interviews. Specifically, a bi-linear relationship between the game and the social contexts of the players was established. Namely, I reflected that the game appears to help foster social relationships and interactions, yet also the existing social phenomena of players' lives seem to determine the interest and playing of the game (e.g., recommendations from friends, playing to become part of the in-group). Particularly in relation to friendships and social interactions, Extract 1 illustrates how Susan reflected on the importance of friendships when playing Football Manager.

Extract 1: Susan

Int: *Is that important when you're playing, in term of playing with your friends?*
Susan: *Yeah.....we often like give each other tips on best players to sign, and the best tactics, and the downloads off the internet to add on to the game and stuff, so yeah, I'd say so.*

The positive experiences of playing Football Manager are therefore framed by the social context of friendships, and their involvement in the gameplay process. Here it is difficult to establish the boundaries of gameplay and the extent to which interactions with friends about Football Manager occurred within gameplay. Indeed, my interpretation of Susan's discourse is that such interactions with friends are not solely based within gameplay, but extend to wider everyday social experiences. Similarly in Extract 2, James describes how social interactions with friends are an important part of playing Football Manager;

Extract 2: James

James: *I do like chatting to people about their experiences I suppose, or their games; I've got a couple of good friends who play quite regularly, so it's always nice to have a good catch up with them and also talk to each other about player signings, good formations, tactics, things like that.*

These case findings present an interesting insight into the social dimensions of the game, and their function in the social discourses of players' lives. These appear to reflect both in-game interactions as well as those operating outside of gameplay. This highlights the role of the game in contributing a focal point of social opportunities for players, which may in turn, be instrumental in enhancing players' social identities with the game and the in-group. This largely reflects the notion proposed by Bender (1978) in suggesting that a sense of identity with a community emerges through social networking.

As well as enhancing social interactions, the interviews revealed wider social impacts for players. That is, it was insightful how frequently participants discussed Football Manager as a "friendship enhancer". This was found to be facilitated through the way in which Football Manager remains a common interest or focal point between friends. This is nicely illustrated in Extract 3, in which Kirk and Peter describe their early encounters in the development of their friendship:

Extract 3: Kirk and Peter

Peter: *It's a topic to talk about as well. I mean me and Kirk are friends at uni, and we met in the first year and it was the kind of thing we talked about, it was something we had in common. When we found out we knew about that...I'm not saying our whole relationship is based on Football Manager, but having something to talk about.....So having a topic to talk about with Kirk, or anyone, I mean Sam, the one in the documentary, he's the first guy I met at university, and one of the first things we talked about was Football Manager. And it's just something to talk about, so the social aspects do come into it, and you don't really think about it.*

Interestingly, these ideas largely reflect the notion of "common interest" which Licklider and Taylor (1968) discuss as being a key requirement for the formation of a sense of community or group identity. Similarly, James (Extract 4 below) describes his experiences with his ex-housemate, and the way in which Football Manager functioned in their friendship and social interactions. Namely, the experiences of playing the game not only provided direct social experiences within gameplay, but also functioned as a focal point in directing their socialisation experiences outside gameplay.

Extract 4: James

James: *I used to play with one of my housemates. We had a massive game actually. It went on basically for the entire time we were at uni, for two years, we played the same game, we got about six seasons in, so that was quite an epic game!*
Int: *Do you think it brought you closer together as friends?*
James: *Oh definitely, definitely, yes. It was a common activity; we played on the same computer or laptop, so we always had to be next to each other or be close to each other to be able to play that game so we were always chatting and socialising outside the game, obviously the game was normally the focal point but we chatted and it definitely brought us closer together as friends.*

These case findings highlight the wider social experiences of Football Manager through its ability to promote enjoyable social experiences, which can be translated to conversations and interactions beyond the boundaries of gameplay. These interactions therefore function within building and enhancing friendships between players, which may not necessarily be enhanced to the same extent without the existence of Football Manager as a common activity. In this way, one common feature which characterises these experiences is the notion of physical proximity between Kirk and Peter, as well as James and his Football Manager friend, in enhancing the development of their friendships. This reflects Jones' (1998) notion of relational development occurring through processes of physical proximity of participants, and the extent to which this functions in their perceptions of in-group membership to the game. This is further reflected in the case interviews which discussed the notion of "in-group affiliation". Extract 5, for example, illustrates Kirk and Peter's lived experiences of being part of the in-group as Football Manager players, and how they distinguish themselves from others who may be perceived as the "out-group". They describe this distinction through their acknowledgement of the importance and meaningfulness of the game, which those in the out-group (non-players) do not appear to appreciate.

Extract 5: Kirk and Peter

Peter: *As much effort as you put into Football Manager, as someone else does, you have that......affiliation with them that's...they understand why you do it, and like my girlfriend..."I don't know why you do it; I don't know why you spend that many hours on it..."*

Kirk: *Yeah, my dad's actually been on at me, saying...like when I was at home, he was like "you spend all your day looking at a computer, are you ever going to get off it" and I'm like "I'm sorting my team here!"*

These experiences reflect the way in which Kirk and Peter have formed a social identity as Football Manager players, which helps consolidate their shared understandings and affiliations which may otherwise be unobtainable. I also observed that this notion of "shared understanding" was apparent within the interview narratives. That is, it became clear that Football Manager players could establish a shared point of reference in their wider social experiences. Extract 6, for example, illustrates the notion of a shared understanding through the affiliation with the Football Manager in-group. In particular, this is shown to be relevant in relation to participants' partners or parents who are described in terms of being an out-group due to not having an understanding of the terminology and workings of the game.

Extract 6: Kirk and Peter

Peter: *Before that [networked gaming] it was just word of mouth, and get together and talk and you'd talk about what you'd done. Like school, I remember talking about formations and stuff, I'd draw it in the back of my diary and stuff, and that's before social networking...*

Kirk: *Your friends telling you tactics and...*

Peter: *Honestly, I could probably get all my old year diaries from school and in the back will be loads of circles and arrows, of where they're going and stuff.... You're there showing your mates, and....*

Kirk: *The good thing about it is that everyone who plays that game will understand everything that you're talking about, whereas an outsider, for instance, will be like; "it's only circles".*

Peter: *It's like a niche in between liking football in general, and liking Football Manager. I mean, like my girlfriend, if I showed her that, she's be like "I don't know what that is", "It's a formation"... "what does that mean?!". So.....even if you showed a football fan, like my dad, if I showed him that still he'd be like "what's that?", you have to be someone who plays the game would recognise* *those arrows and the exact name of the formation, and like the abbreviations for the positions, so it's summat, when you've been playing it for so many years, it's like second nature.*

In a similar way, "shared experience" was a theme which emerged from the interview discussions. This related to the idea of Football Manager providing a shared experience, and this was found to be important for future social experiences. Extract 7, for example, illustrates James' current social experiences with his ex-housemate, with whom he played a lot of Football Manager while they were at university together. Their current social interactions and conversations are described as being framed around their past experiences, highlighting the role of the game in providing social experiences beyond gameplay itself.

Extract 7: James

James: *When I see Bill, I will always talk about our old Aston Villa/Liverpool game, and we can still remember the players, the players that we bought and players that did well, players that didn't do well, and at the end of the day, I can't remember how many days it was...at the start of FM you can get a..it tells you how long you've been playing and how many days..it was like 28 days or something ridiculous that we'd spent on Football Manager from two years at uni, and it's a bit part of that and...erm....if you enjoy it, why not?*

This suggests the importance of the positive within-game experiences, in facilitating subsequent social experiences, which are characterised by reminiscing about the processes of gameplay, through their shared experiences. Additionally, it highlights the way Football Manager was previously intertwined within their daily experiences during the time of play, as well as in their subsequent social encounters with one another. In this way, their shared experiences frame their socialisation both within gameplay and beyond. This reflects Jones' (1998) statement relating to the way in which social meanings associated with spaces emerge through on-going discourses, creating a sense of community or in-group membership.

Through the processes of undertaking and analysing these particular case interviews, I reflected on the overwhelmingly positive nature of the participants' narratives. There was seemingly very little evidence which drew on any negative experiences associated with this game. Indeed, even though some attention was given to competitiveness and "bragging" in relation to the more competitive aspects of the game, this, as well as the other socially-based experiences were largely experienced in a positive way. This was extremely interesting to me, and led to me consider the importance of these findings, in relation to the more typical negative preconceptions about digital gaming as a societal "issue". The fact that these cases highlighted largely positive social impacts for players led me to reflect on the importance of further qualitative work exploring these issues for different types of games. This may help to deconstruct the typical negative conceptions about digital gaming, to demonstrate the way in which it can provide a positive contribution, particularly for social relationships, community and general connectedness.

CURRENT CHALLENGES

Through the processes of the interviews with Football Manager players, it become apparent to me that this game, particularly its social functions, comprises an important contribution to these players' social experiences. This was particularly apparent in Peter and Kirk's interview, in which their narratives largely reflected their shared experiences, both in and beyond gameplay. Similarly, other participants discussed the social utility of the game, for facilitating social interactions, enhancing friendships, and formulating affiliation to the Football Manager in-group, through shared understandings and experiences. This appears to hold a positive function to these players in their social experiences, but it is unclear whether this results in any detriments to friendships and/or social interactions and experiences beyond "in-group" membership. That is, given that these case interviews revealed such a distinction between in- and out-groups, it is conceivable that their perceptions of the out-groups' lack of understanding of the in-group meaning of the game could hold some challenges to effective social integration in particular contexts. For example, in situations where a non-player (out-group member) is amongst players (in-group) discussing Football Manager experiences, this may be detrimental to the out-group member being fully integrated into the social discourse. This notion is discussed by Stivale (1997) and Connery (1997) who suggest that the success of overcoming such challenges are largely dependent on the "dynamic flexibility" of the in-group.

Challenges within the general psychological literature on digital games surround the fact that little acknowledgement is dedicated to examining the role of social contexts on gaming experiences and effects. Although a growing body of research is developing on these issues (e.g., Kaye & Bryce, 2012; in press), within the broader academic context, little theoretical evidence is available to explain the way in which gaming occurs within social contexts, and how these different contexts may be related to different societal outcomes. Through my reading of the psychological literature, it has become clear that there is predominantly a focus on the violent content of games, and its relationship to aggressive attitudes and behaviours. Although this is underpinned by established psychological theory, and supported by a number of empirical studies (Anderson et al., 2010), this remains too restrictive in accounting for a broader range of outcomes. That is, digital games as entities are extremely complex, comprising features which extend beyond violence. Additionally, the activity of gaming itself provides largely positive experiences for players, through the way in which it functions as a form of enjoyable leisure, in reducing stress, providing escapism and enhancing players' emotional states. Therefore, my interpretation of this leads me to challenge the commonly held assumption that digital games are a societal issue. Indeed, I would contend that they can provide players with unlimited opportunities to experience enjoyment, as well as providing players with opportunities to socialise with friends, which should be by no means deemed as a detriment on either an individual or societal level. Although some recognition of this is starting to emerge within the academic literature, this remains a one of my personal challenges, in presenting ideas which more or less "go against the grain". From my perspective, it is disappointing that the mainstream psychological literature on digital gaming is largely preoccupied with violence, aggression and addiction, which, in my opinion, present only a small percentage of a much wider area of enquiry. Challenging these current trends entails presenting a strong contention that the processes of gameplay itself is a varied and dynamic experience, which holds a range of potentially positive and negative outcomes, particularly when considering the dynamic social experiences gameplay can provide. These outcomes are also largely determined by the individual, in relation to their personality traits, and the way in which they may influence gameplay experiences. Given these issues, I would contend that psychological researchers have further investigations to pursue before being able to conclusive claim their understanding of the way in which digital games are affected us.

Further to this, the fact that digital games are largely integrated into players' everyday lives, presents challenges to researchers in this area on how best to study their effects. Existing approaches which utilise experimental methodology to examine effects of digital games cannot account for such issues, and remain too restrictive for the purposes of researching social experiences and impacts of games. Therefore, research which adopts more ethnographic approaches may present an alternative here. This will help develop an alternative to the current research paradigm, to present a stronger case for considering the alternative impacts of digital games for individuals and society.

SOLUTIONS AND RECOMMENDATIONS

Based on the main findings from my case interviews, it appears that the game Football Manager has some utility in enhancing our social lives. Specifically the fact that it was found to be a "friendship enhancer" this may indicate its utility as an intervention for individuals who have difficulty making friends. In particular, this might be an area worth considering for other types of games, in examining the extent to which the development of games-based interventions can be incorporated as a social strategy within organisations (e.g., schools or inclusion programmes) for enhancing the friendship opportunities of individuals. In this way, it is conceivable that this strategy can be applied to other games, beyond the focus of this case.

Through undertaking these case interviews, I have found myself immersed in considering the endless possibilities for further research on this issue. Indeed, these interviews have formed a strong foundation for my future research plans, in which I hope to work collaboratively with Sports Interactive, the game developer of Football Manager. Within this, I hope to explore the way in which the game promotes the development of friendships and social identities, and its contributions to players' self-esteem and overall sense of well-being. This will indicate how Football Manager functions within players' psychological and social experiences of everyday life. Taken together, these findings are intended to provide evidence of the positive social impacts of the game. This will help formulate new understandings of this game, and its role in promoting positive gameplay experiences, as well as its role in the wider contexts of players' lives. This will help move beyond the existing negative preconceptions of digital games being a societal issue, to consider the way in which they may contribute towards positive impacts for players, particularly for formations of friendships, and enhancement of self-esteem, both of which are key to promoting positive overall well-being.

REFERENCES

Abrams, D., & Hogg, M. A. (1988). Comments on the motivational status of self-esteem in social identity and inter-group discrimination. *European Journal of Social Psychology*, *18*(4), 317–334. doi:10.1002/ejsp.2420180403

Agre, P. E. (1998). Designing genres for new media: Social, economic and political contexts. In S. G. Jones (Ed.), *Cybersociety 2.0: Revisting computer-mediated communication and community* (pp. 69–99). London: Sage Publications Inc. doi:10.4135/9781452243689.n3

Anderson, C. A., Shibuya, A., Ihori, N., Swing, E. L., Bushman, B. J., & Sakamoto, A. et al. (2010). Violent video game effects on aggression, empathy, and prosocial behavior in Eastern and Western countries: A meta-analytic review. *Psychological Bulletin, 136*(2), 151–173. doi:10.1037/a0018251 PMID:20192553

Bakardjieve, M. (2007). Virtual togetherness: An everyday-life perspective. In D. Bell, & B. M. Kennedy (Eds.), *The Cybercultures reader* (pp. 236–253). London: Rouledge.

Bender, T. (1978). *Community and social change in America*. New Brunswick, NJ: Rutgers University Press.

Connery, B. A. (1997). IMHO: Authority and egalitarian rhetoric in the virtual coffeehouse. In D. Porter (Ed.), *Internet Culture* (pp. 161–180). New York: Routledge.

Crawford, G., & Gosling, V. K. (2009). More than a game: Sports-themed video games and player narratives. *Sociology of Sport Journal, 26*, 50–66.

Ellemers, N., Haslam, S. A., Platow, M. J., & Knippenberg, D. (2003). Social identity at work: developments, debates and directions. In S. A. Haslam, D. V. Knippenberg, M. J. Platow, & N. Ellemers (Eds.), *Social identity at work: Developing theory for organisational practice* (pp. 3–26). Hove, UK: Taylor and Francis Group.

Harwood, J. (1999). Age identification, social identity gratifications and television viewing. *Journal of Broadcasting & Electronic Media, 43*(1), 123–136. doi:10.1080/08838159909364479

Highet, G. (2003). Cannabis and smoking research: Interviewing young people in self-selected friendship pairs. *Health Education Research: Theory and Practice, 18*(1), 108–118. doi:10.1093/her/18.1.108 PMID:12608688

Hogg, M. A., & Abrams, D. (1990). Social motivation, self-esteem and social identity. In D. Abrams, & M. A. Hogg (Eds.), *Social identity theory: Constructive and critical advances* (pp. 28–47). London: Harvester-Wheatsheaf.

Isiklar, A. (2012). Examining psychological well being and self esteem levels of Turkish students in gaining identity against role during conflict periods. *Journal of Instructional Psychology, 39*(1), 41–50.

Jenkins, R. (2008). *Social identity*. London: Routledge Taylor and Francis Group.

Jones, S. G. (1998). *Cybersociety 2.0: Revisting computer-mediated communication and community*. London: Sage Publications Inc.

Kaye, L. K., & Bryce, J. (2012). Putting the fun factor into gaming: The influence of social contexts on experiences of playing videogames. *International Journal of Internet Science*, *7*(1), 23–37.

Kaye, L. K., & Bryce, J. (in press). Go with the flow: The experiences and affective outcomes of solo versus social gameplay. *Journal of Gaming and Virtual Worlds.*

Kong, F., Zhao, J., & You, X. (2013). Self-esteem as mediator and moderator of relationship between social support and subjective well-being among Chinese university students. *Social Indicators Research*, *112*(1), 151–161. doi:10.1007/s11205-012-0044-6

Lea, M., Spears, R., & Rogers, P. (2003). Social processes in electronic teamwork: The central issue of identity. In S. A. Haslam, D. V. Knippenberg, M. J. Platow, & N. Ellemers (Eds.), *Social identity at Work: Developing Theory for Organisational Practice* (pp. 99–115). Hove, UK: Taylor and Francis Group.

Leonard, D. (2004). High tech Blackface- race, sports video games and becoming the other. *Intelligent Agent, 4* (4).

Licklider, J. C. R., & Taylor, R. W. (1968). The computer as a communication device. *Science and Technology*, *76*, 21–31.

Mauthner, M. (1997). Methodological aspects of collecting data from children: Lessons from three research projects. *Children & Society*, *11*(1), 16–28. doi:10.1111/j.1099-0860.1997.tb00003.x

Mayall, B. (2000). Conversations with children: Working with generational issues. In P. Christensen, & A. James (Eds.), *Research with Children: Perspectives and Practices* (pp. 120–135). London: Falmer Press.

Michell, L. (1997). Pressure groups: Young people's accounts of peer pressure to smoke. *Social Sciences in Health*, *3*, 3–16.

Postmes, T. (2003). A Social identity approach to communication in organisations. In S. A. Haslam, D. V. Knippenberg, M. J. Platow, & N. Ellemers (Eds.), *Social identity at Work: Developing Theory for Organisational Practice* (pp. 81–97). Hove, UK: Taylor and Francis Group.

Reid, S. A., Giles, H., & Abrams, J. R. (2004). A social identity model of media usage and effects. *Zeitschrift für Medienpsychologie*, *16*(1), 17–25. doi:10.1026/1617-6383.16.1.17

Stivale, C. J. (1997). Spam: Heteroglossia and harassment in cyberspace. In D. Porter (Ed.), *Internet culture* (pp. 133–144). New York: Routledge.

Tajfel, H. (1974). Social identity and intergroup behaviour. *Social Sciences Information. Information Sur les Sciences Sociales*, *14*, 101–118. doi:10.1177/053901847501400204

Tajfel, H. (1978). *Differentiation between social groups*. London: Academic Press.

Tajfel, H. (1979). Individuals and groups in social psychology. *The British Journal of Social and Clinical Psychology*, *18*(2), 183–190. doi:10.1111/j.2044-8260.1979.tb00324.x

Tajfel, H., & Turner, J. (1979). An integrative theory of inter-group conflict. In J. A. Williams, & S. Worchel (Eds.), *The social psychology of inter-group relations* (pp. 33–47). Belmont, CA: Wadsworth.

KEY TERMS AND DEFINITIONS

In-Group Affiliation: Sense of meaning or ownership to a specific social group, whereby distinction between in-group (members) and out-groups (non-members) can be made.

In-Group Membership: The belonging to a social group where by a distinction may be made between members (in-group) and non-members (out-group).

Social Identification/Social Identity: The way in which an individual defines themselves in terms of their membership to a certain social group.

Social Identity Theory: A psychological theory which underpins the way in which individuals' self-concept if defined by their membership to social groups.

Social Psychology: A branch of Psychology, which explains human behaviour as a result of interactions with the environment or social context.

Chapter 2
Equality Game, Anxiety Attack, and Misfortune:
A Pedagogical Post-Mortem on Engines, Modding, and the Importance of Player Experience

Victoria McArthur
York University, Canada

EXECUTIVE SUMMARY

In this chapter, I present a post-mortem covering three consecutive offerings of a course on persuasive games at the university level over a three-year period from 2010 – 2013. The course, "Designing Persuasive Games," is part of a larger, multi-disciplinary program on digital media and game design. In this course, students are invited to engage both with theory and praxis, the process of "practicing" theory (Shaffer, 2004), by not only reading and writing about persuasive games but also through the design and development of one. Here, I present the overall design of the course across the three offerings and describe the most significant aspects of the course, from a pedagogical perspective, that I believe to be of value to others designing similar courses. These aspects include choosing a game engine, scaling projects to retain rhetoric, modding as praxis, and player experience testing. A sample grading rubric for persuasive games is also included at the conclusion of this chapter.

DOI: 10.4018/978-1-4666-6206-3.ch002

Copyright ©2014, IGI Global. Copying or distributing in print or electronic forms without written permission of IGI Global is prohibited.

BACKGROUND

The *Designing Persuasive Games* course is a third year undergraduate course offered in a multi-disciplinary media studies programme. Students in this program take a number of core courses together, but are otherwise able to construct their academic path through the selection of electives from a multitude of disciplines, including computer science, visual art, history, classics, and geography. As such, the students who elect to take the *Designing Persuasive Games* course in their third year have extremely varied strengths and skill sets. Despite their unique backgrounds and academic trajectories, they all have one thing in common: an interest in video games and gamer culture. This is also something they have in common with their instructor.

Like my students I have also come from a multidisciplinary background. As an undergraduate I pursued a degree in music while I paid my tuition by building websites. I came to learn how to program as a bit of a hobby during my undergraduate studies, and later found myself teaching web design at the local community college. A lifelong video game enthusiast, I wound up pursuing a master's degree and later a PhD with a focus on game design. My academic work in this field led me to teach courses on game development at a number of institutions, always to a group of (mostly) male programmers looking to break into the industry, but my favourite course to date is and always has been the *Designing Persuasive Games* course.

Whereas the other undergraduate courses I taught focused on programming in *ActionScript* or preparing game design documentation, the *Designing Persuasive Games* course presented a chance for me to design a course where students could engage with the theoretical implications of some of the most influential games scholars through the design of a serious game. For many reasons, each iteration of the course involved a great deal of trial and error as I attempted to find the best mix of theory and praxis, the process of "practicing" theory (Shaffer, 2004). Aside from Ian Bogost, who should they read? As students with a diverse skill set, what kinds of projects can they tackle? Do they work better alone or in groups? How can I guide them so that they are happy with the ways in which their final game conveys their message to future players? With these questions in mind, I present what I believe to be the four most significant aspects of the course design and reflect upon how these pedagogical approaches worked with an interdisciplinary group of students. These are: choosing a game engine, scaling projects to retain rhetoric, modding as praxis, and the importance of player experience testing.

SETTING THE STAGE

The *Designing Persuasive Games* course presently uses Ian Bogost's *Persuasive Games* book (Bogost, 2010) as its primary text. Sections of Flannagan's *Critical Play* (Flanagan, 2009) and a number of articles on game design serve as supplemental texts. I like to begin the class with an activity I borrow from a former instructor in which we play Gonzalo Frasca's game *September 12th,* (Frasca, 2003) and read an interview with Noam Chomsky discussing the aftermath of the September 11th attacks in 2001. Both the game and the interview with Chomsky present the same thesis: a broken logic drives post 9/11 conflicts and that violence begets more violence. By engaging with the same thesis, expressed first by game mechanics and second in written form, students begin to understand how procedural rhetoric works.

We then read Bogost's work and discuss it in seminar. All three iterations of the course have included this text as our primary theoretical source as it is the most prolific and thorough text on the subject. We explore the examples provided in the book, playing the games whenever possible (a handful of games described in the text are no longer hosted online or are rare games for the Atari), and seeking out additional examples of persuasive games. Students are then asked to design a persuasive game of their own. This project represents a significant portion of their final grade and involves the following milestones: a written proposal (500 words), a research poster presentation, a paper (analogue) prototype, a digital prototype, and a playtesting session. While the written proposal and final game are for me, the other milestones present opportunities for students to explore their proposed game in meaningful ways with their peers.

The poster presentation and playtesting session provide opportunities for students to obtain feedback from a variety of sources on their design and efficacy of their game. The paper prototypes give students the opportunity to first explore their mechanics using analogue media, which allows them to explore their game mechanics more freely. The paper prototyping phase did not exist in the first iteration of the course but was added in the second and third offerings. This was largely to address a common issue noted in the first offering: students became focused on what their game engine could do and designed around it, rather than designing a game and then figuring out how to program their games. In offerings where paper prototypes were used, students were able to sidestep the focus on the technological affordances of their game engine and instead spend time exploring procedural representation. Paper prototypes also allow for rapid modifications to gameplay. Overall, students have found these three milestones to be especially meaningful in each offering of the course.

CASE DESCRIPTION

In the following sections I present some of the lessons I have learned after teaching three consecutive offerings of the course. Where possible, each of these lessons is presented in the context of one or more student games. Following this, a sample grading rubric for persuasive games is also included in the Solutions and Recommendations section of this chapter.

Choosing a Game Engine

There are a number of free engines available, each with different strengths and weaknesses. In the first two offerings of the course, students were taught how to use *Game Maker (Overmars)*. This engine works on both Mac and PC platforms and has an online community that hosts a number of tutorials and help sections. The developers of *Game Maker* also have a book called *The Game Maker's Apprentice* (Habgood, Overmars, & Wilson, 2006), which ships with detailed instructions for creating 10 different games. The engine is simple enough to use that it allows students to create simple games without having to learn complex programming languages, but also supports scripting in Game Maker Language (GML) - the engines own proprietary, object-oriented programming language. *Game Maker* also supports exporting games to an executable file - something that is easily shared with others.

Students with little to no programming experience found this engine easy to use, although some felt it was best suited to the development of specific kinds of games (the kinds presented in *The Game Maker's Apprentice*). Games like platform games and maze games were easy to create, but other styles, such as role-playing games (RPGs) were not. Also, while it was easy to create simple games following the tutorials, some students felt as though they were not skilled enough to implement the kinds of sophisticated mechanics they had envisioned when writing a proposal for their persuasive game. The paper prototype was included in order to address some of these issues. Paper prototyping helped the students side-step these issues initially; however, as soon as they started working with the game engine some felt as though they didn't know how to properly translate their mechanics from analogue to digital representation. In the third offering of the course, students were encouraged to research a free engine of their choice.

Other engines that were explored in the third offering include *Game Salad* (GameSalad), *RPG Maker* (Enterbrain), *Adobe Flash* (Adobe Systems), and *UNITY* (Unity Technologies). Students found that *Game Salad* was easier to learn than *Game Maker*, but later discovered that projects could not be easily exported for sharing (and later submitting to the instructor). *Flash* and *UNITY*, while more powerful than *Game Maker*, had a much steeper learning curve with regard to getting a project off the ground. Also, since I was only peripherally knowledgeable in *Flash* and *UNITY*, I was unable to offer sufficient technical support to groups that chose to use (and subsequently learn) these engines. For instructors interested in exploring procedural rhetoric through game design, *Game Maker* is the easiest engine to learn, teach, and use in the absence of a strong technical background.

For students who are able to teach themselves a specialized engine, the results can be very rewarding. One student wanted to allow players to explore the complexities of social stigma and bullying in his project. After spending some time doing *Game Maker* tutorials, the student decided that a role-playing game (RPG) would be better suited to support his game's thesis and approached me about using *RPG Maker* to build his prototype. The student's game was well designed and the student reported back that *RPG Maker* was very easy to learn. A specialized engine such as this is really only capable of making RPGs, but if the genre supports the rhetoric, I will support the engine.

By the time you read this chapter, I imagine that an even better free engine will exist. That being said, the aforementioned points will still hold true – the game engine you choose should be easy to use, be able to create the kinds of games your students want to create, it should have an active community dedicated to its use, and there should be some (well written) tutorials available for free use. The techni-

Figure 1. Misfortune: an RPG game designed to explore the complexities of social stigma and bullying

cal literacy required to learn the engine should also be appropriate to the academic program (and level of education) in which the engine will be used. I have offered a similar course to Engineering students who used *UNITY* to build their games. They came to me already knowing *UNITY*, so I was able to spend instructional time focusing on game design and procedural rhetoric.

It should also be noted that I have used *Game Maker* to teach school aged children (6 years to 17 years) to build games, so don't be afraid to adopt it for use in primary or secondary schools. If, after spending a few weeks learning *Game Maker* you feel as though you're not comfortable enough to teach it in class, consider hiring a student mentor from a computer science department to act as technology facilitator. We have found that having such an aid in the classroom can allow younger students to build more complex projects and also allows the instructor to focus more on pedagogy and less on "technical support."

Scaling Projects to Retain Rhetoric

A common issue in each iteration of the course is scalability. Students are exceptionally ambitious when it comes to the design of their game. This statement holds true for gamers and non-gamers alike. Every effort is made to monitor student projects from the proposal phase through to the prototype phase to ensure that the project is appropriate in scope given the constraints imposed by the academic calendar. Even with this assistance, students may still find that they must simplify their projects at some point in order to ensure that a polished and functioning prototype is complete by the end of the term. Scaling game design projects is tricky enough to begin with, but scaling a game while retaining procedural rhetoric is even trickier.

In some cases, this meant the removal of a fancy game mechanic or proposed levels. With significant parts of the original design gone, many students then had to re-visit the design of their game in order to ensure that it was still persuasive without the missing levels or mechanics.

One such example is the game *Anxiety Attack*, created using *Game Salad.* In the initial proposal, *Anxiety Attack* was actually called *Debt Attack* and was designed to suggest that student debt is inescapable when one must balance finances, education, and the loosely defined "happiness" factor in one's life. The initial proposal was to be an RPG style game where players would participate in various aspects of undergraduate student life. School, work, and leisure activities would be represented as RPG battles, affecting the three variables (education, finances, and happiness). The three variables would be impossible to keep in balance – ultimately one or more of these aspects begins to suffer as the student becomes bogged down with the balancing act of living a productive and happy academic life. The anxiety created by the imbalance of your three "resources" would decrease your effectiveness in battle. Subsequent failures continue to add to the player character's anxiety until, ultimately, the game becomes unwinnable. Here, the students explore what Bogost refers to as "the rhetoric of failure" (Bogost, 2010, page 85).

As the course progressed the original design proved to be too ambitious to be carried out by the students. Through a variety of re-designs, the students arrived at an appropriately scaled version that still procedurally explored their thesis. The final version of *Anxiety Attack* is shown in figures 2 and 3. Here, the player character participates in 3 levels: school, work, and leisure time. In each level, contextually relevant objects (homework, money, and possessions) fly by and must be caught by intercepting the object using movement keys. Successfully collecting as many objects as you can raises your score in the relevant category, but also ultimately raises your anxiety bar. When the anxiety bar is full, the game environment changes to represent an anxiety attack (see figure 3). The backgrounds become harsh, the objects fly by at an impossible pace, the music becomes unpleasant, and the words "ANXIETY ATTACK" flood the screen.

The simplified version of the game not only allowed the students to develop a polished prototype on time, but also allowed them to redirect their focus from asset creation and complex mechanics to ensuring that their mechanics supported their thesis. It is important when proposing simplification to students that they find ways to retain rhetoric in their redesign. Even a simple game can present a powerful message.

Figure 2. Anxiety Attack (final version): a game exploring the relationship between anxiety and student debt. A student tries to collect homework assignments as they approach.

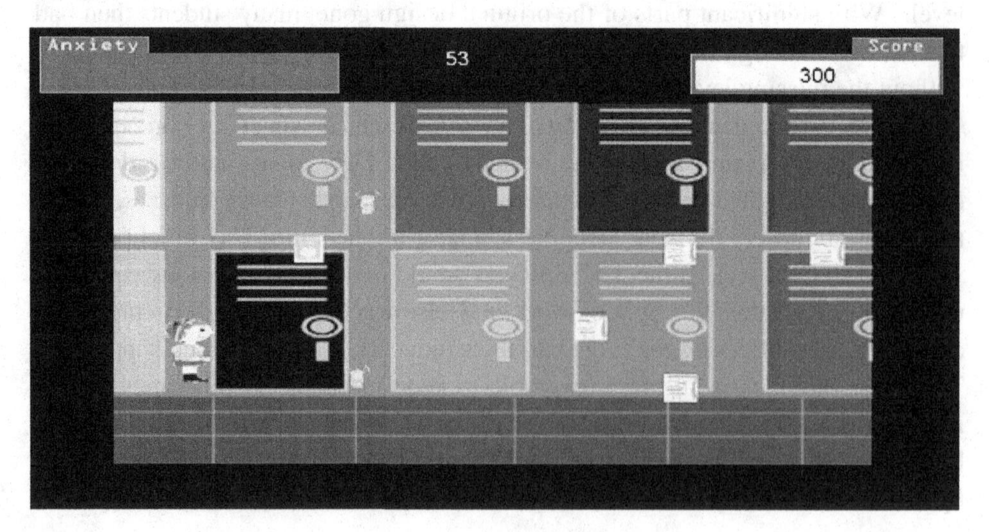

Figure 3. Anxiety Attack (final version): a game exploring the relationship between anxiety and student debt. With a full "anxiety bar" the game environment becomes frantic.

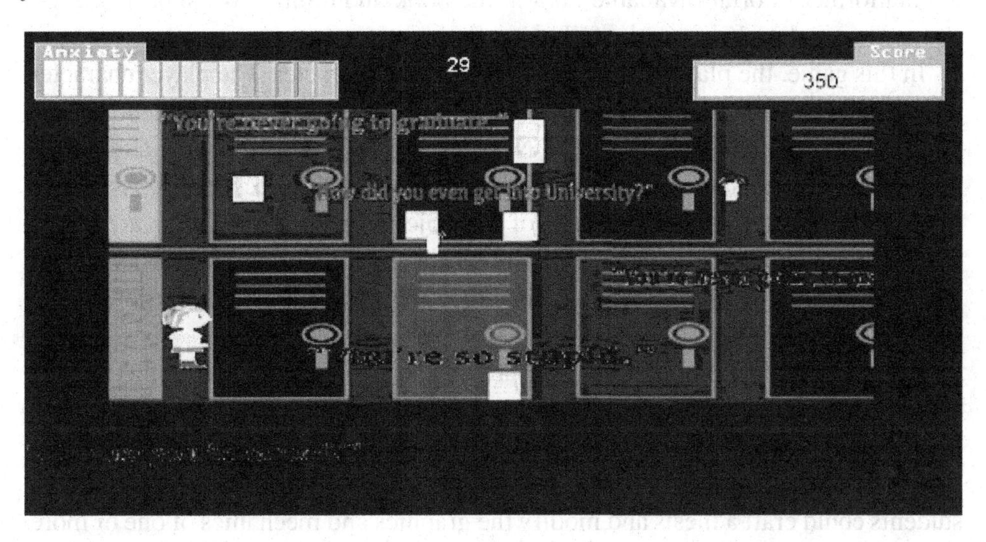

Modding as Praxis

Since technical ability varied greatly between students, it was important to offer a "modding" option in the course. Here, modding refers to the editing of an otherwise functional, pre-existing game. The pedagogical benefits of modding have been presented by others (El-Nasr & Smith, 2006) as a way to allow students to side-step game development barriers in order to gain access to meaningful design activities. This practice may also be particularly beneficial when working with younger students.

Here, after working through a number of tutorials, students may choose a game and modify it in an attempt to explore procedural rhetoric. This usually involves two fundamental steps: re-skinning objects in the game with new sprites, and modifying the game's mechanics. Ian Bogost cautions us that re-skinning alone is a problematic way to explore procedural rhetoric, since, at best we are playing with "graphical logics" (Bogost, 2010, page 89). A change of imagery alone would, in most cases simply be an exploration of visual rhetoric, but through mechanical modding, students who have difficulty developing a game may still actively explore the concept of procedural rhetoric. Once the mechanics are in place, the graphical aspects of the game can easily be redesigned to suit the theme of the new game.

In the first offering of the course, one student, a female athlete, wished to explore gender inequality in sports. As a woman, she was ever aware of the lack of opportunities and unequal pay between the sexes in the athletic profession. She had formed a thesis, but found it difficult to design an original game to present her thesis. As a pedagogical experiment, I invited her to try modding a game. The student chose the platformer tutorial (available both in the book and online) as the basis for her game. Her mod is shown in the following figure.

In this game, the player first plays as the female character. The player character jumps from platform to platform, attempting to collect each of the athletic opportunities (money, endorsements, etc.). Next, the player plays as a male, attempting to collect all of the same objects using the same interaction techniques. As the female, the character cannot reach all of the objects as easily and they are worth fewer points. As the male, there are more objects to collect, they are easier to collect, and they are worth more points than they are when the female sprite collects the exact same objects.

It's a very simple modification and provided a rich writing opportunity for the student. Since she found the idea of creating a game from scratch to be overwhelming, modding afforded her the same academic opportunity: to engage theory in a meaningful way. Modding could also be used as a class-wide activity. As a class, students could craft a thesis and modify the graphics and mechanics of one or more complete games to explore the procedural expression of that thesis. Alternatively, individual students could submit their mods to the class for playtesting and the

Figure 4. Equality Game: a game exploring inequality between the sexes in athletics

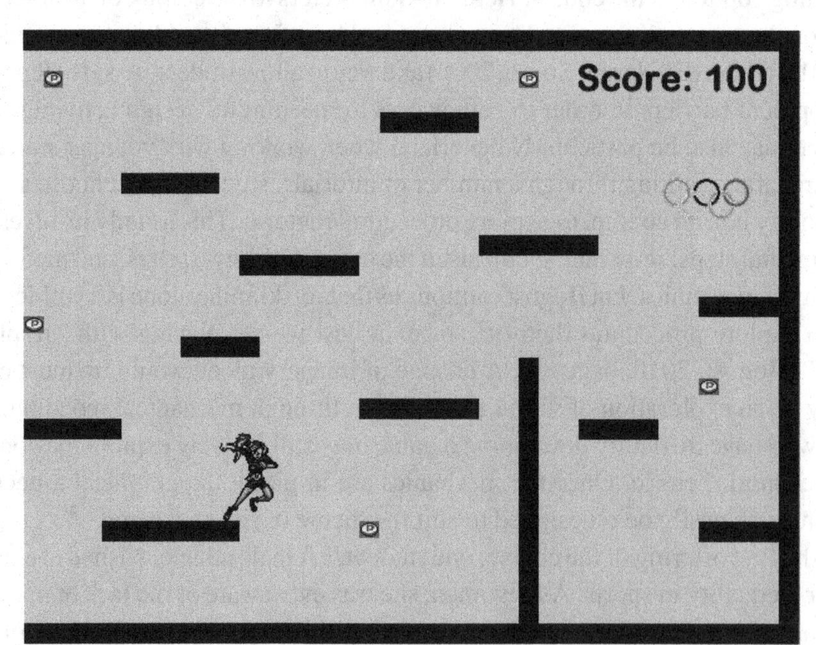

class could then discuss the different mods and how effectively they explored the thesis through mechanics. Playtesting, or player experience testing, is explored in the following section.

Player Experience Testing

Lastly, students were shown how player experience testing is used not only to assess usability and design issues (Fullerton, 2008), but also the "readability" of a students' rhetoric (Flanagan, 2009). Students were shown how to conduct player experience testing, how to collect data from these tests, and how to design the tests so that they could find out whether or not their game's rhetoric was evident to participants.

Early in the semester we explore Mary Flanagan's model for designing for critical play (Flanagan, 2009). Here, Flanagan takes the existing iterative design process and re-frames it in a way that is relevant to the design of a persuasive game. Where the traditional model focuses on the design, prototyping, usability testing, and subsequent re-design of a digital artefact, Flanagan's model includes a an evaluative aspect that encourages critical game design. For example, in her model, one begins with the design of a prototype created with specific critical goals in mind. The artefact is then prototyped and evaluated not only for usability but also the artefact's ability to address its critical agenda. For our purposes, the critical agenda is a thesis or argument. Students propose a game that conveys a thesis through mechanics. They then design their game and test it with a representative player base. Through the act of playtesting, students are not only checking whether or not their game is easy to play or has any inherent usability issues; they are also user testing their game's thesis. At the conclusion of the test, students can then re-visit their designs with the results of these tests in hand. If their thesis was unclear (or misread) by players, they redesign their games and start the cycle over again until satisfied that their design expresses their argument through mechanics.

The importance of player experience testing is also echoed by game designer and educator Tracy Fullerton. In an article contributed to ACM Interactions (Fullerton, 2008), Fullerton shares her original anxieties over turning her games over to usability testing too early, for fear that the results would be counterproductive in the absence of a polished product. When she did allow for her prototype to be evaluated, she found the insights to be invaluable. An impartial third party was able to notice design flaws that weren't visible to someone too close to the project – her. Had she put off usability testing until her game was polished, the issues would have been much harder to fix. By allowing her game to be evaluated in its early stages, she was able to modify the design to address these issues easily during the prototyping phase.

We have had similar experiences in our class as well. Students have worked hard to create meaningful obstacles and puzzles in their games, only to discover during playtesting that these are easily avoided or ignored by players. In many cases these puzzles or obstacles were important parts of their game's rhetoric. Players bring a great deal of variability to the way games are played and subsequently "read" in terms of procedural rhetoric. As the saying goes, "If you can't bring Mohammed to the mountain, you must bring the mountain to Mohammed."

SOLUTIONS AND RECOMMENDATIONS

Each of the aforementioned points presented rich opportunities for pedagogical strategy that I had never experienced before. I suppose this is one of the reasons why the *Designing Persuasive Games* course has been my favourite course to teach to date. Not only has it allowed me to explore game design in a meaningful way – it has also allowed me to explore pedagogy in a meaningful way as well. I will conclude this chapter with what I deem to be the last and most interesting aspect of a course like this, and that would be evaluation.

Students in the humanities are accustomed to working within specific kinds of guidelines when it comes to assignments. Things like page numbers, word counts, and minimum numbers of references provide them with some semblance of both the quantity and quality of material required to submit a satisfactory assignment. Games, as can be expected, are much harder to measure in these terms. In the first offering of the course I struggled with how to set minimum guidelines for student projects. Do pages become levels? Do word counts become gold coins or player objectives? Different genres of games are difficult to compare in such terms. Thus, I have had to work closely with each student in order to ensure that their own game was adequate in scope to submit at the end of the course. Luckily, typical enrolment in my course is anywhere from 10 – 20. For larger classes it might be appropriate to have students work in groups of 4 or 5.

The following rubric was designed for evaluative purposes. I present the rubric to students early on in order to provide them with a reference for how their games will be evaluated. The rubric not only helps them to be mindful of the various aspects of their projects, but also makes it easier for me to be consistent in how I evaluate student games. The rubric was intended to keep students mindful of the most important aspects of the project without constraining the resultant games.

Students taking the course this year benefit greatly from the various leaps of faith I took with previous cohorts. I not only thank my former students profusely for attempting every project and milestone I put before them with the utmost creativity and determination, but also for their feedback on the design of the course and for creating some of the most wonderful projects I have ever had the pleasure of evaluating.

Table 1. Grading rubric for persuasive games

	Excellent	Good	Needs Improvement	Poor	F
Graphical Design /15	Game has a polished graphical design. Graphical elements have a consistent aesthetic (e.g. realism, perspective, etc.). **12 - 15 points**	Overall graphical design is good but some elements are lacking. Graphical elements do not have an overly consistent aesthetic. **8 - 11 points**	Overall graphical design is lacking. Graphical elements are either inappropriate for the genre, or do not present a consistent aesthetic. **5 - 7 points**	Graphical design is severely lacking. Graphical elements are either missing or appear to have been placed without careful consideration for graphical design standards. **2 - 4 points**	There are few or no graphical elements. Inconsistent graphical design makes the game difficult to play. **0 - 1 points**
Interaction Model & Mechanics /15	Game has a polished interaction model. Game mechanics are easy to understand and game play is logical. **12 - 15 points**	Interaction model is good, but some game mechanics are unclear. **8 - 11 points**	Interaction model is inconsistent. Game mechanics are difficult to understand. **5 - 7 points**	Interaction model is problematic. Game mechanics detract from game play. **2 - 4 points**	Game mechanics are broken or too inconsistent to facilitate critical play. **0 - 1 points**

REFERENCES

Adobe Systems. (n.d.). *Adobe Flash*. Retrieved January 26, 2014, from http://www.adobe.com/ca/products/flash.html

Bogost, I. (2010). *Persuasive games: The expressive power of videogames*. Cambridge, MA: MIT Press.

El-Nasr, M., & Smith, B. (2006). Learning through game modding. *Computers in Entertainment, 4*(1).

Enterbrain. (n.d.). *RPG Maker*. Retrieved January 26, 2014, from http://www.rpgmakerweb.com/products/programs/rpg-maker-xp

Flanagan, M. (2009). *Critical play: Radical game design*. Cambridge, MA: MIT Press.

Frasca, G. (2003). *September 12th*. Retrieved January 26, 2014, from www.newsgaming.com/games/index12.htm

Fullerton, T. (2008). Playcentric design. *Interaction, 15*(2), 42–45. doi:10.1145/1340961.1340971

GameSalad. (n.d.). *GameSalad*. Retrieved January 26, 2014, from http://gamesalad.com/

Habgood, J., Overmars, M., & Wilson, P. (2006). *The game maker's apprentice: Game development for beginners*. Berkeley, CA: Apress, Springer.

Overmars, M. (n.d.). *Game Maker*. Retrieved January 26, 2014, from www.yoyogames.com/studio

Shaffer, D. W. (2004). Pedagogical praxis: The professions as models for post-industrial education. *Teachers College Record, 10*(7), 1401–1421. doi:10.1111/j.1467-9620.2004.00383.x

Unity Technologies. (n.d.). *Unity*. Retrieved January 26, 2014, from http://unity3d.com/

KEY TERMS AND DEFINITIONS

Game Engine: A program for developing computer games.

Modding: A slang expression derived from the term modify. In the case of computer games, it refers to the modification of an existing game.

Playtesting: A user-centered approach to testing a game. Playtesting involves inviting players to play the game in early stages. Playtesting can serve to identify technical issues (bugs) or logical issues (poor design).

Post-Mortem: A technical analysis of a finished project.

Praxis: The art of practicing theory.

Prototype: An early sample of a game designed to test its concept. prototypes can be analogue (made of physical materials, such as paper), or digital.

Rubric: A set of criteria for grading assignments.

Serious Game: A game designed for a purpose other than entertainment. Serious games include, but are not limited to, educational games and games for training purposes.

Chapter 3
Games for Top Civil Servants:
An Integrated Approach

Hester Stubbé
TNO, The Netherlands

Josine G. M. van de Ven
TNO, The Netherlands

Micah Hrehovcsik
HKU University of Arts – Utrecht, The Netherlands

EXECUTIVE SUMMARY

In designing De BurgemeesterGame—The Mayor Game—we aimed to develop a game that would be used and appreciated by a target population that was hardly used to being trained and had little affinity with applied gaming: mayors. To make sure that the (learning) goals, the context, the characteristics of the target population, and the creative design were all integrated into the game, we chose to work in a consortium with a focus group. We included engaging elements like simple gameplay based on actual processes, authentic scenarios presented in the way of dilemmas, time pressure, and collaboration. This resulted in a game that was accepted by the target population and has been played by more than half of all mayors in The Netherlands. Mayors feel the game challenges them to explore their decision making during crisis management and stimulates them to discuss this with other mayors.

DOI: 10.4018/978-1-4666-6206-3.ch003

Copyright ©2014, IGI Global. Copying or distributing in print or electronic forms without written permission of IGI Global is prohibited.

BACKGROUND

Our consortium consisted of three main partners, covering the three main roles in game design: (1) Domain experts, (2) Educational experts and (3) Game design experts. (1) Domain experts have expertise in the domain for which the game is designed. They can have knowledge about and contacts in the domain (internal expert) or they can be working in the domain (external expert). In our consortium we used both. (2) Educational experts have expertise in defining competencies and in selecting the didactics needed in the game to achieve the learning goals. (3) Game design experts know how to make a game work and involve players. They also have technical knowledge and experience about possibilities to include certain elements into the game, and how. The three partners and the focus group together have the knowledge needed to develop a serious game that will be accepted in the domain, that will support participants to develop themselves in the competencies decided upon and that they like to work with. Collaboratively we went through the design process, each contributing on the basis of their own expertise.

The companies involved in our consortium were:

- *TNO* is an independent research organization whose expertise and research make an important contribution to the competitiveness of companies and organizations, to the economy and to the quality of society as a whole. TNO's unique position can be attributed to its versatility and capacity to integrate this knowledge. We develop knowledge not for its own sake but for practical application. To create new products that make life more pleasant and valu-

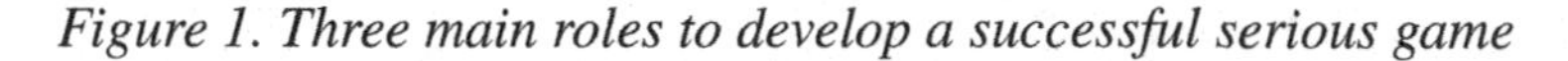

Figure 1. Three main roles to develop a successful serious game

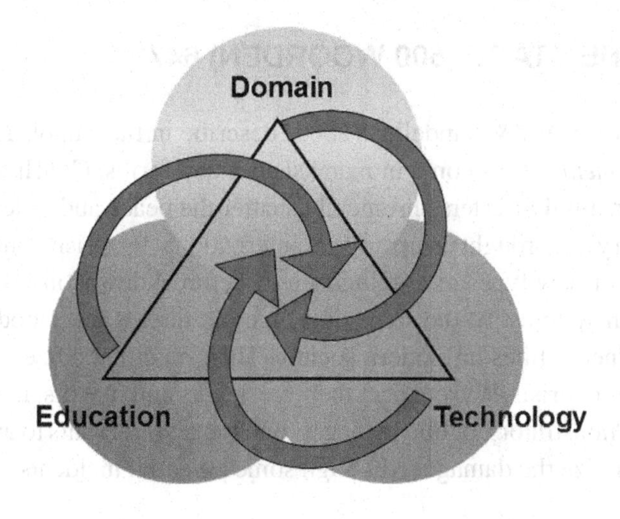

able and help companies innovate. To find creative answers to the questions posed by society. TNO took the role of internal domain expert and educational expert.

- *HKU University of the Arts, Utrecht* is one of the largest schools of art in Europe. Students are offered preparatory courses, bachelor and master programs and research degrees in fine art, design, media, games and interaction, music, theatre and arts management. The HKU's Games and Interaction research group works on innovation in designing games and playful interaction. We ask how design can create more impact in both entertainment and the application in areas such as cultural heritage, learning & inspiration, healthcare & wellbeing, science, ecology and organizational change. The HKU took the role of game design expert.
- *T-Xchange and Thales* designs and develops serious games for public and private organizations. They have in-depth knowledge of using serious games and gaming as facilitating instruments to support time and place invariant human dialog and reasoning during social interactions at individual and group levels, and more importantly, making these sense-making processes transparent during creative explorative stages of problem solving activities. T-Xchange/Thales took the role of game design expert.

Apart from that we had a focus group, consisting of external experts like trainers and coaches of mayors, a former mayor and experts who work with mayors during a crisis. The *Genootschap van Burgemeesters* (Dutch Association of Mayors) has helped us to develop the scenarios for the game, based on authentic incidents and real crises. The Institute for Safety (IFV) has played an important role in making a training, including the game, available for the domain and in the training of trainers.

SETTING THE STAGE (500 WOORDEN) 647

As Boin, 't Hart, Stern & Sundelius (2005) describe in their book *The Politics of Crisis Management*, crises come in many shapes and forms. Conflicts, man-made accidents and natural disasters chronically shatter the peace and order of societies. The new century has brought un upsurge of international terrorism, but also a creeping awareness of new types of contingencies like breakdowns in information and communication systems. At the same time, age-old threats like floods continue to expose the vulnerabilities of modern society. They state that citizens look at their leaders in times of crisis. They expect their presidents and mayors, local politicians and elected administrators, public managers and top civil servants to avert the threat or at least minimize the damage. Although some potential incidents can be averted

Figure 2. Kick-off event with 20 mayors playing the game simultaneous

by adequately assessing and addressing them, crises will occur. When policy makers respond well, the damage is limited; when they fail, the crisis impact increases. In extreme cases this may be the difference between life and death. These are no easy tasks because the management of a crisis is often a big, complex and drawn-out operation with the mass media scrutinizing and assessing leaders and their leadership. In the Netherlands mayors have, by law, a responsibility for crisis management. They are not only required to make resources available for a crisis management organization, they also have an active role during a crisis. To train mayors in crisis management, training focuses on procedures and getting to know the other players in Het Beleidsteam - the Policy Team. These table top exercises usually deal with the first hour after the policy team is assembled during a crisis. The contribution to the professionalization of policy makers in crisis management is limited for several reasons: because of the focus on protocol and cooperation, mayors cannot train the typical dilemmas policy makers face during a crisis. Moreover, the focus on this first hour makes it impossible to train the strategic dilemmas that typically present themselves later on during a crisis and are (mostly) not directly related to the operational activities of first responders.

Trainers find it difficult to develop scenarios that deal with these more strategic dilemmas; their own experience usually lies in operational crisis management. Apart from that, they are less familiar with the training methods needed to train strategic decision making. Another issue is that mayors usually train once to twice a year. Their busy schedules and the logistics needed to arrange table top training (authentic situation) make it hard to increase this frequency. Lastly, mayors are in a position where they need to show they are in control. This attitude often leads to the belief that they do not need more or another training.

Although gaming is a popular trend, at the start of this project it was very difficult to convince policy makers and the people around them that serious gaming could be applicable and interesting to them. Apart from the general reluctance to train at all, they feel they have serious jobs and do not like to waste their time on a game.

The Mayor Game trains strategic dilemmas, thus supporting training needs for policy makers that could not be met before (Bronzwaer, 2011). The use of a game as the training method supports these specific training needs. On top of that, the Mayor game is played in about 60 minutes, including the (group) reflection session afterwards. This makes it possible to increase the frequency of training which helps to achieve learning goals and progress. The didactics used are based on the theory of self-directed learning. This means that the learner is very much in control over his or her own learning. In this way we could deal with the reluctance of mayors to admit that there is room for improvement. Lastly, we gave much attention to the engaging elements of the game. Graphics as well as gameplay were designed to create 'an authentic situation'.

CASE DESCRIPTION

The Mayor game consists of a web-based game (small game) that is integrated in a training session (big game) to stimulate reflection and share experiences. At the moment it contains 12 scenarios that focus on strategic dilemmas.

The training became available through the IFV on 1 April 2012 and can only be used by licensed trainers. Since then 30 trainers have obtained their license. On average, the Mayor game is used every week, with a group of eight people.

Figure 3. Screenshot of dilemma screen

Small Game

The Mayor game is a dilemma trainer. The game itself is web-based and presents a scenario in eight dilemmas. Dilemmas are introduced with a few lines of text and end with a question. Participants need to give a 'yes' or 'no' answer to this question. They work with a Digital Policy Team; the team members sit behind a table and are available for advice and extra information. Once the dilemma has been answered, the participant is asked to record what information items were important to decide on this dilemma. The game is over when either the time limit has been reached (15 minutes) or all dilemmas have been answered.

Immediately after the game, the participant receives feedback on the choices made. This is done in several ways. First of all, they can read a newspaper article that is personalized according to the choices made. This will give the participant an impression of how the press will write about them the day after. Then the participant can click through to statistics about e.g. the time needed to answer dilemmas, the number of dilemmas answered, the number of times the advice of the different team members was recorded as important. Last but not least, the participant's answers are related to the three different roles he or she needs to take during a crisis, (1) 'Boegbeeld': taking a prominent role in the media, (2) 'Bestuurder': focusing on the law and administration side of the situation and (3) 'Burgervader': acting as a father/mother to citizens that suffer from the situation. A graph shows how many of their answers can be related to each of the three roles. The feedback focuses on presenting what the participant has done (mirror). The small game does not provide any interpretation or advice.

This really makes me think!

Big Game

A typical training with the game takes 60 minutes. It starts with a 5 minute introduction on the goals and the use of the game. Then the game is played, which takes 15-20 minutes. The remaining time is spent on discussing the thoughts and considerations the mayors had while reaching their decisions. The actual answers to the dilemmas are not important at all. The responses to real dilemmas can never be right or wrong; only the outcome of the real situation determines, in hindsight, if this was the most effective decision. Therefore, the thoroughness of the decision process is more important. The trainer will stimulate reflection on this process and invite participants to share earlier experiences with similar incidents.

The training takes place in a group of 8 - 15 participants. Each participant individually plays the same, preselected scenario.

Although the Mayor game is designed in a very thorough way, it is our experience that most learning takes place in the discussion/reflection phase at the end of the training. Participants quickly experience what dilemmas can occur in relation to specific scenarios. This experience leads to curiosity and a sense of urgency, which increases the effectiveness of the discussion phase (as compared to a case discussion without the game). To achieve this, trainers need to master the game and know all the scenarios in all their angles. Apart from that they need to be able to stimulate mayors to reflect on their (future) actions and decisions.

Persuasive Game Elements

'The game is not an experience, it only enables the experience' (Schell, 2008). With the Mayor game we aim to create an experience from which mayors will learn. As Blumberg and Ismailer (2009) say, the efficacy of formal games to instruct depends strongly on developmental and cultural appropriateness. To achieve that we have included game elements that will be motivating to our target population. We had Schell's book of lenses (2008) in mind when we looked at the four motivational characteristics of gaming: (1) Immersion, (2) Achievement, (3) Cooperation and (4) Competition. (1) To stimulate immersion the Mayor game represents a simplified version of the real world with authentic dilemmas and time pressure. The simplicity shows in the 2-dimensional game world, the point-and-click system and the rather static background. Though simple, interaction is based on the actual processes in crisis management and scenarios were developed on the basis of real incidents, with experts in the domain. The only aspect of reality we have changed is that participants are forced to give a 'yes' or 'no' answer to dilemmas. In reality, their response would be a shade of grey. This obligation to choose stimulates deep thinking and mental simulation. Sometimes mayors do not like this aspect in the beginning, but admit later that it contributes to the experience. We have added a small engaging element in the facial expression of the members of the policy team: if they agree with the participant's decisions they will smile, if they do not agree, they will look a little angry. They also have a neutral expression. (2) Mayors are deeply motivated by achievement. Therefore, one of the first versions of the game did have a scoring system. However we tried to define right and wrong, it always felt 'unfair' to the participants. Because crisis management is not an exact science, decisions need to be made on the basis of incomplete information and under dynamic circumstances. Only later is it possible to see if a decision turned out well. Therefore, the Mayor game in its present form does not have a scoring system, and there are no right answers. Feedback consists of a summary of the choices made and the relation of these to the roles of a mayor, presented in graphs. (3) The idea of the Mayor game is to stimulate collaboration; 'we all have the same position, and this could happen to me too'. Learning from each other's experiences is very helpful. The element of cooperation was included in the 'big game'. Participants are stimulated to share experiences rather than right answers during the reflection after playing the game. These can be game experiences but also real-life incidents. (4) Competition is based on doing better than somebody else. This automatically leads to a focus on results, which is contrary to our learning goals. We, therefore, do not stimulate competition with our game.

Domain

As described before, we have worked with a focus group (external domain experts) right from the start of the project. This focus group consisted of people with relevant expertise, experience and contacts within the domain of policy makers in crisis management. During the project we have held regular focus group meeting. Apart from that individual focus group members were approached with specific questions between the meetings. Together, we have determined the learning goals and the boundary conditions that should be met for the game to be adopted and successful. Then, in an iterative process, the concept, paper-based design, digital mock-up and digital prototype were discussed and tested with the focus group.

Since the Mayor game addresses new learning goals for policy makers (strategic dilemmas), we had to start by defining those goals. Using the SLIM-method (Specifying Learning means in an Iterative Manner; Verstegen & van der Hulst, 2000) the following learning goals and boundary conditions were identified:

- **Learning Goals:**
  - Recognize policy dilemmas (What is your role and which dilemmas belong to that role?).
  - Decision making under time pressure.
- **Boundary Conditions:**
  - Training session of maximum one hour.
  - Safe and anonymous: no information about performance or scores can become public.
  - 'Serious' enough: not too playful, meaningful context.
  - Mayors must be able to use it: simple Human-Machine Interface.

The advantage of using this method was that we had a good idea of the learning goals and boundary conditions, from the beginning of the project. Because we had consensus in the focus group, these were not changed anymore during the project.

We managed to keep the focus group involved during the whole process, using an iterative approach. That meant that the several aspects of the game that needed to be addressed were discussed in different phases. For example, we started with the learning goals and boundary conditions. In the second meeting we checked if the learning goals chosen were still accepted. At the same time we introduced three possible game concepts, based on those learning goals. At the end of the meeting we had a discussion on the boundary conditions and how we could allow for them.

A typical focus group meeting would consist of the following elements: an update of the status of the project, a check of something discussed before, an (inter)active part of the meeting in which new developed materials were tried out, a discussion/brainstorm about a new aspect. This gave us the opportunity to show that we were taking their feedback seriously, get their feedback on newly developed materials, and obtain new information for the next step. We started and/or ended with lunch or drinks to enable the focus group members to meet and get to know each other and made sure that the methods used, varied during the meetings (presentation, discussion, hands-on try-out).We always kept in mind that these were very busy professional, which meant that our meetings had to be interesting enough for them to join.

When the overall game concept had been chosen, we presented four possible screen representations. Initially, most people chose a representation that showed the situation in all its aspects: a map of the location of the crisis, video images, and the dilemmas. After some discussion, one group member suddenly mentioned that this was too complex for the target population. It would increase the complexity of the game if the necessary information was 'all over the place'. That was when the 'table concept' was chosen.

Our focus group was also very much involved in the design of the graphics. When we showed them the very first concept drawings, they immediately set us right about the uniforms used for the representatives of the police and the fire brigade. Where we had used the prototype policeman and firefighter, they explained to us that those are the people that work outside, on the streets. The representatives in the Policy Team are dressed quite differently. At this stage in the project we could easily change that. Later on they gave us much input on the 'seriousness' of the characters.

Educational Expertise

Our domain experts shared the characteristics of mayors with us. This led to the belief that we should stimulate mayors take control over their own learning process. First of all, mayors are in a position where not many people tell them what to do or give feedback on a not-so-good performance. Apart from that, when people want to learn something/feel the need to learn something, learning will be more effective than when somebody else tells you to. This is why we used the theory of self-directed learning as a basis for the didactics in the Mayor game. This had serious implications for the game play and the game design. In collaboration the elements of self-directed learning were integrated into the game.

Self-Directed Learning

In self-directed learning, the heart of all definitions is that the control over all educational decisions is in the hands of the learner (Percival, 1996). In interaction with the environment, social and physical, the learner decides what he needs to learn and how he can achieve this. It consists of five underlying elements, that interact with each other (Stubbé & Theunissen 2008). (1) Learner control: the learner is in control of his/her own learning process. This means that the learner has control over all educational decisions, but also that the learner can manage his/her own learning. (2) Self-regulating learning strategies: these are a number of skills that support the learner to manage and monitor his/her own learning process. (3) Reflection: this is the combination of self-assessment and self-evaluation on both the performance and the learning process that gives the learner insight in his/her own development. (4) Interaction with the social environment: this can be described as the interaction with others, learners and teacher/coaches, to support learning. (5) Interaction with the physical world: this implies that the learning experience should be set in the 'real world', or a virtual world that is real enough to evoke the real world. The problem, which is the basis for the learning process, should be a 'real-life problem', something the learner could come across in the work situation. Furthermore the learner should be allowed to manipulate the problem and try out possible solutions.

How Did We Do This in the Mayor Game?

- **Learner Control:** In the Mayor game, participants can decide many things for themselves. They can decide whether they answer a dilemma immediately or take a look at the other dilemmas first; whether they ask their Policy team for advice or not; whether they click on and read the extra information or not; and last but not least what they do with the feedback (What do they look at? Do they share their results with others?). They are not obligated to take certain actions and they are not 'punished' if they skip them. The effect of this was that mayors felt invited to play and learn instead of feeling judged and assessed. Their reluctance to training and using a serious game disappeared.
- **Self-Regulating Learning Strategies:** A trainer can invite mayors to set goals for themselves. What would you like to achieve when you play this game? Although all participants play the same scenario, goals can differ: one participant may want to focus on decision making under time pressure,

another may want to know more about a specific scenario and yet another may want to try and be a better father/mother figure for citizens involved. The feedback at the end of the game helps to monitor if these goals have been reached. Trainers are still working in this, it has not been fully implemented at this stage.

- **Reflection:** There are two ways in which the Mayor game stimulates reflection. First of all, the feedback given at the end of the game stimulates the participants to think about the scenario, dilemmas and his or her own decisions. It is like taking a look in the mirror: this is what you did. Precisely because the game does not offer any interpretation or advice, it stimulates participants to think about the meaning of the statistics. This can be used in the group discussion. Furthermore, the game is always played in a training session with a licensed trainer. We know that some people find it difficult to reflect by themselves. This trainer will ask questions that stimulate reflection, and invite participants to share their experience. Because participants do not feel judged, there is hardly any resistance towards the game. The effect is that most participants feel involved and actively participate in the reflection phase.
- **Social Interaction:** One of the reasons to embed the Mayor game in a training session is to ensure that they can interact with others playing the same game. During the reflection phase there is ample opportunity for participants to give help or feedback to others, or ask for it. On top of that there is a trainer who actively supports the participants in their learning process and can be asked for specific advice.
- **Physical Interaction:** All scenarios are derived from the real world. This is what mayors experience in their work setting. Also the graphics of the game were designed to match the 'seriousness' of a mayor's job (well-dressed people, the right uniforms, modest wallpaper). The effect of this is that participants are willing to play the game. 'This is what it is really about!'

Game Design

What is the starting point for a game designer when designing a serious game? On one hand it is very important to get familiar with the domain in which the game will be played, on the other hand there is always the danger of being distracted from designing the game as the expectations and learning goals add up. To avoid this loss of perspective, the AGD framework was used (Hrehovcsik, 2010). The framework asks the game designer simple questions of who, where, when, how, why and what; these in turn provide clues to the kind of needs expected from the design. Using these questions helps to create a quick-scan analysis of the context, content and

Figure 4. Game design model

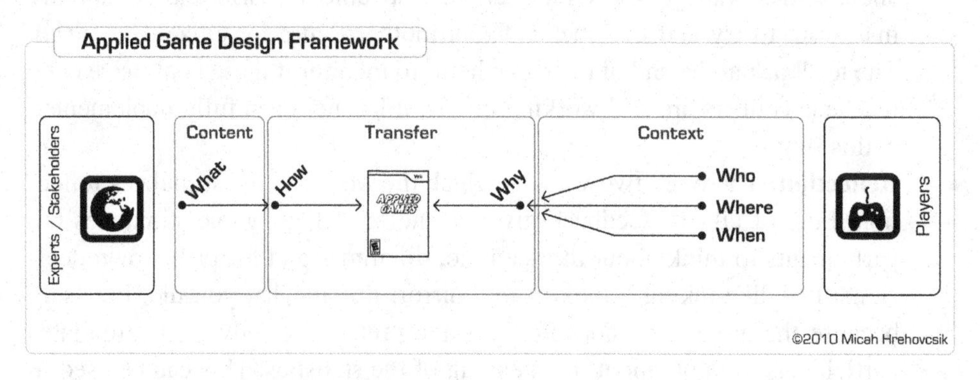

transfer of expectations placed on the game. Working from this perspective, game concepts can be selected on their merits of best fulfilling the quick-scan's expectations. The AGD framework provides the creative space to explore possible alternative directions for the game's design, instead of pre-selecting a game genre or fulfilling a list of requirements.

Making a good game is hard. Making a good serious game is even harder. (Winn, 2009)

This choice for the AGD framework automatically implies that the partners who can answer the questions posed must be involved right from the beginning of the design process. Our integrated approach with domain experts, educational experts and game design experts fulfilled this need. In practice, external domain experts are not always able to explain what is needed to fulfill the end-users' needs. Therefore we used a design and development process in four phases: concept, paper-based design, digital mock-up and digital prototype. This gave our focus group the possibility to try out and experience early in the design process instead of only talking about what the game could be like. This provided valuable feedback. Throughout the process we went through cycles of iteration on the design and development by creation, play-testing, evaluation and ideation. The most significant changes to the design occurred during the first phases and became minimal in the later phases.

The concept process started with brainstorm sessions and resulted in more than six game concepts. These were then synthesized into three game concepts that focused on different aspects of a mayor's role in crisis management. After pitching the three concepts in a focus group meeting, one concept was chosen. We then created a paper-based game of this concept. At the time, using paper-based prototyping for what would become a digital game was innovative for digital serious game design.

Paper-based games had been used as serious games before (Abt, 1970) but using paper-based games as a prototyping tool for digital entertainment games had only been a recent suggestion (Fullerton, 2004).

Using paper-based games to design serious games offers advantages such as:

- Quicker cycles of iterations;
- Play-testing occurs earlier in the development process;
- External experts can better understand the game's systems.

The paper-based game underwent more than eleven iterations, based on the feedback from play-testing within the design team, with the project team and with the focus group. The result provided the foundations for what would become the Mayor game, in which participants are challenged to manage a crisis in 15 minutes.

The next step was to translate the paper-based game into a digital mock-up. This meant we had to decide on workflow and interaction between participant and game. Our focus group members were leading in the choice for the workflow: it should be recognizable for the target population and recognizable as a game. In the end the workflow chosen reflected the mayor's perspective when dealing with a crisis most. The best way to develop a 'bad game' is to have poor interaction. A game

Figure 5. First session with digital mock-up

must 'feel right' and react the way participants expect. Designing good interaction is again an iterative process where testing with participants is essential. The digital mock-up game was so successful that one of our focus group members even thought that it was the final product.

The quality of visual graphics is another common point of critique. Although attention to graphics is enormously important, there are numerous examples where developers aimed at 3D realism according to entertainment industry standards, but only achieved uncanny visuals. For the mock-up we used rudimentary graphics, most of which had been taken from the paper-based game. One of our focus group members thought this would do: since the Mayor game was meant to be a serious game, the visual quality of the graphics did not need attention. Once the final graphics were shown, he was impressed and then admitted that it did make a difference to the gameplay experience. In an effort to determine the tastes of our future players, we attempted to gauge their tastes by running tests. The preliminary test showed that a realistic style would most likely be preferred by our target population. In the next test our focus group were shown 3D graphics with variations of one character: more or less deformed and more or less detail. A rather stylized character type with some detail was chosen. This determined the final style for the game.

When we started the project, serious games had the reputation of being ugly, and not fun to play. Usually they resembled simulations more than games. On top of that the target population are typically non-gamers, serious, computer illiterate and with no time to play. We had the feeling we were embarking on 'Mission Impossible'. And it may have been, except for the collaboration between game design, domain experts and educational experts. During a workshop session with mayors and similar city officials at the end of the project, there were many laughs of surprise and smiles of knowing. Soon after the game session many of the participants began discussing their experience. Our target population, who were supposed to be too serious to play games, had no time for games and was computer illiterate, were now acting like typical gamers. Mission accomplished!

CURRENT CHALLENGES

During and after the project we encountered several challenges. In this paragraph we will share the most important ones.

First of all, the project was part of the GATE research program. The aim was to prove that it was possible to develop a game for strategic decision making that would be played by mayors. We had not really spent any thoughts on actual use in training programs and implementation of the game. After playing the digital game, our focus group thought it would be a pity if the Mayor game was not used in actual training. As the project team consisted of research partners only, we could not achieve this ourselves.

Another challenge we faced during development of the game was security. For mayors it would be absolutely unacceptable if their training results became public, in whatever way. We discussed various methods to ensure this safety. We worked with unique accounts, that do not trace back to names and people. Still, we felt we could not guarantee total security. In the end we chose to remove all results at the end of every day. From an educational point of view, this is not ideal. Comparing your results to those of earlier training sessions, helps to understand and learn. We did explore some technical solutions, but these were rather expensive.

The game has been used for some time now, and we have receive many requests if the concept of the game can be used in different contexts as well. We were invited to try out for a training with a Dutch Ministry, involving ministers and top civil servants in crisis management. The game was used to prepare the civil servants for a large scale exercise. Although they thought the game helped them in their preparation, we received feedback that the current way of providing feedback is not optimal for ministers and other top civil servants. They have even less time for training than mayors. On top of that they are not really convinced they need to be trained at all. Feedback for this target population should, therefore, support individual learning even more.

The mayor game is a dilemma trainer. The whole learning experience is based on the fact that a dilemma cannot be answered by a simple 'yes' or 'no', which stimulates participants to consider various perspectives and angles of the situation. This has led to two different challenges: (1) we have noticed that it is difficult for trainers to use our game and come up with new scenarios themselves. The most important issue here is that it is rather difficult to describe real dilemmas. (2) for the use of the mayor game in other contexts we need to check thoroughly if the learning goals involve real dilemmas or are actually about legislation or protocol.

Apart from this we experienced the dilemma of making choices: our thorough, three-way analysis (domain, educational and game-design) gave us many ideas of how to develop the perfect game. It was not possible to work out all our ideas, because of limitations in budget and time. We did share all ideas with our focus group and together we chose which ones would and which ones would not be integrated in the game. We have not added to the features of the game since the end of the project, but would like to do so.

SOLUTIONS AND RECOMMENDATIONS (500 WOORDEN)

We were able to solve some of the challenges mentioned in the previous paragraph during the projects. Others were taken up immediately after the project ended and some still have to be solved.

In this project we were taken by surprise by the demand for implementation. As a result it took quite some time to find the right partner and come to an agreement. We have learned that prototypes of research projects can be very interesting for implementing partners, and that we should involve them at an early stage of the project.

We still have not found a working solution for the security issue. Some solutions are too expensive, others complicate the use of the game too much. As security is a real risk for mayors and thus for the success of the game, we have decided not to take any chances and to err on the side of caution. All results are still deleted at the end of every day. On top of that we still use the anonymous, randomly generated accounts.

The need for other types of feedback coincides with our dilemma of choosing which elements we could implement in the game at this stage, and which ones had to wait. At a concept level, more ways of providing feedback had been thought out. The remarks of ministers and top civil servants show that we do need to implement these to support individual participants to get the most out of the game experience.

The most difficult element in using the Mayor game in its most effective way has proved to be the development of dilemmas. Experienced trainers of mayors as well as trainers in new contexts find it difficult to identify the dilemmas in a situation. They tend to describe more basic, operational issues. So far we have solved this by staying involved in the development of the scenarios. We hope that the trainers will learn from this cooperation and develop the skills to do it themselves in the future. In the meantime, we guarantee the quality of the scenarios that are used with our game.

On the whole our consortium was very well equipped to solve issues and challenges dealing with the Small game. Together we were able to define the learning goals, assess what would really benefit mayors and design a game that was culturally appropriate. Because of that, the Mayor game is successful in a context where nobody believed it would be. Involving our target population from the beginning has given us unique information and support from within. Seeing the expression on their faces, hearing their thoughts during the game play and talking with them afterwards ensured us that we had the right angle and that we could succeed in our mission. This worked even better than we had anticipated, resulting in a request for implementation of the game in actual training and application in other domains. To incorporate appropriate persuasive elements in a serious game it is of the utmost importance to understand and collaborate with domain, educational and game design experts.

ACKNOWLEDGMENT

We would like to thank the ambassadors and early adopters of the game and training who have given their time and effort to (the results of) this project. Their support was our motivation to continue.

This project was part of the GATE project (Game Research for Training and Entertainment, 2007-2012). The aim of GATE was to develop awareness, insight and technology that can be used both in the entertainment gaming industry and in the serious gaming industry, with a special focus on learning and training experiences.

ICTRegie is a compact, independent organization consisting of a Supervisory Board, an Advisory Council, a director, and a bureau. The Minister of Economic Affairs, and the Minister of Education, Culture, and Science bear the political responsibility for ICTRegie. The organization is supported by the Netherlands Organization for Scientific Research (NWO) and SenterNovem. (GATE was funded by ICTregie)

REFERENCES

Abt, C. C. (1970). *Serious games*. New York, NY: The Viking Press.

Blumberg, F. C., & Ismailer, S. S. (2009). What do children learn from playing digital games? In *Serious games – Mechanisms and effects*. New York: Routledge, Taylor and Francis.

Boin, R. A., 't Hart, P., Stern, E., & Sundelius, B. (2005). *The politics of crisis management: Public leadership under pressure*. Cambridge, UK: Cambridge University Press. doi:10.1017/CBO9780511490880

Bronzwaer, S. (2011, August 11). Extreem-rechts wil BBQ, wat nu?. *NRC Handelsblad*.

Fullerton, T., Swain, C., & Hoffman, S. (2004). *Game design workshop: Designing, prototyping, & playtesting games*. San Francisco, CA: CMP Books.

Hrehovcsik, M. (2010). *An applied game design framework: Prioritizing game design in serious game development* [Web log message]. Retrieved September 09, 2013, from http://gamedesigntools.blogspot.kr/2013/05/2cat-framework-for-applied-game-design.html

Percival, A. (1996). Invited reaction: An adult educator responds. *Human Resource Development Quarterly*, *7*(2), 131–139. doi:10.1002/hrdq.3920070204

Schell, J. (2008). *The art of game design – A book of lenses.* Boca Raton, FL: CRC Press Taylor & Francis Group.

Stubbé, H. E., & Theunissen, N. C. M. (2008, June). Self-directed adult learning in a ubiquitous learning environment: A meta-review. In *Proceedings - Special Track on Technology Support for Self-Organised Learners during 4th EduMedia Conference 2008 Self-Organised Learning in the Interactive Web - A Change in Learning Culture?* Salzburg, Austria: EduMedia.

Verstegen, D.M.L., & Hulst van der, A.H. (2000, November). *Standardized development of a needs statement for advanced training means.* Paper presented at the Interservice/Industry Training, Simulation and Education Conference (I/ITSEC). Orlando, FL.

Winn, B. M. (2009). The design, play, and experience framework. In R. E. Ferdig (Ed.), *Handbook of research on effective electronic gaming in education* (pp. 1010–1024). Hershey, PA: IGI Global.

KEY TERMS AND DEFINITIONS

Crisis Management: Is the process used by an organization to deal with a major event that threatens to harm the organization, its stakeholders, or the general public.

Decision Making: Can be regarded as the cognitive process resulting in the selection of a belief and/or a course of action among several alternative possibilities.

Reflection (Reflective Learning): is the process of internally examining and exploring an issue of concern, triggered by an experience, which creates and clarifies meaning in terms of self, and which results in a changed conceptual perspective.

Self-Directed Learning: The learner takes responsibility for his or her own learning and collaborates with others, as an active participant in his or her learning process.

Strategic Decision Making: Is the process regarding long-term decisions, which in a crisis situation is usually 12 hours and longer.

Chapter 4
Six Factors That Determine the Conceptualization of Persuasive Strategies for Advergames:
The Case Study of "Tem de Tank"

Teresa de la Hera Conde-Pumpido
Utrecht University, The Netherlands

EXECUTIVE SUMMARY

In this chapter, I define six factors that determine the conceptualization of persuasive strategies for advergames. Advergames are understood here as "digital games specifically designed for a brand with the aim of conveying an advertising message" (De la Hera Conde-Pumpido, In Press). These six factors have been used for the analysis of the advergame Tem de Tank (DDB Amsterdam & Flavour, 2010), which was launched in 2010 by Volkswagen to introduce the Volkswagen Polo BlueMotion. The reason for selecting this game as a case study for this chapter is that, although the advergame's goals were properly defined, the game contains, in my opinion, a series of problems in terms of persuasion. Therefore, this game is a perfect case study to exemplify how the factors presented here can be useful to identify problems in the persuasive strategy of an advergame.

DOI: 10.4018/978-1-4666-6206-3.ch004

Copyright ©2014, IGI Global. Copying or distributing in print or electronic forms without written permission of IGI Global is prohibited.

BACKGROUND

The case study presented in this chapter is part of the PhD research project that I have conducted as a member of the Center for the Study of Digital Games and Play at Utrecht University (The Netherlands). The Center for the Study of Digital Games (GAP) is focused on the examination of digital games as well as the role of play in our contemporary culture. Studying these games, the way they are played, as well as the culture that formed around them, the GAP provides insight in past and contemporary media use.

In my PhD thesis "Persuasive Structures in Advergames", that aims to broaden the understanding of how advertising messages can be conveyed through digital games, I have proposed a theoretical model for the study of persuasive structures in advergames that addresses one main questions: (1) how can an advertising message be conveyed through a digital game and (2) the factors that determine the conceptualization of advertising strategies for advergames. This chapter applies the part of the theoretical model of my PhD thesis that answers to the second question. This part consists of six factors that determine persuasive communication through advergames.

Due to the interactive nature of digital games, the factors that determine the conceptualization of advertising strategies for advergames are related not only to the way the advertising message should be conveyed but also to how the game motivates players to play it and how it engages them to keep playing. I have thus identified six factors that should influence advertisers' decisions: advergames' objectives, products in advergames' integration, advergames' target, advergames' visibility, advergames' credibility and advergames' playability. In the following sections I explain how these factors influence the process of decision-making.

SETTING THE STAGE

The theoretical model presented in this chapter has been applied for a content analysis of the advergame *Tem de Tank* (DDB Amsterdam & Flavour, 2010). I have also conducted an in-depth interview with Jain van Nigtevegt, Creative Director at Flavour, the game company that has developed the advergame. In this interview I have collected specific information about the objectives of the advergame, the conditions of the development process and the difficulties they have encounter within it. In this interview, I have also discussed the results of the analysis of the advergame, which has served to adjust some of my appreciations about the content, and therefore the results of the analysis.

Based on the results of the analysis together with the feedback obtained during the interview, I have proposed a series of alternatives and solutions to improve the quality of the game in terms of persuasion. The solutions that I propose here would have not supposed changes in the budget of the advergame or the time dedicated for its development, because the objective was to propose solutions that could be implemented taking into considerations the constrictions of the real case.

CASE DESCRIPTION

The Game

The game selected as game as a case study for this chapter is *Tem de Tank*[1] (DDB Amsterdam & Flavour, 2010), launched in 2010 by Volkswagen as part of a 360 degree integrated campaign to introduce the Volkswagen Polo BlueMotion. The Polo BlueMotion was designed to be environmental friendly and was Volkswagen's response to customers' beliefs that the most important issue for the auto industry was tackling environmental issues. The car had lower fuel consumption and CO2 emissions. The Polo BlueMotion was the first step of the brand to become the "greenest automaker globally" (Volkswagen Group, 2011, p. 5-8).

As part of the campaign to introduce this new version of the Polo, Volkswagen launched *Tem de Tank* in Holland aiming to make Dutch players aware about the energy efficiency of the car and also to create awareness of the influence drivers have on fuel consumption. In order to convey the advertising message, the designers created a game in which the player's goal was to drive a Volkswagen Polo BlueMotion from Amsterdam to Milan using only one tank of fuel and arriving to the final destination with as little combustible material left as possible. In order to do this, the player needed to make a series of decisions that influenced the fuel consumption of the car.

At the beginning of the game, players needed to decide upon the number of passengers in the car and the number of pieces of luggage. Players had the possibility to login with their Hyves[2] account and to choose a maximum of four friends to become passengers on the trip. Furthermore, during the trip, they were able to collect or to get rid of passengers or luggage. These features were designed to make them aware about the influence the weight transported has in the consumption of fuel.

Besides this, players could collect weather icons that resulted in changes in the meteorology, also influencing fuel consumption. During the trip players could also control other features that had an influence on fuel consumption: they could decide whether to use the air conditioning, whether to open or close the windows, and about using the cruise control. Furthermore, at certain points of the journey, players were encouraged to choose between two different routes. The decision made about the route had also consequences for fuel consumption.

During the trip, players were able to check the amount of fuel left in the tank, and use that information to make decisions to try to arrive to Milan with as little fuel left in the tank as possible. The game was available to be played during one month, and after that month, the three players with the best performance were invited to participate in a real event, in which they had to drive a real Polo BlueMotion from Amsterdam to Milan with the same objective. The one that arrived to Milan using only one tank of fuel, and having the least fuel left won a Polo BlueMotion.

Six Factors that Determine the Conceptualization of Persuasive Strategies for Advergames

The game *Tem de Tank* is analyzed in this chapter attending to six factors that determine the conceptualization of persuasive strategies for advergames. As stated before, advergames are understood here as "digital games specifically designed for a brand with the aim of conveying an advertising message" (De la Hera Conde-Pumpido, In Press, p. 40). This relation of advergames with advertising purposes makes them differ from other type of persuasive games such as political games or educational games. It is important to underline that advertising messages are usually unwanted communication (Messaris, 1997). Therefore, in order to avoid players' resistance, the persuasive intentions of an advergame should not be as obvious as the persuasive intentions of an educational game, for example. Furthermore, due to the interactive nature of digital games, the factors that determine persuasive communication within advergames are not only related with the way the advertising message should be conveyed, but also with how the game motivates players to play it and how it engages them to keep playing.

Taking the above into consideration, I have identified six factors that should influence advergames' designers' decisions in order to ensure the efficiency of the game in terms of persuasion. These six factors are: advergames' objectives, integration of products in advergames, advergames' target, advergames' visibility, advergames' credibility and advergames' playability.

The first factor that should be considered when conceptualizing persuasive strategies for advergames are the objectives of the creative strategy. When I say the objectives of the creative strategy, I not only mean the advertising objectives, but also the game objectives. As previously stated, advergames are a form of advertising that because of its interactive nature requires from active players, who voluntarily approach the game and play it. Therefore, the objectives of a creative strategy for advergames should take into consideration not only how to convey the advertising message, but also how to motivate players to play the game and how to engage them to keep playing.

The second factor to consider when conceptualizing a persuasive strategy for an advergame is how to integrate advertised products within the game. The characteristics of the product and the strategy chosen to advertise it should guide the decision of how to represent the product within the advergame, and in which manner the player is going to be able to interact with it. In their research report on advergames, Chen and Ringel (2001) identified three levels of product-game integration: associative, illustrative and demonstrative. Associative advergames corresponds to those in which the product advertised is not present at all, therefore those that present the lowest level of product-game integration. At the other extreme, demonstrative advergames correspond to those that are simulations of products or services, and therefore, those that present the highest level of product-game integration. Between both extremes, illustrative advergames are those that present the product but are not a simulation of it.

The third factor that can influence advergames' persuasive strategy is the target of the advertising campaign. Three are the strands of advergames' target that are important in the process of conceptualization of creative strategies: (1) their demographic, psychographic, and geographic characteristics; (2) their attitude toward the game; and (3) their attitude toward the advertising message.

The forth factor that can influence advergames' persuasive strategy is visibility. The goal of advergames' visibility is to make players aware of the existence of the advergame, which is the first step to make players want to play it. Advergames' visibility depends on where the advergame is placed, and when it is available to be played. The evolution of technologies has enabled the proliferation of different types of advergames that can be played in multiple platforms, such as mobile advergames, online advergames or console advergames; and which are also placed in many different contexts, such as game portals, app stores, microsites or banners.

The fifth factor that can influence advergames' persuasive strategy is credibility. Advergames' credibility is defined as the extent to which the player perceives claims made about the product or brand within it to be truthful and believable (MacKenzie & Lutz, 1989). Advergames' credibility is useful to avoid players' disagreement (Messaris, 1997) and it is specially important when an advergame calls for a substantial change in consumers' behavior (O'Keefe, 1990). The credibility of the claims made within an advergame is a multidimensional construct that depends (1) on the perceived truthfulness or honesty they already have about the brand, (2) on advertising credibility, namely players' perceptions of the truthfulness and believability of advertising in general and (3) on the perceived discrepancy players can have of its claims (MacKenzie & Lutz, 1989).

The sixth and last factor that can influence advergames' persuasive strategy is playability. What makes players to start playing, keep playing or quit playing advergames is related with all the experiences they feel when interacting with the game system. The term playability has been defined to identify and analyze all the attributes of digital games that interfere in this process. The term playability is related to the degree to which a game is fun to play and is usable, with an emphasis on the interaction style and plot-quality of the game (González Sánchez, Gutiérrez Vela, Montero Simarro, & Padilla-Zea, 2012). Advergames' playability can be measured by the "degree to which specific users can achieve specific goals with effectiveness, efficiency and, especially, satisfaction and fun in a playable context of use" (González Sánchez et al., 2012, p 1037). The attributes of digital games that influence on advergames' playability are: motivation, learnability, memorability, efficiency, incorporation and satisfaction.

ANALYSIS AND RESULTS

Taking into consideration the six factors explained in the previous section, I have conducted and in-depth content analysis of the advergame *Tem de Tank*. In this analysis I have paid attention to the 18 variables included in the table below with the objective of identifying problems in the persuasive strategy of the game and proposing solutions attending to the objectives of the campaign (see Table 1). In this section I present the most salient results of this analysis.

Table 1. Variables for the analysis of the persuasive strategy

Advergame's Objectives	Product Integration	Advergame's Target
• Advertising Goals • Game Goals	• Associative Advergame • Demonstrative Advergame • Illustrative Advergame	• Demographic/ Psychographic/ Geographic Characteristics • Attitude towards the Game • Attitude towards the Advertising Message
Advergame's Visibility	**Advergame's Credibility**	**Advergame's Playability**
■ Where is it Placed? ■ When is it Placed? ■ Complementary actions	• Perceived Truthfulness • Perceived Discrepancies	• Motivation • Learnability • Memorability • Incorporation • Satisfaction

Tem de Tank's Advergame's Objectives

As explained above, on the one hand the advertising goals of *Tem de Tank* were to make players aware about the energy efficiency of the car and also to create awareness of the influence drivers have on fuel consumption. On the other hand, the game goals were to drive a Volkswagen Polo BlueMotion from Amsterdam to Milan using only one tank of gasoline and arriving to the final destination with as little fuel left as possible.

In this case, designers decided to create an advergame in which the advertising goals and the game goals overlap. The first advertising goal, which is to make players aware about the energy efficiency of the car, overlaps with the first game goal, that is to arrive to Milan using only one tank of fuel. Similarly the second advertising goal, which is to create awareness of the influence drivers have on fuel consumption, overlaps with the second game goal, which is to arrive to the final destination with as little fuel left in the tank as possible. Therefore, in order to succeed in the game players need to understand the advertising message.

However, the selection of the advertising goals might be distant to the campaign's objectives. As stated before, the BlueMotion line of Volkswagen was launched with the aim of making VW the greenest automaker globally. Volkswagen's objective is to communicate that BlueMotion cars are environmentally friendly, and the advertising goals of the game are focused on communicating that the Polo is a low fuel consumption car. Yet I cannot find any claim in the game that using less fuel is good for the environment. Even though there is an implicit relation between the low use of combustible fuel and respect for the environment, the latter is not the focus of the advertising message conveyed throughout the advergame. Nevertheless, regardless of whether there is a distance between the campaign's goals and the advergame's goals, the advergame's goals are still related with the benefits of the car, and are coherent with the brand and the product advertised.

Tem de Tank's Product Integration

In this section, I analyze how the Volkswagen Polo BlueMotion was integrated in the game. As stated before, one of the goals of the game is to drive a Volkswagen Polo BlueMotion from Amsterdam to Milan. In order to do this, the player has to control a photographic representation of the vehicle along a graphic representation of the road between Amsterdam and Milan. To drive the car, the player needs to use the arrow keys of the keyboard which allow him to accelerate or to turn left or right. Furthermore, as stated above, using two selectors on the interface players can also decide to use the air conditioning or to open the windows. Besides this, by pressing the space bar players can activate the cruise control to control the speed of the vehicle. Despite all these features, the behavior of the car within the game world is not realistic. Therefore, the player is not looking at a simulation of the car.

It follows that in this case the advergame is in between an illustrative advergame and a demonstrative advergame, because although the car is integrated in the game in its natural context, a road, the game does not allow players to interact with it in the way they would do it in real life. Furthermore, the game supposedly provides a realistic reference of the consumption of fuel in relation to the mileage, the use of the air conditioning, the use of the windows, the number of passengers, the kilograms of luggage and the weather. Therefore, there is an association that can be established between the way the player has to interact with the car in the game and the experience of fuel consumption in real life.

On the basis of the above, I claim that in this case the game is in between an illustrative and a demonstrative advergame, and it takes advantage of both forms of advertising. Demonstrative advertising provides direct information about the nature of a product. Example of demonstrative advertising in the game is the fact that the player needs to complete a driving route using only one tank of fuel and making decisions related with that action in real life. The fact that the game is not a realistic simulation of driving the car introduces elements into the game experience that help to convey the advertising message, such as objects that players need to collect on their way. These elements could not be used in a realistic environment.

In summary, in my opinion, the decisions made regardless of the integration of the product in the game are appropriate attending to the characteristics of the product, and help to convey the advertising message. However, as I will explain in depth in the following sections, there are a series of features within the gameplay that might undermine the credibility of the supposedly realistic fuel consumption of the car for the journey between Amsterdam and Milan in the game, which might have had consequences in the strategy followed regarding to the product integration.

Tem de Tank's Advergame's Target

In this section I focus on analyzing the adequacy of the advergame in taking into consideration its target. For this purpose, I analyze three strands of advergames' target that influence how players receive the advertising message conveyed throughout the advergame: (1) their demographic, psychographic, and geographic characteristics; (2) their attitude toward the game; and (3) their attitude toward the advertising message.

According to Jain van Nigtevegt, Creative Director at Flavor, the target of the advergame in this case was young people, especially men, between 25-35 years old with an interest in environmentally friendly and fashionable cars with low taxes (personal communication, January 2013). According to the Interactive Software Federation of Europe, 8 out of 10 Dutch males between 16 and 35 years old are gamers, and 55% of them usually play online games (ISFE, 2012). If compared to other age groups, the target group of the advergame analyzed here is the group that plays more games online (ISFE, 2012). Therefore, the strategy to select a game to advertise the Polo BlueMotion is in my opinion a good one attending to the age and the gender of the target group. It follows that a positive attitude of this target group towards a game as a form of advertising can be expected. However, as I will explain in depth later, I consider that the game presents flaws in terms of playability that might have negative consequences on the final attitude of players towards the advertising message.

Furthermore, even though the designer's target group are young men with an interest in environmentally friendly cars, as I explained before, the game does not include any clear argument related to the benefits of the use of the car for the environment. On the contrary, the advergame is only focused on the low fuel consumption of the car, which is only interesting for those who want to save money on fuel. As stated before, although there is an indirect relationship between low fuel consumption and low CO2-emissions, the latter is not the focus of the advertising message conveyed throughout the advergame.

In sum, although the strategy of using an advergame to convey the advertising message seems to fit well attending to the target group, I consider that there are flaws in the playability of the game that might cause a negative reaction of players towards the advertising message. Furthermore, there is a distance between the definition of the advergame's goals and the definition of the target group.

Tem de Tank's Advergame's Visibility

In this section I analyze the decisions made concerning the visibility of the advergame *Tem de Tank*. Furthermore, I analyze these decisions attending to the advergame's goals. In this case the brand decided to design an online advergame and place it on a microsite under the domain of Volkswagen. Furthermore, the brand opted to launch their game focused on the Dutch market. For that reason, the decision was to launch the game only in Dutch. This decision left out non-Dutch speakers, who were not going to play the game even if they had access to it due to language limitations. Furthermore, the game was only available during the one month of the campaign, and it was removed from Volkswagen's domain when the month was over. Apart from that, the game took advantage of the total control of the content that a microsite provides, and designed a game totally adapted to the advertising goals.

The placement of the advergame on a microsite was however a disadvantage concerning visibility, and necessitated an extra effort to make players know about the existence of the game. In order to overcome this disadvantage, the brand decided to publicize the game on the radio and to make use of viral marketing through Hyves and blogs. In less than one month the game was played 170,000 times by more than 100,000 unique players (Jain van Nigtevegt, Creative Director at Flavor, personal communication, January 2013). Therefore, it can be concluded that in terms of visibility the work done was in consonance with the objectives of the campaign.

Tem de Tank's Advergame's Credibility

In this section I analyze *Tem de Tank* (DDB Amsterdam & Flavour, 2010) in relation to the factors that might have affected its credibility, namely the extent to which players perceived claims made about the Polo BlueMotion within it to be truthful and believable. For that purpose, I examine (1) the perceived truthfulness or honesty players had about the brand, and (2) the perceived discrepancy players might have had of Tem de Tank's claims.

Firstly, Volkswagen is a brand with a high reputation. In 2012 VW was in the top three in the rank of brands with the highest reputation in the world according to the Reputation Institute, the world's leading reputation management consultancy (2012). It follows that players would have been expected to have a high level of trust in Volkswagen. Furthermore, although the BlueMotion was a new version, the Polo, as a model, already existed for a long time and it could be expected that its reputation would be linked to the new version of the model. In that sense, the advergame did not need to make a big effort in generating familiarity and loyalty for the brand.

The game promised players that it is possible to get from Amsterdam to Milan on only one tank of fuel by making efficient use of the Polo BlueMotion. Furthermore, the game supposedly provides a driving experience in which the consumption of fuel is realistic and it is calculated according to the performance of the player. However, there are a series of features in the gameplay that are not realistic and that in my opinion might lower the credibility of the experience. At the beginning of the game, for example, the player needs to decide with how many passengers he wants to travel with to Milan. That decision has influence in the consumption of fuel in the journey. However, during the trip, the player can collect or get rid of luggage and passengers with the intention of arriving with as little fuel as possible to the final destination. This is not realistic at all, because when people do long trips, they do not use to collect or get rid of passengers indiscriminately. Furthermore, whilst the player can collect icons that correspond to weather conditions, thereby influencing fuel consumption, in real life the weather is not something that can be changed by a driver during a trip. Therefore, I consider that the possibility given to the player to control the consumption of fuel with these actions lacks credibility to the claim that is possible to arrive from Amsterdam to Milan with one tank of fuel, meaning it becomes a promise more than a realistic experience.

In sum, although the game starts from a good position in terms of credibility, I claim that some decisions made related to its playability do not help to reinforce it. Small changes that are proposed in the next section could help to overcome this and to reinforce the credibility of the advergame's claims.

Tem de Tank's Advergame's Playability

In this section I evaluate the playability of *Tem de Tank*, namely the degree to which the game is fun to play and is usable. Therefore, I analyze the game attending to the following attributes: motivation, learnability, memorability, efficiency and satisfaction.

Motivation is the capacity of digital games to encourage players to undertake specific actions and continue undertaking them until they are completed (González Sánchez et al., 2012, p. 1040). In terms of motivation *Tem de Tank* (DDB Amsterdam & Flavour, 2010) used an appealing strategy to encourage players to join the game. This was the final prize which, as mentioned before, allowed the three players who performed best during the month the game was available on the microsite to participate in a real event in which the best of the three won a Polo BlueMotion. Undoubtedly, the possibility to win a car was an important motivation for most of the players not only to play the game for the first time, but also to keep playing to improve their performance. Furthermore, the fact that the final goal of the real event was the same as the digital game, might have also served to motivate players to pay attention to the indications given throughout the game to drive efficiently.

However, in my opinion there are some flaws in terms of learnability that might have affected the understanding of the challenge and therefore, players' motivation. According to the researchers in interaction design Yvonne Rogers, Helen Sharp, Jenny Preece, learnability is related to players' capacity to understand what they have to do in the game and understand how to do it (2011, p. 21). The first flaw is that the challenge of the game is not properly presented to the player before playing it. On the first screen of the game the player reads the following message: "With the Polo BlueMotion you can drive with one tank [of fuel] to Milan. Indeed, arriving on empty is a quite a challenge". With this text the player might have understood that the goal of the game is to arrive to Milan without fuel, because that action is presented as a challenge. However, right after the text continues saying: "Tame de tank and get to Milan with as little fuel as possible left". It follows that the player might have been confused about the real objective.

Furthermore, the text explaining the challenge appears in a really small size on the upper left corner of the screen. Therefore, it is possible that some players directly played the game and did not read the challenge of the game. Since *Tem de Tank* (DDB Amsterdam & Flavour, 2010) looks like a racing game, the lack of presentation of the challenge can make players understand that the objective of the game is to arrive to the final destination as soon as possible. And in fact, it is possible to arrive to Milan without consuming the tank of fuel with this objective in mind. It follows that it is possible to finish the game without understanding the objective of the game. Moreover, at the end of the game there is not clear feedback about the result of the performance. This means that is possible to finish the game without understanding its goal and thinking that the performance was correct. This flaw might result in players who did not get the advertising message, even after playing the game more than once.

Furthermore, there are a series of elements in the interface of the game that need to be understood and controlled in order to perform well, but also in order to get the advertising message. The presentation of those elements and what they are useful for is made in a small slideshow on a loading screen before the game starts. This slideshow runs too quickly and the size of the texts of the explanation is too small. Therefore, it is really difficult to read and understand the explanations to control those elements. Again, this might have cause problems to understand how to perform in the game, which also has consequences for the transmission of the advertising message. I would suggest increasing the size of the slideshow and making it run slower to give the player time to read the texts.

The elements that need to be controlled in the game are the gas gauge, the buttons to use the air conditioning and to open the windows, the elements that can be collected during the journey and the "actieradius meter". The "actieradius meter" is an indicator of the performance of the player regarding the efficiency of the car that supposedly helps him to have an idea about if he is wasting too much fuel or if he is not consuming enough fuel according to the challenge. However, there is no explanation in the game that helps to understand how this indicator works. Therefore, it becomes a distraction in the game.

Besides this, in my opinion there are also flaws in the game related to efficiency, namely the way the game supports players in achieving the objectives and reaching the final goal. On the one hand I think that the goal of the game is well chosen, because even if the game is easy to play, the challenge of arriving to the final destination on as little fuel as possible serves to balance the game. Furthermore, the challenge also encourages the player to keep playing once he has finished, with the purpose of improving his result. However, the problems with the presentation of the challenge might have resulted in problems of efficiency.

Moreover, I also consider that there is a flaw related to the way the player has to interact with the game, which also might result in problems of efficiency. As stated before, the player needs to control the car using the arrow keys on his keyboard. However, there are other elements in the game that need to be controlled by the player at the same time that he is driving the car. For example, the player is encouraged to open or close the windows, to use the air-conditioning or to choose between different routes during the journey. Surprisingly, the player has to use the mouse to control those elements in the game. I say surprisingly, because players usually control the arrow keys with the same hand they control the mouse. Therefore, the game is asking the player to lose control of the car if he wants to control the rest of the elements in the interface. In my opinion this is a flaw in terms of efficiency, leading players to focus on control of the car and forgetting about the other options. Consequently, the efficiency of the transmission of the advertising message is reduced, because the player does not experience the consequences on fuel consumption of using these features.

These flaws might also have consequences for the advergame's utility, namely what allows players to carry out the tasks they have to complete in the way they want. Utility is related with players' freedom within the advergame. In this case although players' freedom is limited, there are a series of decisions that they can make and that have consequences in the final outcome. However, if the player does not understand what he has to do in the game, the utility provided can become meaningless.

Finally, regarding satisfaction, namely the pleasure derived from playing the advergame, it is a subjective property really difficult to evaluate in this analysis. Nevertheless, the strategy of using the real challenge as a prize for the three best players might have generated extrinsic pleasures for those players who were able to be on the leaderboard during the month the game was online, and for those who finally were offered to face the challenge. Therefore, in my opinion the strategy of the game was able to generate pleasure that could be experienced after playing the game, and that might have had positive consequences on player's lives which they might have linked to the brand.

SOLUTIONS AND RECOMMENDATIONS

Paying attention to the results of the in-depth content analysis of *Tem de Tank* I conclude that although the challenge of the game is well designed and there is a balance between the difficulty of controlling the game and the difficulty of mastering the game, in my opinion the game presents flaws that might have affected its playability and that might have had consequences in the transmission of the advertising message. In this section I propose a series of solutions for those flaws. The solutions that I propose here would have not supposed changes in the budget of the advergame or the time dedicated for its development.

In the previous section I have identified a flaw related with the visual representation of the "actieradius meter"[3]. As explained before, the "actieradius meter" was designed to serve as an indicator of the performance of the player regarding fuel consumption, supposedly helping him to have an idea if he is wasting too much fuel or if he is not consuming enough fuel according to the challenge. However, the visual representation of this meter makes it difficult for players to decipher the information provided by it. The meter has an arched representation with two extremes. At the two extremes of the meter there are two indicators: one that says "+100", that supposedly warns the player that with the fuel left he can ride 100 kilometers after Milan; and another one that says and "-100", that supposedly means that he is going to run out of fuel 100 kilometers before he arrives to Milan. Players are supposed to keep the pointer of the meter in the middle point, represented in blue, to accomplish the best performance, but it is quite difficult to decipher what is going on when the player is not in the blue zone. The use of the same color, the red, on both sides of the meter, does not allow the player to get quick feedback on what is going on.

A solution for a representation of the "actieradius meter" could be to create a vertical visual design in which the center represents arriving into Milan with as little fuel remaining as possible (see Figure 1). In this representation players could see how the level of fuel goes up in a green color when they drive with too much fuel

Figure 1. Design proposal for the actieradius meter

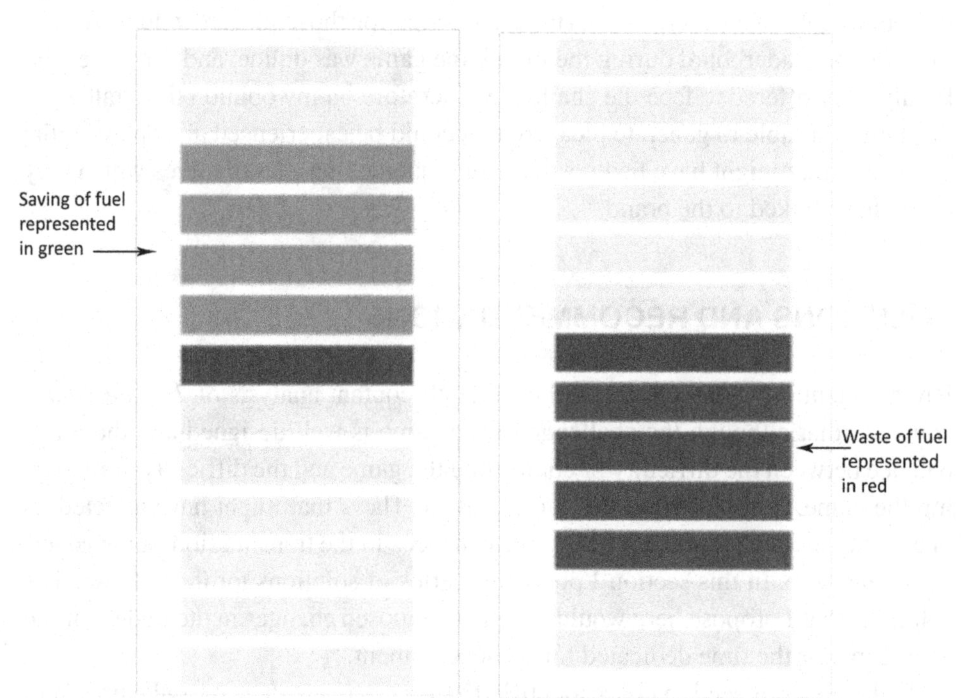

left and the level of fuel goes down into a red color when they are wasting too much fuel. The colors would increase and decrease in a way that when the players sees the red color he understands that he is wasting too much and when he sees the green color he understands that he has a lot of fuel left. This representation would help players to quickly understand the meter just by identification of the colors. At the same time, it also would be helpful to link it with the advertising message, because the green color would mean getting to Milan on only one tank on fuel. This means that even the player would not be in the highscore table, he is able to finish the game.

Moreover, this new representation of the "actieradius meter" would give sense to the color code used in the items the player needs to collect during the journey. Those items are represented in two colors, green and red, green meaning that they are saving fuel when collecting them and red that they are wasting more fuel when collecting them. With this representation players would also understand that collecting green items increases the level of the "actieradius meter" and collecting red items decreases the level of the "actieradius meter". Therefore, the color would be not only linked to the advertising message, but also with the game goal.

Another flaw that I have previously mentioned is that the items that the passenger can collect during the journey do not help to reinforce the message that it is possible to arrive to Milan on one tank of fuel. I have argued that collecting passengers and weather icons is not realistic. The solution I would suggest to solve this problem is to change the type of items the player can collect during the trip. Firstly, since the player can decide at the beginning of the trip how many passengers and suitcases he wants to travel with, I would suggest removing the possibility of collecting these elements during the trip. Instead, I would substitute those elements with other signs that represent the windows and the air conditioning. Therefore, the experience would be more realistic, because switching the air conditioning or and off and opening and closing the windows several times during a journey is realistic and also has consequences on fuel consumption. At the same time, I would remove the actual controllers for the air conditioning and the windows, thus improving playability.

Furthermore, the weather is also something that cannot be controlled by players. For this reason and in order to improve the credibility of the advergame, I would suggest removing the possibility of collecting weather signs during the trip. I have also noticed that although at some points during the journey the roads splits and the player has to decide which way to follow, there are no direct consequences of choosing one route or the other. At least, there is no feedback about the consequences. Taking this into consideration, I would suggest adding signs indicating the weather forecast for the different routes the player can choose during the journey. Even if it is not normal to find weather signs on the road, the forecast for different routes is something that can be checked by drivers. Therefore, this solution is more realistic than the actual one, and could serve to give credibility to the advertising message.

CONCLUSION

After the in-depth content analysis of the advergame *Tem de Tank* I would conclude that most of the flaws identified in the game come from a strategy that establishes a clear relationship between game goals and advertising goals, but that ultimately does not result in a gameplay in which all the elements and the relationships between them attend to both sets of goals at the same time. In my opinion, this might have been the result of thinking about the design of those elements in isolation and not thinking about the possible meanings of the relationships between them.

I consider that the result is an advergame which had the great potential to become really effective in terms of persuasion but that might have not always conveyed the advertising message in the way pretended. When I say that it had great potential to become effective I mean that the starting points were all in favor of the brand. VW's credibility is high, the target is the ideal one to whom to convey an advertising message using a digital game, and the advergame's goals were perfectly designed to convey the advertising message. Furthermore, the challenge of the game of reaching Milan with as little fuel as possible left in the tank made this advergame unique, which was good reason for players to play it instead of another racing game. Besides this, the quality of the visual design of the game might have also aroused positive feelings toward it. Finally, the possibility of winning a real Polo BlueMotion was the perfect motivation to attract players to play the game.

However, I have identified some flaws in the design that in my opinion have resulted in a game that might not have always achieved the purpose of conveying to players the influences drivers' performance has on fuel consumption. I also consider that there is a chance that some players might not even have realized about the energy efficiency of the car. The advergame might have been beneficial for VW because it might have served to increase recall and some players might have had a positive memory of the game thanks to the real challenge which came out of it. However, in my opinion advergames should work without any prize promised after the game, because having fun playing them should be enough reward for players. I consider that if players play the advergame unwillingly with the only purpose of getting a final prize, the persuasive potential of digital games is being disregarded.

REFERENCES

Chen, J., & Ringel, M. (2001). *Can advergaming be the future of interactive advertising?* Retrieved from http://www.kpe.com/ourwork/pdf/advergaming.pdf

DDB Amsterdam, & Flavour (Producer). (2010). *Tem de tank*. Retrieved from http://www.flavour.nl/volkswagen/

De la Hera Conde-Pumpido, T. (In Press). *Persuasive structures in advergames.* Utrecht, The Netherlands: Utrecht University.

González Sánchez, J. L., Gutiérrez Vela, F. L., Montero Simarro, F., & Padilla-Zea, N. (2012). Playability: Analysing user experience in video games. *Behaviour & Information Technology, 31*(10), 1033–1054. doi:10.1080/0144929X.2012.710648

ISFE. (2012). *Videogames in Europe: Consumer study*. The Media, Condent and Technology Research Specialists.

MacKenzie, S. B., & Lutz, R. J. (1989). An empirical examination of the structural antecedents of attitude toward the ad in an advertising pretesting context. *Journal of Marketing, 53*(2), 48–65. doi:10.2307/1251413

Messaris, P. (1997). *Visual persuasion*. London: SAGE Publications.

O'Keefe, D. J. (1990). *Persuasion: Theory and research*. Newbury Park, CA: Sage.

Reputation Institute. (2012). *2011 global rep trak: Results and report*. Reputation Institute.

Rogers, Y., Sharp, H., & Preece, J. (2011). *Interaction design: Beyond human-computer interaction*. Chichester, UK: John Wiley & Sons.

Volkswagen Group. (2011). *Green machine*. Author.

ADDITIONAL READING

Armstrong, S. J. (2010). *Persuasive Advertising*. New York: Palgrave MacMillan. doi:10.1057/9780230285804

Bogost, I. (2007). *Persuasive Games: The Expressive Power of Videogames*. Cambridge: MIT.

Cauberghe, V., & De Pelsmacker, P. (2010). Advergames. The Impact of Brand Prominence and Game Repetition on Brand Responses. *Journal of Advertising, 39*(1), 5–18. doi:10.2753/JOA0091-3367390101

Challapalli, S. (2008). Brands take to mobile advergaming. *Business Line*. http://www.thehindubusinessline.com/todays-paper/tp-marketing/article1612184.ece

Chang, Y., Yan, J., Zhang, J., & Luo, J. (2010). Online in-game Advertising Effect: Examining the Influence of a Match between Games and Advertising. *Journal of Interactive Advertising, 11*(1), 63–73. doi:10.1080/15252019.2010.10722178

Chen, J., & Ringel, M. (2001). Can Advergaming be the Future of Interactive Advertising? http://www.kpe.com/ourwork/pdf/advergaming.pdf

Consalvo, M., & Dutton, N. (2006). Game Analysis: Developing a Methodological Toolkit for the Qualitative Study of Games. *Game Studies, 6*(1).

Dal Cin, S., Zanna, M. P., & Fong, G. T. (2004). Narrative Persuasion and Overcoming Resistance. In E. S. Knowels, & J. A. Linn (Eds.), *Resistance and Persuasion* (pp. 175–192). New York: Psychology Press.

Davidson, D. (2003). Games and Rhetoric: a rhetorical look at gameplay. *The Ivory Tower, August*(2003). http://www.igda.og/coluns/ivorytower/ivory_Aug03.php

De la Hera, T. (2013). *A Conceptual Model for the Study of Persuasive Games*. Paper presented at the DiGRA, Atlanta.

Deal, D. (2005). *The Abilty of Branded Online Games to Build Brand Equity: An Exploratory Study*. Paper presented at the DiGRA 2005 Conference: Changing Views-Worlds in Play, Vancouver.

Eliseev, S. (2010). Advertising and Games: pros and cons. *All News in One*. http://yqyq.net/

Entertainment Software Association. (2011). Entertainment Software In-Game Advertising.

Eriksen, E. H., & Abdymomunov, A. (2010). Angry Birds will be bigger than Mickey Mouse and Mario. Is there a success formula for apps? *Entrepreneurship Review*. http://miter.mit.edu/article/angry-birds-will-be-bigger-mickey-mouse-and-mario-there-success-formula-apps

Ferrari, S. (2010). *The Judgment of Procedural Rhetoric. (Master of Digital Media)*. Georgia: Georgia Institute of Technology.

Frasca, G. (2007). *Play the message. Play, Game and Videogame Rhetoric. (Ph.D.)*. Copenhagen: IT University Copenhagen.

Glass, Z. (2007). The Effectiveness of Product Placement in Video Games. *Journal of Interactive Advertising*, *8*(1), 22–32. doi:10.1080/15252019.2007.10722134

Grace, L. D. (2012). *A Topographical Study of Persuasive Play in Digital Games*. Paper presented at the Proceeding of the 16th International Academic MindTrek Conference, Tampere, Finland. doi:10.1145/2393132.2393149

Grundey, D. (2008). Experiential Marketing vs. Traditional Marketing: Creating Rational and Emotional Liaisons with Consumers. *The Romanian Economic Journal, 29*(3).

Heide, J., & Nørholm Just, S. (2009). Playful persuasion. The Rhetorical Potential of Advergames. *Nordicom Review*, *30*(2), 53–68.

Juul, J. (2010). *A Casual Revolution: Reinventing Video Games and their Players*. Cambridge, Mass.: MIT Press.

Lee, K. M., Seung-Jin, A., Park, N., & Kang, S. (2009). Effect of Narrative on the Feelings of Presence in Computer-Game Playing. http://www.allacademic.com/meta/p13584_index.html

Lombard, M., & Synder-Duch, J. (2001). Interactive Advertising and Presence: A Framework. *Journal of Interactive Advertising*, *1*(2), 56–65. doi:10.1080/15252019.2001.10722051

McGonigal, J. (2011). *Reality is broken*. New York.

MCInsight. (2008). *Advertising and Video Games* (M. Contacts, Ed.).

KEY TERMS AND DEFINITIONS

Advergame: Digital games specifically designed for a brand with the aim of conveying an advertising message.

Credibility: The extent to which the player of an advergame perceives claims made about the product or brand within it to be truthful and believable.

Learnability: Players' capacity to understand what they have to do in the game and understand how to do it.

Persuasive Strategy: The creative strategy of an advergame determines what the advertising message will say and how the strategy will be executed.

Playability: The degree to which a game is fun to play and is usable, with an emphasis on the interaction style and plot-quality of the game.

Product Integration: The representation of a product within an advergame, and the manner in which the player is able to interact with it.

Visibility: Is the part of the persuasive strategy of an advergame focused on making players aware of the existence of the advergame.

ENDNOTES

1. To see pictures of the game go to http://www.flavour.nl/#/Projects/VW.
2. Hyves is the most popular social network in Holland.
3. To see pictures of the "actieradius meter" go to http://www.flavour.nl/#/Projects/VW.

Chapter 5
Game-Based Learning as a Promoter for Positive Health Behaviours in Young People

Andrew Sean Wilson
Birmingham City University, UK

EXECUTIVE SUMMARY

Maintaining the healthcare of young people living with long-term medical conditions is dependent upon them acquiring a range of self-care skills. Encouraging them to attain these as well as assessing their competency in them beyond the health-care setting is challenging. The development of educational computer games like Health Heroes, Re-Mission, and Sparx have been shown to successfully improve self-care, communication, and adherence to medicines in young patients. Therefore, this medium might be an alternative means for delivery of healthcare information. In this chapter, we propose that by encapsulating healthcare processes in Game-Based Learning (GBL) either by computer games or by applying the principles of gamification, a more fun, structured, and objective process would be created, one to which young people can relate. The framework we suggest will provide doctors with an insight into how GBL could be used positively in a healthcare setting as well as provide a basis for application to other disciplines where knowledge and skill acquisition can be challenging.

DOI: 10.4018/978-1-4666-6206-3.ch005

Copyright ©2014, IGI Global. Copying or distributing in print or electronic forms without written permission of IGI Global is prohibited.

BACKGROUND

Many people are now expected to live longer than in previous generations. With this increase in life expectancy the incidence of age-associated long term illness is also predicted to rise (WHO, 2013). Nearly 133 million Americans (Bodenheimer, Chen, & Bennett, 2009) and 15 million United Kingdom (UK) citizens (DoH, 2011) live with a long term medical condition. Two thirds of all deaths in the USA are attributable to conditions such as cancer, diabetes, lung disease, heart disease or stroke ("Tackling the burden of chronic diseases in the USA", 2009). In the UK mental and behavioural disorders (including substance abuse) and musculoskeletal conditions are some of the major causes of long term disability. Long term medical conditions are not restricted to the elderly; young people may also experience conditions such as cancer, asthma, diabetes, cystic fibrosis, heart disease or (juvenile) arthritis. Approximately 14% of young people in the UK live with a long term medical condition (Hagell, Coleman, & Brooks, 2013).

As young people grow older patterns of health promoting as well as health risk behaviours are established which can be maintained throughout their life (Sawyer, Drew, Yeo, & Britto, 2007). Activities such as unhealthy diet, drugs and alcohol misuse, smoking, physical inactivity, and unsafe sexual practices can lead to poor health outcomes (Murray et al., 2013). If a young person is already living with a long term illness engaging in health risk behaviours can further complicate their condition. Smoking can accelerate the development of heart disease in diabetics (Grundy et al., 1999); alcohol can enhance the toxicity of medications (Nash, Britto, Lovell, Passo, & Rosenthal, 1998) whereas drug therapies can affect fertility and can potentially cause birth defects (Janssen & Genta, 2000). Research has shown that health risk behaviours can be more common in young people with long term conditions with greater potential for adverse health outcomes in them (Sawyer at al., 2007; Suris, Michaud, Akre, & Sawyer, 2008; Philpott, 2011).

Care for a young person with a long term medical condition is on-going and ideally will adapt to their needs as they grow older. If they are diagnosed during childhood their care will be in an environment which is focused towards theirs and their family's needs. Their condition may persist over time and even follow them into their adult life. This means that they will eventually have to transfer to the adult healthcare system which is very different from that which they were used to as a child. In the UK this normally occurs between 16 and 18 years of age but is more dependent upon local policy rather than the young person's confidence in being able to cope in the new setting. Transfer can be a cause of anxiety to both the young person and their family as they will have to experience a new healthcare environment. If the transfer is not managed appropriately there is an increased chance that the young person will "drop out" of the healthcare system (Wacker et al., 2005;

Yeung, Kay, Roosevelt, Brandon, & Yetman, 2008), risking poorer health outcomes (Annunziato et al., 2007) and subsequently be re-admitted to hospital (Yeung et al., 2008; Nakhla, Daneman, To, Paradis, & Guttmann, 2009).

In order to maintain the continuity of a young person's care careful and purposeful planned preparation for transfer is important (McDonagh, 2007), this is known as transitional care. It addresses the medical, psychosocial, educational and vocational needs of the young person in an age and developmentally appropriate manner. As the young person matures the responsibilities for their care moves away from the parent requiring the young person to acquire a range of knowledge and skills that are important for them to function independently in the adult healthcare system. These include understanding of health and condition specific issues, self-management skills, effective information-seeking skills, managing psychological and general health, effectively utilising the healthcare available to them, coping with social issues, maintaining their education and vocation aspirations as well being able to live independently.

Adolescence (10-19 years of age (WHO, 2001)) is considered to be one of the most sensitive developmental periods next to early childhood (Viner et al., 2012) Therefore this is a time when positive health behaviours can be influenced and where preventative healthcare strategies should be focused (Catalano et al., 2012). Educational resources to support the needs of young people's healthcare are very important but many patients and their families find it difficult to get the information they need and it is not always in appropriate style for them (Shaw, Southwood, & McDonagh, 2004). There is interest in how information can be presented in a youth friendly way and how technology can be used to support this (Stinson et al., 2008). It has been proposed that new forms of communication which are used and enjoyed by young people such as social marketing, creative design and information technology should be taken advantage of to assist them with their varying developmental and disease-related needs (Sawyer et al. 2007; Sawyer et al., 2012). The Internet has established itself as a major source of health information and a medium that young people often use in order to answer their queries (Suris, Akré, Berchtold, Bélanger, & Michaud, 2010). Their involvement has seen the development of dedicated youth-centred online healthcare programmes. These have included interactive activities, self-monitoring and reflection, youth-to-youth information sharing and social support, and accurate, accessible and developmentally targeted health specific information (Scal, Garwick, & Horvath, 2010).

In 2012 a report by the entertainment software association (ESA, 2013) showed that 32% of Americans who played video games were less than 18 years of age and in Europe 26% of gamers were under 24 years of age (Interactive Software Federation of Europe, 2012). Computer games therefore form an important part of young people's lives and could be used as a way of engaging them in learning how to take care of themselves.

The use of computer games as an educational medium is of increasing interest because their inherent characteristics give players unique personal perspectives and experiences. The mechanics that define how the game works and is played, for example use of goals, challenges, rewards, competition and feedback, can be used to positively reinforce learning. Therefore computer games offer the potential for creating educational environments that are fun and can facilitate active participation in learning as the players can construct knowledge at their own pace and under their control. However negative connotations have been associated with playing computer games for example increasing sedentary behaviour and their potential to increase antisocial and aggressive behaviours in young people (Janssen, Boyce, & Pickett, 2012; Ferguson, Miguel, Garza, & Jerabeck, 2012).

Despite this computer games have been used to educate young people about healthcare issues but there is very little work in its application to issues that are associated with transitional care. In this chapter we will discuss how I (as the technologist) and my colleague Dr. Janet McDonagh (a clinician and specialist in adolescent care) from Birmingham Children's Hospital (BCH) in the UK have been exploring how game based learning (GBL) could be used to provide important healthcare information to young people.

In this chapter we will discuss the research we have been conducting in how GBL could be used in healthcare for young people. This includes a review of some of the computer games which have been designed to help young people with long term medical conditions. We will also discuss the importance of using appropriate assessments as a way of measuring health outcomes as well as how game mechanics could be used to support the acquisition of self-care and self-management skills. We will also highlight the issues and concerns by the healthcare services associated with young people using technology and computer games.

SETTING THE STAGE

Young people may be diagnosed with a long term medical condition early in their life. As they grow older their healthcare needs will change. Gauging when a young person is ready to transfer their healthcare depends upon the young person's self-belief, involvement of their parents and the healthcare team. This approach is subjective which has resulted in the development of several assessment tools aimed at helping measure the range of skills which have been identified as being important for successful transfer of care (Sawicki et al., 2011; Ferris et al., 2012). Many of the items in these assessments are common to one another for example understanding the condition and its treatment, how to self-care, how to communicate with the healthcare team and understanding how to use support services. However these assessments may be disease specific and have aspects that are more relevant to their country of origin for example dealing with medical insurance and the financial aspects of healthcare.

A series of healthcare plans and transitional readiness checklists which reflect the developmental stages of adolescence, ("Getting Ready" – early adolescence, "Moving Along" – mid adolescence and "Moving Up" – late adolescence) have been developed by BCH's Adolescent Rheumatology Team and are more appropriate to the needs of young people in the UK. These were initially developed as part of a national multi-centre transitional care research project (McDonagh, Southwood, & Shaw, 2006) and although they are not formally validated (a measure of dependability) they continue to be used with all young people attending the adolescent rheumatology service from age 11 years irrespective of condition. The checklists have evolved in response to the young people's requests which include the need to include statements regarding pain, fatigue and anger management. The checklists have eight domains which encompass knowledge, self-advocacy (speaking up for oneself), transferring to the adult healthcare system, health and lifestyle, activities of daily living, school and vocation, leisure as well as managing emotions. Figure 1 provides examples of the early and late adolescent checklist which the young person completes based upon their confidence in being able to perform the given task. The level of required knowledge and understanding that the plans assess increases as the young person matures.

Complementing the patient checklists is a doctor's checklist (Figure 2). This records the young person's progress in acquiring these skills as they grow older. The views of the young person always take precedence over the views of the clinical team.

CASE DESCRIPTION

Review of Existing Healthcare Games for Young People

Our initial research focussed on assessing existing games for young people with long term medical conditions. Websites such as social impact games (http://www.socialimpactgames.com) and games for change (http://www.gamesforchange.org) indicated that several had been produced however there was little evidence that the games had been evaluated for their effectiveness. Subsequently a review of the literature was conducted using Thomson Reuters' Web of Knowledge (http://wok.mimas.ac.uk) and Elsvier's Science Direct (http://www.sciencedirect.com). The search terms used included serious games, computer games for health, computer games and adolescence, computer games and young people. The results indicated that

Figure 1. BCH Adolescent Rheumatology Team transitional readiness checklists for young people. The checklists show a range of tasks within the domains of knowledge, health and lifestyle for early and late adolescence. The items are used to gauge the level of increasing knowledge, understanding and skill acquisition as the young person matures.

Getting Ready!
(Early Adolescent Plan)

Knowledge and Skills	Yes I can do this on my own and don't feel I need any extra advice	I would like some extra advice/ help with this	Date/ discussed/leaflet
KNOWLEDGE			
I can describe my condition			
I know my medication regime – names, doses, how often etc			
I know who's who in the rheumatology team			
HEALTH & LIFESTYLE			
I am able to manage my fatigue (tiredness)			
I am able to manage my pain			
I understand what healthy eating means for both my general health and my condition			
I understand the importance of exercise/activity for both my general health and my condition			
I am aware that my condition can affect how I develop e.g. puberty			
I understand the risks of alcohol and drugs to my health			

Moving Up!
(Late Adolescent Plan)

Knowledge and Skills	Yes I can do this on my own and don't feel I need any extra advice	I would like some extra advice/ help with this	Date/ discussed/leaflet
KNOWLEDGE			
I am confident in my knowledge about my condition and it's therapy			
I understand what is likely to happen with my condition when I am an adult			
I order and collect my repeat prescriptions and book my own appointments			
I call the hospital myself if there is a query about my condition and/or therapy			
HEALTH & LIFESTYLE			
I understand the effect of smoking, drugs or alcohol on my condition and general health			
I know where and how I can access providers of reliable accurate information about sexual health			
I understand the implications of my condition and drug therapy on pregnancy/parenting (if applicable)			

Figure 2. BCH Adolescent Rheumatology Team transitional readiness checklists for doctors. The checklist illustrates four of the domains (knowledge, self-advocacy, transfer to adult care and health and lifestyle).The doctor will record the progress of acquisition of knowledge and skills as the young person develops through early (E), mid (M) and late (L) adolescence.

(K) KNOWLEDGE		E	M	L
1.	Describes condition	☐	☐	☐
2.	Understands effects on body	☐	☐	☐
3.	Knows medicines / regimes	☐	☐	☐
4.	Understands team roles	☐	☐	☐
5.	Understands medical terms	☐	☐	☐
6.	Understands side effects of medicine	☐	☐	☐
7.	Understands prognosis		☐	☐

(S) SELF-ADVOCACY		E	M	L
1.	Asks own questions	☐	☐	☐
2.	Preparation for independent visits	☐	☐	☐
3.	Understanding rights to confidentiality	☐	☐	☐
4.	Independent visit		☐	☐
5.	Preparation for joint injections without GA		☐	☐
6.	Has JI without GA		☐	☐
7.	Self medication / self injection		☐	☐
8.	Calls hospital with queries		☐	☐
9.	Orders repeat prescriptions			☐

(T) TRANSFER TO ADULT CARE		E	M	L
1.	Meaning of transition	☐	☐	☐
2.	Difference between paediatric / adult		☐	☐
3.	Independent GP visits		☐	☐
4.	Transfer plan			☐
5.	Prep visit to adult provider			☐
6.	Able to use other adult services			☐

(H) HEALTH AND LIFESTYLE		E	M	L
1.	Pain management	☐	☐	☐
2.	General exercise	☐	☐	☐
3.	Healthy eating	☐	☐	☐
4.	Dental care	☐	☐	☐
5.	Smoking	☐	☐	☐
6.	Alcohol	☐	☐	☐
7.	Substance use	☐	☐	☐
8.	Puberty	☐	☐	☐
9.	Sleep / fatigue	☐	☐	☐
10.	Sexual health issues		☐	☐

several games had been devised to help young people in a range of health issues and they had been evaluated to assess their effectiveness. The Health Hero (Bronkie the Bronchiasaurus, Captain Novolin, Packy & Marlon and Rex Ronan - Experimental surgeon) series of games were developed in the late 1990's. Bronkie the Bronchiasaurus is an adventure game for young people with asthma. The premise of the game is for players to keep two dinosaur characters (Bronkie and Trakie) in good health by monitoring their breathing, making sure that they take their medicines, correctly use their inhalers and avoid environmental triggers of asthma attacks. In Captain Novolin the player controls the aforementioned hero who must defeat alien invaders (in the form of junk food) and rescue the diabetic mayor in the game. During the game Captain Novolin must eat healthy meals in order to maintain safe blood glucose levels otherwise he may die. Bonus points can be earned by players if they correctly answer questions about diabetes. Packy & Marlon was designed to help young people improve their self-management of diabetes. The adventure game involves playing Packy & Marlon, two diabetic elephants, who must find missing food and diabetic supplies which were scattered through the game by the enemy characters (rats and mice). The players must keep Packy & Marlon's diabetes under control by managing their blood glucose levels, taking their medication (insulin) as well maintaining a balanced diet. Rex Ronan - Experimental surgeon (smoking prevention) involves taking control of a character that has been miniaturised and entered into a human body. By using his weapons (laser scalpel) he must remove the adverse effects of smoking (tar, phlegm, plaque, debris, and precancerous cells) from the patient. During the game the player must answer questions about the side effects of tobacco smoking. Research into the use of these games showed that young people who played them improved their knowledge, self-management, perceived self-efficacy and communication with family and friends. These games were developed for the Super Nintendo Entertainment System (SNES) consoles and Microsoft Windows 95 but are no longer available.

More recent games such as Re-Mission (http://www.re-mission.net) and Re-Mission 2 (http://www.re-mission2.org) have been designed to help young people understand childhood cancers and their treatment (Figure 3).

In Re-Mission the player controls a nanobot character called Roxxi who has been injected into a patient's body. The player must monitor the patient's health while trying to defeat various forms of cancer. The game is split into levels which have information on drug treatments, how they work and the importance of adhering to them. Re-Mission2 is a series of free online games (Nanobot's revenge, Stem Cell Defender, Nano Dropbot, Leukemia, Feeding Frenzy and Special Ops) which deal with a range of issues associated with childhood cancers and their treatment.

Figure 3. Screenshots of the original Re-Mission [A] game and NanoBotsRevenge from Re-Mission2 [B] (©2013 HopeLab. Used with permission)

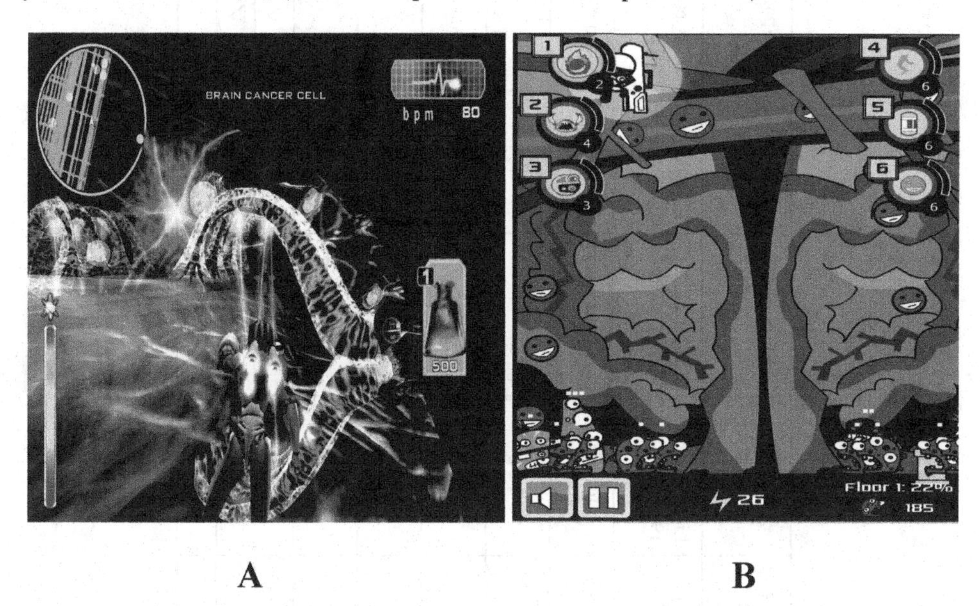

Sparx (http://sparx.org.nz) is a fantasy game designed for cognitive behavioural therapy (a psychological technique aimed at helping people change their behaviours) in order to help young people who are suffering from depression. The 3D game requires the player to take control of an avatar and compete in a series of challenges designed to restore the balance in the game world which is dominated by gloomy negative automatic thoughts (GNATs).

Both Re-Mission and Sparx have been the subject of clinical trials where their effectiveness has been tested with young people who were living with the conditions. Re-Mission has been shown to improve cancer related knowledge, self-efficacy and adherence to drug treatments (Beale, Kato, Marin-Bowling, Guthrie, & Cole, 2007; Kato, Cole, Bradlyn, & Pollock, 2008) whereas Sparx reduced depression, anxiety, hopelessness and improved the quality of life (Merry et al., 2012). Innovative games such as Creep Frontier, a game to help young people with cystic fibrosis, incorporates the use of a spirometer (a tool to measure the volume of air that a person breathes in and out) into the game in order to train the young player how to use it (Bingham, Bates, Thompson-Figueroa, & Lahiri, 2010). Table 1 summarises games that have been produced and formally evaluated with indication of their outcomes.

The Health Hero games were innovative in their approach to educating young people about their health by taking advantage of new and emerging recreational media and applying it to serious issues. The games allowed the player to explore

Table 1. Examples of computer games for healthcare in young people

	Rex Ronan - Experimental Surgeon *	Bronkie the Bronchiasaurus *	Packy & Marlon *	Captain Novolin *	Re-Mission	Creep Frontier	Sparx
Study author(s)	Lieberman & Brown, 1995	Lieberman, 1997	Brown et al., 1997	Lieberman, 1997	Beale et al., 2007; Kato et al., 2008	Bingham et al., 2010	Merry et al., 2012
Aspect of health covered in game	Smoking prevention	Asthma	Diabetes	Diabetes	Cancer	Cystic Fibrosis	Depression
Improved knowledge about subject	✓	✓			✓	✓	
Improved self-efficacy and self-management		✓	✓		✓		
Improved communication skills		✓	✓	✓			
Improved adherence to drug treatments.					✓		
Improved health outcomes							✓
Improved quality of life							✓
Decreased visits to doctor			✓				

* Information about game can be found at http://www.socialimpactgames.com

worlds and interact with them; tested their understanding of their medical condition and provided a way to discuss issues with friends, family and the healthcare team (Lieberman, 1997). The principles (game mechanics) used in them occur in the other games we have review too. The mechanics used include goals, feedback, achievements and rewards, challenges and curiosity. These combine to give the young person incentives to work towards attaining knowledge about their health and self-care skills. They can also measure how well they are attaining the skills and highlights their achievements. As a result the games' designs encourage young people to participate in learning about their healthcare.

The improvements in knowledge, self-efficacy, self-management skills and adherence to treatment (Table 1) form important parts of transitional care. However the games reviewed are often disease specific (asthma, cancer and diabetes), focus on a set of required behaviours for that condition and may target particular age groups. Consequently they cannot be generalised to a range of long term medical conditions. Developing a game for transitional care is further complicated as the behaviours and levels of knowledge and skills required need to take into account the different stages of adolescence and the maturity of the young person.

The challenge of developing a game for healthcare involves designing it so that it is fun to play and engages its audience. It also has to increase the player's knowledge and positively influence their behaviours. Whether these have been achieved depends upon assessing the outcomes. Assessments may be embedded within the game which provides real-time feedback but this subsequently restricts the flexibility of the type of assessment and also limits its modification or adaptation as the assessment is locked into the game. Assessment may also be implemented outside of the game for example in the form of satisfaction surveys or knowledge quizzes. The latter often use pre and a post-test evaluation where the player is asked to complete a quiz prior to playing the game and then they complete another one after playing it. This is used to indicate whether the intervention (the game) has improved the person's understanding of the subject. Ideally the quiz would be repeated at a later date to see if the knowledge is retained over time.

Trying to embed all the requirements for a good healthcare game could be a challenge and time consuming process. However game mechanics can be applied in non-game contexts which is known as gamification.

Game Mechanics and Gamification

Gamification has been used in training, education, project management, social networks, health and wellness and in the commercial sector where it has been used in marketing and promotion with a view to engaging customers in product, service and brand loyalty. Examples include IBM connections where gamification has been posited as having the capability to improve organizational performance through a process of improving business collaborations (Royle, 2012). RedCritter Tracker (http://www.redcritter.com/company.aspx) is a project management tool for agile software development where team members can unlock badges and earn points for completing challenging tasks which can then be traded in for rewards in virtual stores. This approach is thought to improve both individual and team motivation and productivity. Nike Training club (http://www.nike.com/us/en_us/c/womens-training/apps/nike-training-club) is a mobile app which uses gamification to promote both physical activity and healthy lifestyle.

The BCH Adolescent Rheumatology Team healthcare checklists monitor progress from both the young person's and doctor's perspectives but is qualitative in nature. The process also relies on the intrinsic motivation of the young person to complete the tasks. Using the principles of gamification a quantitative measure of achievement and progress can be implemented and provides a mechanism for feedback. Figure 4 provides a conceptual framework where leader boards and rewards are integrated into the doctor's checklist.

Each domain in the BCH Adolescent Rheumatology Team doctor's checklist (Figure 2) has a series of items associated with it and the response to each may be more demanding based upon the stage of adolescence that the young person is in. The response to each item at each stage of adolescence would have a score associated with it. In Figure 4 there are seven items based upon knowledge and up to three different levels based upon the stage of adolescence (early, mid or late). An example of how gamification may be applied is that each of the responses may have 100 points for each successfully completed item (a total of 2000 points for the domain). If the player completes all items associate with early adolescence they would receive 600 points and they would gain a level. More points and higher levels would be achieved as more knowledge is accumulated as the young person matures through the stages of adolescence. When all items within a domain are completed (2000 points and achieving level 3) a further reward can be earned signifying mastery of that domain. Some of the items in the checklists (Figure 2) lend themselves to aggregation, for example the young person being able to accurately describe their condition as well as the effects it has on their body or that the young person understands their medication and the potential side effects of it. Acquiring these skills in combination lend themselves to the assignment of bonuses points for successfully achieving both of them. Failure to attend clinical appointments is a common problem. Awarding bonus points for successfully attending them would help to reinforce their importance and encourage the young person to stay in the healthcare system.

Figure 4. Leader board and rewards. By completing items within adolescent stages i.e. early (E), Mid (M) or Late (L) the player would gain points and badges for achievements which signify levels of experience. Trophies signify successful mastery of the domain. The knowledge leader board measures progress relative to that domain and feeds into the main leader board.

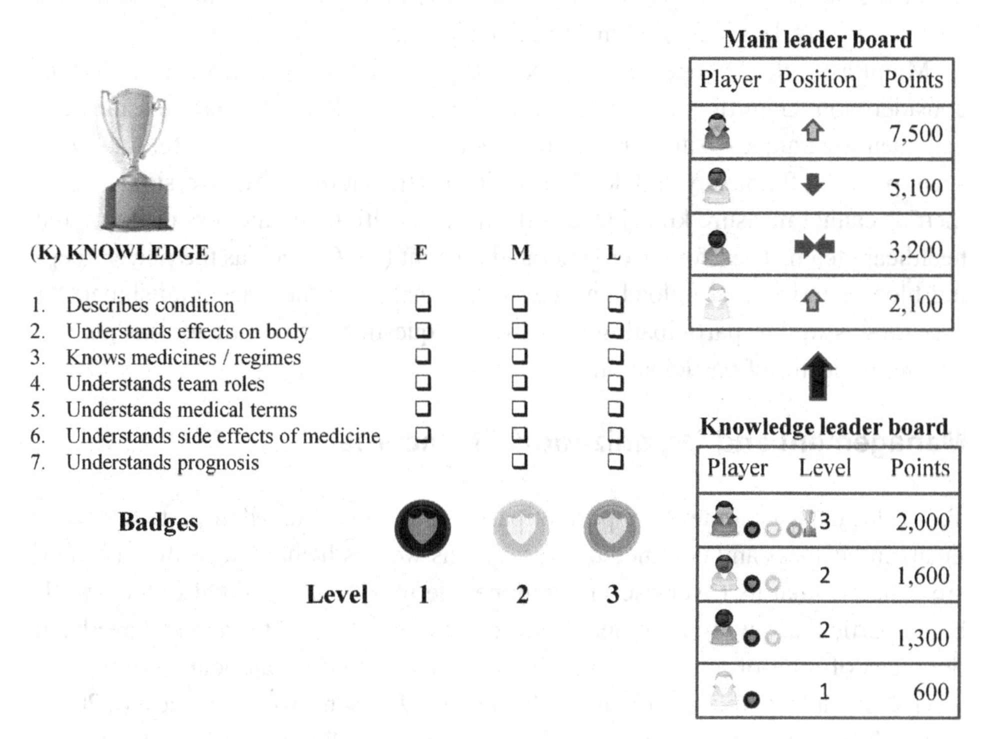

Gamification uses rewards as a way of recognising achievements however it is important not to over use them as this could potentially devalue them conversely if insufficient are used the incentive to engage with the process can be lost. Reward schedule and ratios are used to manage this process with randomness and surprises enhancing the effectiveness.

Highlighting a player's achievement can be managed by recording their scores on a leader board which will also provide them with feedback on their overall progress relative to their peers. Leader boards have the potential to motivate or de-motivate the player if they feel the goals are unachievable. Marczewski (2013) suggests the use of both relative and absolute leader boards. Relative leader boards would indicate the players progress against their peers, within a domain, where as the absolute leader board would indicate overall performance, relative to their peers, in acquiring all the skills associated with successful transition.

Status is an important mark of achievement in games and also provides an opportunity to identify where new players can find help. In our system this provides an opportunity to identify mentors who can provide advice and support. The right to privacy and confidentiality is important when dealing with young people with medical conditions. Anonymity can be preserved by the use of online characters (avatars) that reflect their persona but hide their identity.

Monitoring the engagement of young people in this system is an important consideration. Gamification often uses web based applications and measures engagement by unique visitors, page views per visitor, time spent on site, total time spent per user, frequency and depth of visits, participation and conversions. These metrics cannot measure knowledge and skill acquisition and are less useful to the healthcare team. Therefore user generated content (UGC) such as the player keeping blogs, diaries, and uploading videos demonstrating their accomplishment of activities as well as participating in answering questions in on line forums may be a better indicator of participation.

Management and Organizational Concerns

Technology has become an important part of our world contributing to advancements in business and commerce, education as well as healthcare. However there are concerns that its (over)use in young people may be detrimental to their well-being particularly within the healthcare community. Computer games have been the focus of a lot of attention regarding negative behaviour associated with their use for example its relation to violent behaviour (Janssen, Boyce, & Pickett, 2012) although research in this area highlights quite contradictory findings (Ferguson et al., 2012). One study has associated video game playing with less time being devoted to academic studies however the authors highlight this cannot be taken as an indicator of the person's academic achievements (Cummings & Vandewater, 2007). Gentile et al., (2011) indicate that excessive video gaming can be related to poor psychosocial outcomes in some young people whereas earlier studies found no such effects the authors suggesting that video game playing is a healthy part of adolescence (Durkin & Barber, 2002). A review by Hagell (2012) on the health implications of new technology describes both positive and negative aspects of its use with the suggestion that new technologies should form an important component in health promotion particularly in hard-to-reach groups. New and innovative ways of using technology should be found as a way of communicating with young people and delivering their services; focus being on taking full advantage of this media rather than avoiding it.

Promoting and demonstrating what can be achieved with technology has been one way that we have found that helps to break down the barriers and resistance to its adoption by the healthcare community. Ultimately we have to take into consideration the views of the young people and which media they would like to use which also means recognising that not all young people will want to use technology or computer games as their preferred source of information. Therefore alternatives need to be made available. Where computer games are to be considered as a means for educating young people sensible guidelines for their use should be provided which would allay some of the concerns of the doctors.

CURRENT CHALLENGES

Transitional care is a complex process involving young people acquiring and maintaining a range of knowledge and skills which will vary with the age and maturity. Maintaining a young person's engagement in their healthcare is challenging as other factors increasingly become prominent in their (adolescent) lives. However the implications of not taking care of themselves are a poorer quality of life and delayed transfer to the adult healthcare system. Creating a process that allows both the young person and the doctor to monitor progress and create incentives for maintaining care in a fun way and which the young person can relate to would be advantageous. Using GBL could be one solution to this.

Inherently all games should be fun to play but in terms of GBL there is the requirement for sound pedagogy that will result in measurable and discernable learning outcomes. For GBL to be successful in the healthcare setting the game should also result in positive health outcomes too. Our initial research indicated that not all games being developed were following this approach with little evidence for evaluation of their effectiveness.

Designing games is an art rather than resulting from a prescribed formula. Our research highlighted the lack of any guidelines for developing healthcare games requiring us to work towards devising our own framework for game(s) for transitional care.

The resources for developing games are also a factor that needs to be taken into consideration. Computer games take time, money and skilled technical individuals in order to make them. The Health Hero games were early examples of how new technologies (games consoles) were being used to provide healthcare information in a format that young people could relate to and engage with. As technology moves very quickly there is the danger that it soon becomes out of date and while the information Health Heroes contains is relevant it is in a format that is no longer readily accessible.

When considering the format of GBL its cost - benefit and longevity needs to be taken into account. This led us to consider of the use of gamification where the fun elements of games and their mechanics may be used to enhance young people's engagement with their healthcare but where the technology required is not as sophisticated as would be required with the development of a computer game. Services such as Badgeville (http://badgeville.com/) and gamificationU (http://gamificationu.com/) already provide resources to start creating gamified systems.

It is not to be underestimated that generating a system based upon gamification would still be a challenging and time consuming process. This would result from the implementation and refinement of a range of complex mechanics and balancing them to in order to make the game fair and fun to play with the added necessity to educate the player and positively influence their behaviour. There would also be the requirement to make sure appropriate assessments were incorporated in order to provide meaningful feedback to the players (patients) and doctors. By using this approach to develop a GBL system those game mechanics which are found to be effective in engaging the players in their healthcare process could then be candidate mechanics for incorporation into subsequent computer game(s).

Engagement in GBL may be dependent upon age and stage of development with younger people more likely to enjoy and participate in them. There is also the element of novelty, the challenges faced in order to maintain that game's uniqueness and how different genders perceive and react to it. Therefore there is the possibility it may only be effective within a certain time frame. As the system is a game and based upon achievements there is always the chance that people will try to cheat in order to gain the rewards which would be counter productive to the requirement that the young people are learning a set of behaviours which are important to their healthcare.

SOLUTIONS AND RECOMMENDATIONS

It is now accepted that young people are a distinct group with specialist healthcare needs which are quite distinct from children and adults. Medicine is a discipline which is keen to embrace technology to assist in patient care but may not be as fast to adopt as other sectors. Therefore our work is in its early stages.

We feel that designing GBL to support young people in transitional care will be dependent on understanding their needs and requirements for example acceptance of its use by them, what ages, stages of adolescent development and genders it will appeal to as well as acceptance by their family and the clinical team. It will be very important to include the young people when designing and developing any GBL system as involvement and inclusion is an important consideration that is advocated in healthcare for young people.

Our future work will progress by focussing upon requirements gathering from relevant stakeholders and will involve the following aspects:

- Assessing the acceptance of GBL in stake holders i.e. the healthcare team, patients and family. What do games mean to them and how do they see it helping and benefiting their healthcare process?
- Identifying any issues or barriers to adoption of GBL e.g. different ages and stages of adolescent development, gender and type of long term illness. How do the clinical teams feel about using GBL given the on-going debate about concerns of overuse of technology by young people?
- Identifying the requirements for GBL to assist in transitional care. What do they young people want to see?
- Identifying appropriate educational and psychological measurements to formally measure education and health outcomes as a result of using GBL.

There will be obvious limitations to the effectiveness of GBL in healthcare for young people. However the previous studies we have discussed have shown that games can be effective in engaging young people in learning to care for themselves. By their nature games are wilfully entered (Schell, 2008) and subsequently people want to participate in them. Therefore we would anticipate that by using GBL we would persuade young people to actively engage in their healthcare and build their confidence in wanting to learn to acquire the necessary skills to look after themselves.

In this chapter we have made some suggestions for frameworks for GBL which will provide clinicians with an insight into how it could be used positively in a healthcare setting and provide a basis for application to other disciplines where engagement with knowledge and skills acquisition can be challenging.

REFERENCES

Annunziato, R. A., Emre, S., Shneider, B., Barton, C., Dugan, C. A., & Shemesh, E. (2007). Adherence and medical outcomes in pediatric liver transplant recipients who transition to adult services. *Pediatric Transplantation*, *11*(6), 608–614. doi:10.1111/j.1399-3046.2007.00689.x PMID:17663682

Beale, I. L., Kato, P. M., Marin-Bowling, V. M., Guthrie, N., & Cole, S. W. (2007). Improvement in cancer-related knowledge following use of a psychoeducational video game for adolescents and young adults with cancer. *The Journal of Adolescent Health*, *41*(3), 263–270. doi:10.1016/j.jadohealth.2007.04.006 PMID:17707296

Bingham, P. M., Bates, J. H. T., Thompson-Figueroa, J., & Lahiri, T. (2010). A breath biofeedback computer game for children with cystic fibrosis. *Clinical Pediatrics*, *49*(4), 337–342. doi:10.1177/0009922809348022 PMID:20118101

Bodenheimer, T., Chen, E., & Bennett, H. D. (2009). Confronting the growing burden of chronic disease: Can the U.S. health care workforce do the job? *Health Affairs*, *28*(1), 64–74. doi:10.1377/hlthaff.28.1.64 PMID:19124856

Brown, S. J., Lieberman, D. A., Gemeny, B. A., Fan, Y. C., Wilson, D. M., & Pasta, D. J. (1997). Educational video game for juvenile diabetes: Results of a controlled trial. *Medical Informatics*, *22*(1), 77–89. doi:10.3109/14639239709089835 PMID:9183781

Catalano, R. F., Fagan, A. A., Gavin, L. E., Greenberg, M. T., Irwin, C. E. Jr, Ross, D. A., & Shek, D. T. (2012). Worldwide application of prevention science in adolescent health. *Lancet*, *379*(9826), 1653–1664. doi:10.1016/S0140-6736(12)60238-4 PMID:22538180

Cummings, H. M., & Vandewater, E. A. (2007). Relation of adolescent video game play to time spent in other activities. *Archives of Pediatrics & Adolescent Medicine*, *161*(7), 684–689. doi:10.1001/archpedi.161.7.684 PMID:17606832

Department of Health. (2011). *Ten things you need to know about long term conditions*. Retrieved from http://webarchive.nationalarchives.gov.uk/+/www.dh.gov.uk/en/Healthcare/Longtermconditions/tenthingsyouneedtoknow/index.htm

Durkin, K., & Barber, B. (2002). Not so doomed: Computer game play and positive adolescent development. *Applied Developmental Psychology*, *23*(4), 373–392. doi:10.1016/S0193-3973(02)00124-7

Entertainment Software Association. (2012). *Sales, demographics and usage data: Essential facts about the computer and video game industry*. Retrieved from http://www.theesa.com/facts/pdfs/ESA_EF_2013.pdf

Ferguson, C. J., Miguel, C. S., Garza, A., & Jerabeck, J. M. (2012). A longitudinal test of video game violence influences on dating and aggression: A 3-year longitudinal study of adolescents. *Journal of Psychiatric Research*, *46*(2), 141–146. doi:10.1016/j.jpsychires.2011.10.014 PMID:22099867

Ferris, M. E., Harward, D. H., Bickford, K., Layton, J. B., Ferris, M. T., & Hogan, S. L. et al. (2012). A clinical tool to measure the components of health-care transition from pediatric care to adult care: The UNC TR(x)ANSITION scale. *Renal Failure*, *34*(6), 744–753. doi:10.3109/0886022X.2012.678171 PMID:22583152

Gentile, D. A., Choo, H., Liau, A., Sim, T., Li, D. D., Fung, D., & Khoo, A. (2011). Pathological video game use among youths: A two-year longitudinal stud. *Pediatrics*, *127*(2), E319–E329. doi:10.1542/peds.2010-1353 PMID:21242221

Grundy, S. M., Benjamin, I. J., Burke, G. L., Chait, A., Eckel, R. H., & Howard, B. V. et al. (1999). Diabetes and cardiovascular disease: A statement for healthcare professionals from the American Heart Association. *Circulation*, *100*(10), 1134–1146. doi:10.1161/01.CIR.100.10.1134 PMID:10477542

Hagell, A. (2012). *Health implications of new technology*. Association for Young People's Health (AYPH). Retrieved from http://www.ayph.org.uk/publications/296_RU11%20New%20technology%20summary.pdf

Hagell, A., Coleman, J., & Brooks, F. (2013). *Key data on adolescence 2013: The latest information and statistics about young people today*. Association for Young People's Health.

Interactive Software Federation of Europe. (2012). *Industry facts*. Retrieved from http://www.isfe.eu/industry-facts

Janssen, I., Boyce, W. F., & Pickett, W. (2012). Screen time and physical violence in 10 to 16-year-old Canadian youth. *International Journal of Public Health*, *57*(2), 325–331. doi:10.1007/s00038-010-0221-9 PMID:21110059

Janssen, N. M., & Genta, M. S. (2000). The effects of immunosuppressive and anti-inflammatory medications on fertility, pregnancy, and lactation. *Archives of Internal Medicine*, *160*(5), 610–619. doi:10.1001/archinte.160.5.610 PMID:10724046

Kato, P. M., Cole, S. W., Bradlyn, A. S., & Pollock, B. H. (2008). A video game improves behavioral outcomes in adolescents and young adults with cancer: A randomized trial. *Pediatrics*, *122*(2), E305–E317. doi:10.1542/peds.2007-3134 PMID:18676516

Lieberman, D. (1997). Interactive video games for health promotion: Effects on knowledge, self-efficacy, social support and health. In R. L. Street, W. R. Gold, & T. R. Manning (Eds.), *Health promotion and interactive technology: Theoretical applications and future directions* (pp. 103–120). New York, NY: Routledge.

Lieberman, D. A., & Brown, S. J. (1995). Designing interactive video games for children's health education. In *Interactive technology and the new paradigm for healthcare* (pp. 201–210). Amsterdam: IOS Press.

Marczewski, A. (2013). *Gamification a little on leaderboards*. Retrieved from http://marczewski.me.uk/2013/01/21/gamification-a-little-on-leaderboards/

McDonagh, J. E. (2007). Transition of care: How should we do it? *Paediatrics and Child Health (Oxford)*, *17*(12), 480–484. doi:10.1016/j.paed.2007.09.007 PMID:17715444

McDonagh, J. E., Southwood, T. R., & Shaw, K. L. (2006). Growing up and moving on in rheumatology: Development and preliminary evaluation of a transitional care programme for a multicentre cohort of adolescents with juvenile idiopathic arthritis. *Journal of Child Health Care*, *10*(1), 22–42. doi:10.1177/1367493506060203 PMID:16464931

Merry, S. N., Stasiak, K., Shepherd, M., Frampton, C., Fleming, T., & Lucassen, M. F. (2012). The effectiveness of SPARX, a computerised self help intervention for adolescents seeking help for depression: Randomised controlled non-inferiority trial. *British Medical Journal*, *18*(344), e2598. doi:10.1136/bmj.e2598 PMID:22517917

Murray, C. J., Richards, M. A., Newton, J. N., Fenton, K. A., Anderson, H. R., & Atkinson, C. et al. (2013). UK health performance: Findings of the global burden of disease study 2010. *Lancet*, *381*(9871), 997–1020. doi:10.1016/S0140-6736(13)60355-4 PMID:23668584

Nakhla, M., Daneman, D., To, T., Paradis, G., & Guttmann, A. (2009). Transition to adult care for youths with diabetes mellitus: Findings from a Universal Health Care System. *Pediatrics*, *124*(6), 1134–1141. doi:10.1542/peds.2009-0041 PMID:19933731

Nash, A. A., Britto, M. T., Lovell, D. J., Passo, M. H., & Rosenthal, S. L. (1998). Substance use among adolescents with juvenile rheumatoid arthritis. *Arthritis Care and Research*, *11*(5), 391–396. doi:10.1002/art.1790110510 PMID:9830883

Philpott, J. R. (2011). Transitional care in inflammatory bowel disease. *Gastroenterologia y Hepatologia*, *7*(1), 26–32. PMID:21346849

Royle, T. (2012). *Gamification: Unlocking hidden collaboration potential*. Retrieved from https://www-304.ibm.com/connections/blogs/socialbusiness

Sawicki, G. S., Lukens-Bull, K., Yin, X., Demars, N., Huang, I. C., & Livingood, W. et al. (2011). Measuring the transition readiness of youth with special healthcare needs: Validation of the TRAQ—Transition readiness assessment questionnaire. *Journal of Pediatric Psychology*, *36*(2), 160–171. doi:10.1093/jpepsy/jsp128 PMID:20040605

Sawyer, S. M., Afifi, R. A., Bearinger, L. H., Blakemore, S. J., Dick, B., Ezeh, A. C., & Patton, G. C. (2012). Adolescence: A foundation for future health. *Lancet*, *379*(9826), 1630–1640. doi:10.1016/S0140-6736(12)60072-5 PMID:22538178

Sawyer, S. M., Drew, S., Yeo, M. S., & Britto, M. T. (2007). Adolescents with a chronic condition: Challenges living, challenges treating. *Lancet*, *369*(9571), 1481–1489. doi:10.1016/S0140-6736(07)60370-5 PMID:17467519

Scal, P., Garwick, A. W., & Horvath, K. J. (2010). Making rheumtogrow: The rationale and framework for an internet-based health care transition intervention. *International Journal of Child and Adolescent Health*, *3*(4), 451–461.

Schell, J. (2008). *The art of game design: A book of lenses*. Boca Raton, FL: CRC Press.

Shaw, K. L., Southwood, T. R., & McDonagh, J. E. (2004). User perspectives of transitional care for adolescents with juvenile idiopathic arthritis. *Rheumatology*, *43*(6), 770–778. doi:10.1093/rheumatology/keh175 PMID:15039498

Stinson, J. N., Toomey, P. C., Stevens, B. J., Kagan, S., Duffy, C. M., & Huber, A. et al. (2008). Asking the experts: Exploring the self-management needs of adolescents with arthritis. *Arthritis and Rheumatism*, *59*(1), 65–72. doi:10.1002/art.23244 PMID:18163408

Suris, J. C., Akré, C., Berchtold, A., Bélanger, R. E., & Michaud, P. A. (2010). Chronically connected? Internet use among adolescents with chronic conditions. *The Journal of Adolescent Health*, *46*(2), 200–202. doi:10.1016/j.jadohealth.2009.07.008 PMID:20113927

Suris, J. C., Michaud, P. A., Akre, C., & Sawyer, S. M. (2008). Health risk behaviors in adolescents with chronic conditions. *Paediatrics*, *122*(5), e1113–e1118. doi:10.1542/peds.2008-1479 PMID:18977960

Tackling the burden of chronic diseases in the USA. (2009). *Lancet*, *373*(9659), 185. doi: 10.1016/S0140-6736(09)60048-9

Viner, R. M., Ozer, E. M., Denny, S., Marmot, M., Resnick, M., Fatusi, A., & Currie, C. (2012). Adolescence and the social determinants of health. *Lancet*, *379*(9826), 1641–1652. doi:10.1016/S0140-6736(12)60149-4 PMID:22538179

Wacker, A., Kaemmerer, H., Hollweck, R., Hauser, M., Deutsch, M. A., & Brodherr-Heberlein, S. et al. (2005). Outcome of operated and unoperated adults with congenital cardiac disease lost to follow-up for more than five years. *The American Journal of Cardiology*, *95*(6), 776–779. doi:10.1016/j.amjcard.2004.11.036 PMID:15757611

World Health Organisation. (2001). *The second decade: improving adolescent health and development*. Geneva: World Health Organization.

World Health Organisation. (2013). *10 facts on the state of global health*. Retrieved from http://www.who.int/features/factfiles/global_burden/facts/en/index.html

Yeung, E., Kay, J., Roosevelt, G. E., Brandon, M., & Yetman, A. T. (2008). Lapse of care as a predictor for morbidity in adults with congenital heart disease. *International Journal of Cardiology*, *125*(1), 62–65. doi:10.1016/j.ijcard.2007.02.023 PMID:17442438

KEY TERMS AND DEFINITIONS

Adherence: Complying with the clinical teams prescribed therapies for example taking the right medications at the right time.

Adolescent: A young person in the age range of 10-19 years.

Birmingham Children's Hospital Adolescent Rheumatology Team Transitional Readiness Checklists: A series of healthcare plans which were designed to assess a young patient's knowledge and acquisition of skills which are deemed important for successful transfer of their care to the adult healthcare system.

Chronic Medical Conditions: Medical conditions that may persist for long periods of time and even through out a person's life.

Game Based Learning: The applications of the principles of games (fun and play) in education. In this chapter we use the term game based learning to encapsulate both the use of educational or serious computer games together with the processes used in gamification and how these can be applied to educational and healthcare settings.

Gamification: The application of game dynamics (progression, feedback and behaviours) and mechanics (the rules and actions that constitute a game) to non-game contexts.

Self-Care: The ability to look after ones own healthcare.

Self-Efficacy: The belief in one's ability to carry out a particular task.

Self-Management: The ability to manage ones daily living and affairs that are not directly related to therapy or treatment of their medical conditions.

Transitional Care: Transition addresses the medical, psychosocial as well as educational and vocational needs of the young person in an age and developmentally appropriate manner.

Chapter 6
The Travelling Rose:
A Persuasive Game to Stimulate Walking Behaviour of Older Persons in Low SES Neighbourhoods

Valentijn Visch
Delft University of Technology, The Netherlands

Wessel Bos
Delft University of Technology, The Netherlands

Ingrid Mulder
Delft University of Technology, The Netherlands

Richard Prins
Erasmus Medical Center, The Netherlands

EXECUTIVE SUMMARY

The persuasive game, Travelling Rose, aims to enhance walking activities of elderly residents living in neighbourhoods characterized by a low socioeconomic status. The game consists of a wooden pass-on box containing user-reports and instructions on how to find a companion, how to generate a surprise together, how to give this surprise a fellow neighbourhood member, and how to pass on the Travelling Rose box. The persuasive catch of the Travelling Rose is present when the companions are generating the surprise and are instructed to take a walk in the neighbourhood for easing social communication and flourishing creative ideas. In this chapter, the design process leading to the final version of the Travelling Rose is described, involving user studies, concept testing, and iterative prototyping. Secondly, the final prototype is presented and framed using the theoretical Persuasive Game Design model.

DOI: 10.4018/978-1-4666-6206-3.ch006

Copyright ©2014, IGI Global. Copying or distributing in print or electronic forms without written permission of IGI Global is prohibited.

THE TRAVELLING ROSE PROJECT

Introduction to its Authors

The Travelling Rose project is initiated as a collaboration between the faculty of Industrial Design Engineering of the Delft University of Technology and the Department of Public Health of the Erasmus Medical Center in Rotterdam. The Travelling Rose is embedded in the larger research project 'NEW.ROADS', coordinated by the fourth author Richard (Rick) Prins. This encompassing project aims to enhance walking behaviour of elderly by social and physical interventions in four neighbourhoods of Rotterdam, which are characterized by a low socioeconomic status (SES). The Erasmus Medical Center already developed some interventions, but Rick was also interested in seeking collaboration with designers, which is the reason he contacted the faculty of Industrial Design Engineering. In this faculty we not only focus on product engineering but also on designing products and services for specific user experiences, emotions, or behavioural change. For instance, the second author, Ingrid Mulder, is working in the field of design for social interactions and societal impact, and the first author Valentijn Visch is working in the area of persuasive game design. To design a creative intervention for the NEW.ROADS project, we searched for a good graduate student who was found in the person of the third author, Wessel Bos. Despite the interest and energy of the authors on developing the Travelling Rose, the project would never have been succeeded without the interest, support and collaboration of its end users, i.e., the residents of the Rotterdam Afrikaanderwijk.

SETTING THE STAGE

The Importance of Physical Activity

Being physically active is considered by the World Health Organisation to be a main factor for healthy aging. Physical activity is positively associated with the experienced quality of life in terms of self-efficacy, physical-, mental-, and social health. This effect is even enhanced when performed in companionship (e.g. sport clubs, friends). In terms of its societal impact, being physically inactive has been estimated to cause a burden of disease, which is as high as the burden of disease caused by smoking (Lee, Shiroma, Lobelo, Puska, Blair & Katzmarzyk, 2012). Although the benefits of physical activity are clear for individuals, healthcare professionals, and society, not everyone is physically active. Social healthcare research among densely populated neighbourhoods showed that especially elderly (55+) living in low SES neighbourhoods are not physically active enough. The aim of the NEW.ROADS project is to motivate these groups to be more physically active by increasing their walking behaviour for at least 30 minutes a week.

Problem with Low SES and Physical Activity

Public health literature clearly describes the socio-economic disparities with regard to health. One of the factors contributing to these disparities are the differences in physical activity levels among the various SES groups. Studies show that people having a lower SES typically possess a less physically active lifestyle (Droomers, et al., 2001; Giles-Corti & Donovan, 2002) than people having a higher SES. It is hard to reach and motivate low SES groups to increase their physical activity. Despite various attempts, at least in the Netherlands, there are no lifestyle interventions which have proven to be effective for these groups (Busch & Schrijvers, 2010). Hence, there is a clear need for a new approach to target people living in lower SES neighbourhoods to promote their physical activity levels.

Persuasive Game Design Model

One of the new approaches to motivate users for behavioural change is by means of persuasive game design. In our Persuasive Game Design model (Visch, Vegt, Anderiesen, vanderKooij, 2013) we aim for behavioural change by adjusting the user's real world experience into game world experience. Users do not always need intentionally designed game products by external game designers to experience a game. For instance, children can perfectly play a hide and seek game in natural surroundings such as forest. In contrast, some intentionally designed game products such as chess may not always lead to game experiences as enjoyment or engagement but may as well lead to boredom or distraction. Therefore, we propose that games ultimately exist in the experience of the user. However, game designers can try to seduce and motivate the user by gamification design to enter a game world experience. Various game elements, such as rewards, challenges, phantasy, or social dependency rules, proved to be effective in motivating users to play and have a game world experience. In our model, gamification design is defined as a design which applies game-elements to motivate a user to shift or transport from a real world experience to a game world experience. Such a game world experience is typically engaging, enjoyable, safe, free and providing direct feedback. These experiential qualities can be best described as *symptoms* being not exclusively restricted to game world experience since they might appear in real world experiences as well, albeit often at a less intense experience level or in combination with non-game world experiential qualities. A game world experience and a real world experience should be considered as two extremes between which users hover. Only scarcely, users experience a real world without any game world qualities. Even the humdrum of taking out your garbage may provide some game-related experiences such as satisfaction due to succeeded challenge). In the same vein, a game world will almost never be experienced without any relation to the real world. For instance, during a capturing 3D video fighting game, a player stills knows that her virtual death will have less impact than her death in the real world. Gamification design thus aims to change, or *transport* (c.f. Green, Brock, & Kaufman, 2004), a user from a real world experience towards a game world experience. This is however not where the aims of Persuasive Game Design end. Persuasive game design differs from entertainment game design in that the user's game world experience is not the end goal but serves as a means to achieve specific user behaviour in the real world. Often, this behavioural aim is something the user wants to achieve but experiences difficulty in achieving it – for instance quitting smoking, learning math tables, being therapy adherent, or being physically active. Persuasive game design facilitates reaching such real world goals by motivating the user by means of a gamification into a game world experience that in turn motivates the user for specific real world behaviour. Therefore, the final aim for persuasive game design is to realize specific behavioural transfer effects for the game world experience to the real world. In the present case, the aimed transfer effect consists of a walking trip of elderly in low SES neighbourhoods.

Design Process

To design a fitting gamification in persuasive game design, it is needed to include in the design process both the transfer effect as well as the user's motivation with regard to the current real-world behavior, to the game-elements and to the aimed for transfer effect. The aimed transfer effect is provided by the Erasmus Medical Center (i.e., increase walking behaviour of elderly in SES neighbourhoods with at least 30 minutes per week). In order to achieve this transfer effect by persuasive game design an experienced game world is needed that motivates the user for walking. A straightforward design strategy is to integrate the aimed transfer effect into the game world experience by making it a requirement for the game world experience. This type of gamification is often used in persuasive games aimed at physical training transfer goals, such as exergames (c.f. Staiano & Calvert, 2011). An alternative gamification design strategy to achieve aimed transfer effects is by letting the game world experience not *require* the complete transfer goals in order arrive at full game experience, but by letting the game world *allude* to the transfer goal. Even alluding to a part of the transfer goal might be sufficient for achieving the aimed complete goal as can be shown in the classic serious game Re-Mission

(Kato et al., 2008). In this first person shooter game for (young) cancer patients, the players gain knowledge about their disease and medications during playing. This knowledge in turn significantly increased the medicine adherence of these patients.

In persuasive game design it is thus needed to motivate the user by means of the game world for the aimed transfer behaviour. As shown above, this might be achieved by making the game world experience require the transfer behaviour or by letting the game world allude to a part of the transfer behaviour. Often gamification occurs somewhere in between these extremes. For instance, in the game Fish'n'Steps (Lin et al., 2006) the transfer goal consists of increasing physical activity of its players. In the game, the player's daily pedometer counts were linked to the growth of a virtual fish. In this way, the playing the game required executing the aimed transfer behaviour. However, the main goal of the game was not only to let the fish grow by taking the required amount of steps, but also to motivate the players for physical activity after they play the game. The game was designed as a catalyst, in which the game world alluded to the final transfer goal of an attitudinal change in the real world.

In addition to including the transfer effect in gamification design strategy, it is necessary to include user motivation as well with regard to the following three respects. Firstly, the game-elements need to be motivational for the user. If the user is not motivated by the game-elements used, the user will never experience a game world. Hence, a strategy based game element, such as chess, will not motivate toddles to play the game in the desired way. Secondly, in order to motivate the user for a transfer goal, the transfer goal should be attainable and (potentially) desirable for the user. It will be very difficult to motivate a large person to grow shorter, since he will know this goal is unattainable. Effective persuasive games do often not only include goals that are attainable for the user but also goals for which the user already possess a motivation. For instance, it will be almost impossible to motivate players to quit smoking when they never had any motivation for quitting. Most persuasive games do not create a new motivation for a transfer goal, but modify an existing motivation for the transfer goal. Thirdly, a gamification design has to transport the user's experience from a real world to a game world. In order to achieve this, the gamification design has to relate in some way to the current real-world experience of the user. For a designer this kind of user real-world behaviour and motivation is essential to know in order to select the motivational game-elements for the gamification design. For instance, when a gamification is designed which requires the user to spend daily two hours of gameplay, the designers need to be sure that their users have this time available. In sum, the selection and design of the game elements that will be used in the gamification are dependent on (1) the user's motivation for the transfer effect *to make the transfer effect attainable*, (2) the user's motivation for chosen game-elements *to let the game-elements motivate the user to transport towards a game world*, and (3) the user's motivation and behaviour in the current non-gamified real world situation *to implement the game in the user's real world context*.

THE TRAVELLING ROSE: USER STUDY, CONCEPT TESTING, PROTOTYPE

User Research: Understanding the Users and their Context

For the present research we investigated a) whether the users of our low SES target neighbourhood (Afrikaanderwijk in Rotterdam, the Netherlands) had indeed low levels of physical activity and b) whether we could address causes for this behaviour. To this end Wessel walked along the streets of Afrikaanderwijk and rang doorbells to conduct a short interviews about the residents' interests, experience of neighbourhood, walking behaviour, and life in general. Six people opened their doors and were happy to participate in the interview. In addition to this, three people were interviewed in the neighbourhood centre 'Het Anker'. The ages of the participants ranged from 54 to 86 years and their backgrounds showed a diversity of nationalities (Suriname, Dutch, Cape Verde, Afghanistan) characteristic of the neighbourhood. The nine interviewed people have been living for 5 to 52 years in the neighbourhood. Textbox 1 shows a description of one of the interviews.

Almost all interviewed persons reported to be physically active on a daily basis for instance though walking (as sole activity for an hour a day), shopping (mostly with walking frame), or gymnastics (in front of television, in neighbourhood centre, or in fitness centre). The neighbourhood was generally seen as 'lively', both in a positive and in a negative sense. Positive were the amounts of children, the large market, the social contacts, BBQs and the communal garden. However, the liveliness of the neighbourhood was mostly seen as negative and unsafe due to the factors like burglary, robbery, noise, disrespectful youth, drunks, or lack of car-parking place. Being unsafe and dangerous were dominant negative reported impressions, which was the reason for our researcher Wessel to investigate if this was true. His experience of this field research is presented in Textbox 2.

Our formative qualitative research might be slightly biased since the participants who were willing to be interviewed might be socially engaged-, and active people in general. Moreover, the reported physical activity might be exaggerated as an effect of the social desirability to be physically active (Adams et al., 2005). Both factors can explain the found difference between our study on reported physical activity and the large-scale research of Droomers et al. (2001) and Giles-Corti & Donovan (2002). Moreover, the reported duration of physical activity did not always reach the minimum standard for healthy activity (30 minutes of moderate-to-vigorous intensity activity for at least 5 days per week). A commonly shared motivation among the participants in favour of their neighbourhood was the reported value of social relationships with neighbours – which was explicitly mentioned by 4 of the 6 interviewed. Again, this effect might be biased, since the people who were willing to partake in the interview could be more socially interested than people who did not want to participate.

The Walk as Social Tool

The results of the user study suggested in terms of motivation that the target group was not unmotivated to perform physical activity. Our target group was motivated for the transfer goal of walking. However, the social situation of the neighbourhood was not experienced as safe or welcoming to perform actual walking. A commonly shared motivator for the inhabitants was social relatedness: the majority of the inhabitant placed social connection in their top list of positive motivators for life in general. This observation led to a design vision in which the positive motivator of social relatedness is used to stimulate walking.

Box 1. Interview excerpt during user research. Translated by the authors.

Interview with Mrs S. (72 year old)
22 years ago mrs. S. arrived in the Netherlands. Since then she has lived in Rotterdam and since 16 years in Afrikaanderwijk. She is living in a senior apartment because she cannot walk the stairs anymore. Until recently she was often together with her mother. In August her mother has passed away at an age of 91 years. It makes her sad. Luckily her children live nearby. They visit her almost every day. She has enough to do: maintaining the house and baby-sitting among other things. Her grandchild (2 years old) comes to visit her for two or three times a week for half a day.
Interests
Flowers, arranging flowers, making clothes, liveliness.
What about the neighbourhood?
The neighbourhood is pretty and lively. After having lived here for some time it is nice to know the people around. Sadly the neighbourhood is not quite safe; a few months ago she was mugged at the market. There are also many cases of burglary. If anything could be changed, it should be safety so she would not be anxious anymore.
What about walking and PA?
She tries to walk every day, for 1 or 2 kilometres. She likes to walk with her children. She will also get some groceries because she likes it, and to get some things with discount. Nothing much, because you shouldn't carry around more than € 5 in the first place. Cycling is too dangerous because her knees do not work so well anymore. Standing for longer than 15 minutes is also painful because of back issues. Still she is doing everything herself. She doesn't need help yet and she will do the housekeeping for as long as she can. Given the choice she would rather walk in nature because it is green and healthy.
What is important in life?
Peaceful relations. Giving and receiving love.

Box 2. Context night research by Wessel

Nightly Interviews Triggered by all the stories about nightly clamour in the neighbourhood I decided to go and have a look. I left my phone and wallet at home and set out to Afrikaanderwijk. I was determined to speak openly to even the most dangerous assemblies and deep down I was hoping for some adventure, just to experience what it must be to live in Afrikaanderwijk. After walking the streets for an hour I finally found two groups of people. Talking to both of them cured me of the illusion that Afrikaanderwijk at night is a special sort of Wild-West experience, but yielded an important insight. Afrikaanderwijk at night is like every other neighbourhood: Most of the time sleepy, quiet, and not threatening at all. A group of 10 Moroccan adolescents I met seemed a bit offended when even the chance existed that they were associated with unsafety. Which is understandable. Immediately after the conversations my perception of the situation began to change, which was a powerful experience. Many of the conversations that were playing in my head when walking around appeared to be nonsense, the constantly developing scenarios turned out to be useless. It felt as if the lights went on and the ghosts disappeared. I hadn't even been afraid and yet I was led by expectations that distorted my perception of reality such a great deal, just because of a few stories and a lot of imagination. The important insight is that mental images can be pretty powerful.

Figure 1. Impression of Afrikaanderwijk, Rotterdam, the Netherlands. Photo: Wessel Bos.

Achieving social relatedness is one of the three major needs guiding human behaviour (Ryan & Deci, 2000). Throughout life, people strive to satisfy this need by actively engage in social networks of school, study, social clubs, work, and on a more intimate level by finding a partner and raising a family. The effects of social relatedness on mental- (for a review see Baumeister & Leary, 1995) and physical-health (for a review see Seeman, 1996) are uncontested. For elderly however, the frequency of social interactions declines as they grow older (Carstensen, 1993). For instance, retirements reduce work-related social interactions, physical limitations reduce outgoing and social networking activities, and the number social contacts decline due to illness or mortality. However, although the quantity of social interactions is reduced at old age, the valued quality of the interactions increases by age (Carstensen, 1993). This increase can be explained by the nature of the interactions. Superfluous work-related interactions are removed from the amount of social situations and relation with siblings or children are intensified. In addition, the increase might be the result from elderly-typical emotion regulation strategies to appraise conflicts more positive than younger people (Diehl, Coyle, & Labouvie-Vief, 1996) and damp negative affects (Riedinger, Schmiedek, Wagner, Lindenberger, 2009). Despite the eventual increased emotion positive regulation systems, the reduced social network of elderly may cause negative emotions such as loneliness (Tomaka, Thompson & Palacios, 2006; Victor, Scambler, Bond, & Bowling, 2000). Based on our user-study and literature, social-relatedness will be used as a motivating game-element in the gamification design.

The objective of our study, walking, perfectly lends itself for social interaction. Although there has not been much scientific evidence on the social aspects of walking, daily life experiences show that walking is often used as an activity to intensify social relationships – for instance by families on Sunday strolls or even by Dutch television shows such as 'De Wandeling' ("The Walk" broadcasted by the KRO). The producer of the latter show suggested that interviewing during walking stimulated reflection since people are not facing each other continually and because the talking happens alongside a physical activity (vanderWijst, 2011). These two effects (not staring at each other and being engaged in a physical activity) are recognized by some therapists as well (cf. walkandtalk.com) who use walking as a setting for therapy. In a similar vein, walking the dog provides opportunities for social encounters (Johnson & Meadows, 2010; Rogers, Hart, & Boltz, 1993). At a more general level of motivation for walking, social companionship is seen as an important determinant for physical activity (c.f., Stralen et al, 2009; Wendel-Vos et al., 2007 for a review).

In conclusion, the motivation to walk and the motivation for social relationships can mutually reinforce each other. Our user research showed that the transfer goal walking can be believed to *attainable* for our target group, the game element of social *engagement* is motivating for our target group, and a social walk could be *implemented* in the real world context. Concept testing has to show what detailed gamification design and specific game elements are motivating and can be implemented. Prototype testing will show if the aimed transfer goal is reached.

Concept Testing: The Pass-On Booklet

In order to investigate the relationship between the social and the walking motivation in our target group, a first concept was designed: the pass-on booklet (see Figure 2). For this concept inspiration was found in product service systems that connect strangers. Similar examples are the classic chain letter or, in the product domain, 'Herman Friendship Cake'. This cake is baked after 9 days of unrefridged rising of the dough. However, before baking the cake, the dough of Herman is divided into four parts. The owner uses one part to bake a cake, whereas the other three parts are given to friends with the same baking instructions. These instructions tell each owner to put in ingredients and wait for another 9 days, split, pass-on, etc.

The design of our pass-on booklet was as follows: The first participant had to contact someone above the age of 55, who lived in the neighbourhood, and invite him/her for a walk. After this walk, the first participant wrote down his/her experi-

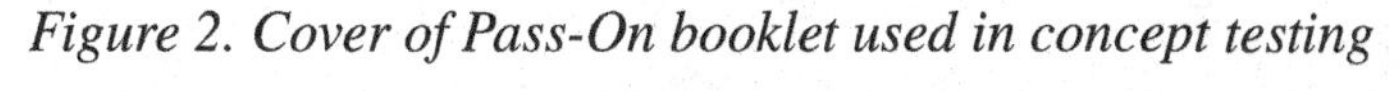

Figure 2. Cover of Pass-On booklet used in concept testing

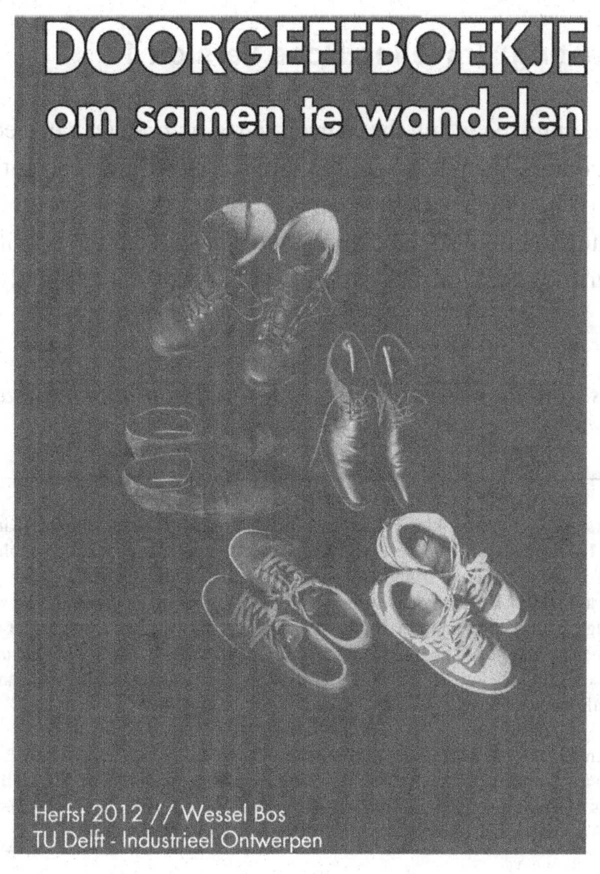

ences and gave the booklet to the second participant who in turn contacted and invited a third person for a walk, wrote down experience and passed the booklet on. The pass-on chain was limited to a maximum of 7 walks.

In our concept testing, four of these booklets were given to four of the people who participated in our former user research. The study showed that the amount of walks made per booklet varied from none to five, including a total of 13 participants. A record of this first walk can be found in textbox 3.

Analyses of the returned booklets showed that people liked the walk ("I think this walk is a good idea"), and people especially liked the social aspect of the walk (e.g., "We got to know each other better"). Often the desire was mentioned to walk together on a more regular basis ("I hope we'll do this more often"). This shows that the expected motivating power of social relatedness to initiate the transfer goal behaviour "walking" was found. The reports in the booklet also showed that inviting someone else for a walk was experienced as a inhibiting social barrier – leading to termination of the pass-on cycle (e.g. "It is difficult to find someone to join, so I give up"). For our next design we therefore had to focus on facilitating (new) social contacts.

Prototype Testing: Traveling Rose Version 1

Based on our findings regarding the concept *pass-on booklet* we needed to develop strategies to encourage users to make the initial contact with each other for a walk. Remember than when the contact was made, the product seemed to work well. In order to facilitate the initial social contact, we investigated the possibility to generate a shared extrinsic goal for the participants that was not explicitly stating social

Box 3. Excerpts of notations made by users in pass-on booklet. Translated by the authors.

Notations Walk 1
I started with the first walk of this booklet. I did find it difficult to invite someone older than 55 years for the walk. Although I live across an old people's home, I didn't dare to contact the old people since I didn't know them. Instead, I contacted an old friend of me to take a walk, but due to bad weather we cancelled our first appointment. Finally on a sunny day we went for a walk at 10 in the morning. We walked via the Rijnharbourbridge, Katendrecht and Maasharbour to Afirkaanderwijk. We talked about work, health, behaviour of other people and how other people reacted to small incidents. My walking companion hadn't walked the Rijnharbourbridge before, so that was very exciting. Our walk ended in Afrikaanderwijk. We agreed to walk together more often.
Notations Walk 2
Subjects: youth, animals (dogs and cats), our neighbourhood.
It wasn't difficult to ask someone to join for a walk – she is one of my neighbours. The walking was nice. We walked in the streets of another neighbourhood, where I normally don't walk. We have been walking almost an hour. I liked it. We got to know each other better. I hope we'll do this more often.

contact and walking, but which implicitly required social contact and walking. To generate such an extrinsic goal was inspired by our assumption that it is easier to motivate people to walk when the walking serves an extrinsic goal, such as shopping or feeding the ducks, than when walking serves an intrinsic goal, such as just the pleasure of walking. Our requirements were therefore to design a goal for the users, facilitating social contact as well as our target behaviour *walking*. In terms of the Persuasive Game Design model, we could not directly use social relatedness as a motivating game element but needed another game element to facilitate achieving social relatedness.

Figure 3 depicts the usage of our first prototype. For this prototype we made use of the game elements *mystery, creativity,* and *social surprise*. The game element of social surprise is used to facilitate the initial social contact between the two participants who both have to invent a surprise for a third participant. We supposed that this shared extrinsic goal facilitates inviting each other for participation. The game element of creativity is used to motivate two participants for a walk and social integration. Finally, the game element of mystery was used as design context for the gamification prototype. This latter element was applied in the design of the box, the logo, and the 'secret society' website where the participants can upload their experiences.

For our first test, seven of Travelling Rose version 1 boxes were sent out to people aged above 55 years. The boxes contained instruction for use as well as a question to email their experiences to the project website. None of the boxes returned and we received 6 mails from 3 of the 7 boxes. In textbox 4, three mails relating to box 6 are presented. The e-mails showed that the surprises varied from parties,

Figure 3. The Travelling Rose version 1

Box 4. User mails during Travelling Rose version 1

Box #6
Mail 1: Dear people of Travelling Rose, […] we organised a birth day surprise party for a lady of 96 years. She was suffering from the effects of a burglary for over a year and we therefore like to give her this party. […] Keep surprising people since many deserve it!! Let this box travel further, and take care it doesn't fall silent ! Success !!! *Mail 2:* […] I wanted to thank my husband for taking care of the household since I'm not so mobile anymore. We took a breakfast at the HEMA [large Dutch department store]. Last week we did this together with our son and grandson. For them it was a complete new experience. The box is ready for the next candidate. It wasn't difficult to find him. *Mail 3:* […our surprise] will be a spring greeting for a friend who faced an unhappy year and who deserve a surprise! We'll celebrate the assignment in the near future at our home by drinking coffee and some drinks with the new receiver of the box. (translated by the authors).

birdfeed houses and chocolates to breakfast and drinks. The user responses also indicated that some parts of the Travelling Rose design still needed improvement. Participation with the Travelling Rose seemed to require social courage, which not all participant possessed. For instance, as was mailed by a lady aged 71: *"Old people are afraid, suspicious and lacking trust"* and someone else *"I haven't slept a wink last night'.* It seemed that, although people like to participate in the Travelling Rose, participation was still experienced as fearful and difficult. For our next and final prototype, we aim to lower the social thresholds to participate even more and change the mystery theme – which might have a discouraging, instead of an encouraging, effect.

The Final Prototype: Travelling Rose Version 2

For this final version we dealt with the design issues that were raised by our former study. Firstly, the game element of mystery was removed from the design by giving the rose an open smiling face and by removing the Secret Society of Heroes. This framing narrative of the website, which was mainly putting the participants in an invitation-only club, was now changed into a narrative frame foregrounding the contribution of the Travelling Rose to the neighbourhood. Secondly, a box Warden is appointed to prevent disappearance of the boxes and to make sure that the boxes' travel activities cannot be stopped by a participant who doesn't want or dare to participate. This Warden enthuses the participants, explains the system, hands out and receives the boxes, maintains a list of willing partners, and helps people to connect with each other. The Warden function will be typically performed by a social worker who is already present in the low SES neighbourhood.

Another new addition to the final prototype of the Travelling Rose are the *Rambler Roses.* These are people who have indicated to the Warden that they are willing to be contacted as partner for the Travelling Rose. If someone has, for one reason or another, difficulty in finding or inviting a partner for the Travelling Rose, he or she can ask the Warden for a Rambler Rose. The Rambler Roses may as well start up a new cycle of the Travelling Rose.

A manual that clearly explains and facilitates the use of the Travelling Rose now accompanies the box. This procedure will make it easier for users to invite someone else to accompany him or her during the walk. Especially, since the manual also presents a list of suggestions for surprises, it is expected to decrease the fear of having to be creative during the walk. The suggestions are grouped to themes and to degree of required energy for giving the surprise. The themes are *complements and attention* (e.g., saying something nice – little energy), *helping* (e.g., helping in the garden – much energy), *giving a present* (cookies – medium energy), *being together* (e.g., going shopping together – medium).

In the previous version, we used a member website to collect all the experiences of the Travelling Rose. For user convenience and to maintain a direct connection to the Travelling Rose, we now asked the users to write their experiences into the non-digital *guestbook* inside the Travelling Rose. In this book, the subjects are asked to introduce themselves by name or initials, to describe their walking experience, to tell why they think walking is important, to describe their surprise and surprise reception, and to provide tips and tricks for a next user. The guestbook aims to functions as a catalyst for the new receiver of the Travelling Rose by making him or her curious and challenge him/her to join. The guestbook also serves to monitor the experiential effects caused by the Travelling Rose. To monitor the realisation of the aimed transfer effect of the Travelling Rose on walking, a pedometer is added to each box. The users are asked to write down the number of steps in the guestbook in order to show the travel distance of the Travelling Rose (see Figure 4).

Overview of Travelling Rose in relation to Persuasive Game Design Model

The aimed transfer effect of the Travelling Rose consisted of increased walking behaviour. In order to achieve the proper game-element to realize this behaviour, a user research was conducted. The results of the user research revealed firstly that our target group, residents of low SES neighbourhoods, were motivated for walking and that some of them were already engaged in some physical activity. Secondly, the user research showed that social characteristics of the neighbourhood were sometimes experienced to increase the motivation for physical activity (lively neighbourhood & going out with friends) and sometimes to decrease the motivation

Figure 4. Overview of interaction in The Travelling Rose version 2

for physical activity (unsafe neighbourhood & no direct social contacts). Based on these results we were choose to apply social relatedness as a general motivational game-element in the gamification design. Our first tested concept showed however that users experienced social barriers when inviting each other for a walk. This inspired us to include a shared extrinsic goal for the participants consisting of creating a surprise together for a third person. In sum, in order to motivate our target group for a walk, we used social relatedness, but instead of using this directly as a game-element, an additional element (shared surprise goal) was needed to facilitate realise its motivational value.

The game world experience consisted of the experiences during the social creative walk and during the performance of the surprise. The aimed for transfer goal, walking, was only marginally present in the game world experience, but it was incorporated in the design as a requirement. We hope that the transferred behavioural effects of the Travelling Rose consist not only of compliance to the walking instructions for the Travelling Rose, but that it also works as a catalyst for increased social walks – just as some of our subjects wished for. As such, the game design of the Travelling Rose persuades its users to comply with the game and make a walk, and it aims at a larger scale to generate an attitudinal change by generating socially engaging walking experiences (Figure 5).

Figure 5. Travelling Rose in persuasive game model

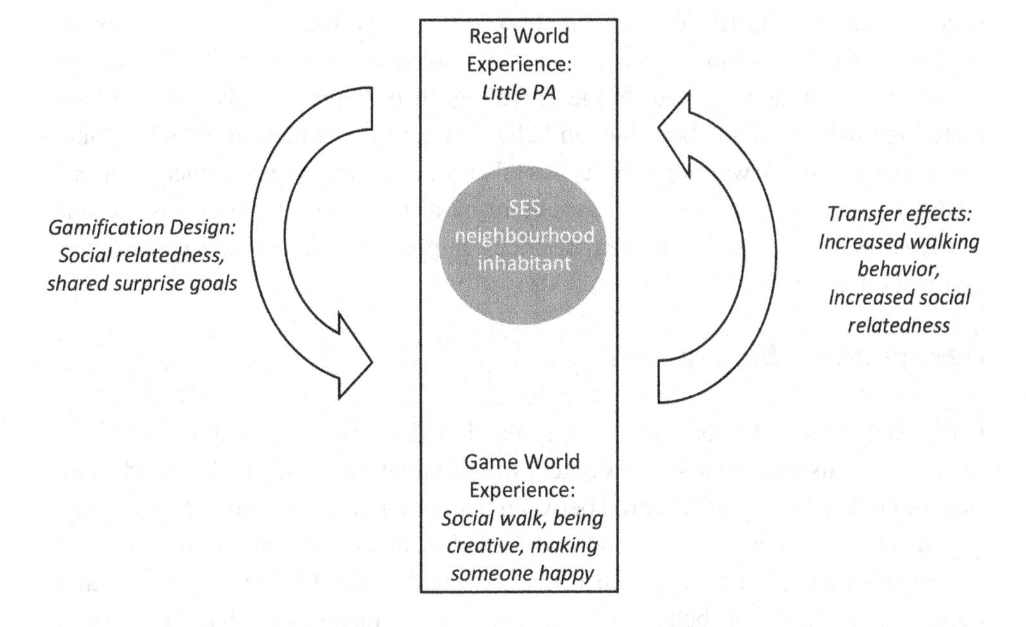

CURRENT CHALLENGES

Involvement of the Target Group

Although we applied iterative design processes involving the end-users at various stages, we still do not know whether we were able to recruit and motivate the typical members of our target group who have low physical activity. The people who participated in our design process were enthusiastic about the project as a whole, but we are unable to state that these people represent the people the Traveling Rose was designed for. However, by involving the participating and motivated people to the Travelling Rose project, a contagion effect might be expected since the Travelling Rose is based on a pass-on structure in the neighbourhood. Moreover, the contagion effect might even be increased by means of the involved social worker functioning as Warden. When testing the concept of the *Pass-On* booklet, we already observed the motivating power of a supporting 'mother organization' functioning as a Warden. It was much harder for us as researchers to motivate people use the Travelling Rose, than it was for *Afri Verbindt*, a local organization in the target neighbourhood. Having a mother organization for the continuation of the game seems to be crucial for achieving the sustainable societal impact.

Integrating natural organisers from the target group into the design, such as the social worker having the Warden function in the Travelling Rose, opens ways to monitor elderly's walking behaviour in a less obtrusive way. Moreover, emerging technologies could bring value to the Traveling Rose. Think of tools that automatically log objective data about human behaviour and the context in which it takes place, and how much walking activity has taken place. By using experience sampling or diary techniques, researchers can obtain more insight in the relationships and interactions between users, their experiences, human behaviour, social context, and application usage (Mulder & Kort, 2008).

Designing for Motivations

Our design method to conduct a user research and user concept test to investigate the motivations and behavioural context of the target group seemed to work well. Interestingly, a large gap appeared between being motivated for social contacts and physical activities on the one hand and for behaving accordingly on the other. In literature on cognitive factors that influence health-related behaviour, this is also known as the "intention-behaviour" gap. It was for instance found that 30% of positive intentions to engage in physical activity were actually transferred into physical activity (Godin & Kok, 1996; Hagger, Chatzisarantis, & Biddle, 2002). It is exactly at this point that persuasive game design can help people realize the motivation they already possess. A large drawback in realizing motivations with the Travelling Rose was the fear for social contact initiation. Obviously, the inhibitive effect of social fear is studied throughout social psychology in diverse experiments. It is however important for game designers to firstly know exactly at which part of the design these inhibitive effects appear. Secondly, game designers should test the effect of their game elements in overcoming these inhibitive effects. During the process of the Travelling Rose, we introduced a shared goal to facilitate social contact making and we facilitated the ease of goal attainment by our surprise suggestions in the Travelling Rose manual.

Our aimed for transfer effect was reached by motivating people using social motivators that require walking. From the perspective of the Travelling Rose user, walking is just one of the processes within the Travelling Rose activities. As such, the design of the Travelling Rose aimed to reach this original target behaviour by only indirectly alluding to the general behavioural aim. It can be reached just by the compliance to the Travelling Rose procedure, but the mutual motivating effect of walking and talking (physical activity and social relatedness), might as well catalyst an attitudinal change (c.f., Oinas-Kukkonen, 2013) leading to a more long-term increase of physical activity.

SOLUTIONS AND RECOMMENDATIONS

The Travelling Rose is a persuasive game design for behavioural change. In this paper we mainly focused at the design process, the prototype itself and the relation to the Persuasive Game Design model. Our main finding are (1) that the gamification design should be aligned to user research results on motivation for transfer goals, motivation for game element and motivations in the real world context, (2) that motivational game-elements (i.e. social relatedness) might not work directly but might need a supportive game-element (i.e. shared procedural goals), and (3) that the intended transfer effect need not to be experienced as the core of the game by its participants.

We think that especially the design for transfer effects should be studied more. In the present case, we did not yet study the effect of the Travelling Rose on walking behaviour at length and cannot, yet, make claims about such effects. Evidence-based knowledge is needed in order to validate the game at stake and to generate knowledge on design principles for transfer effects. For instance, in the Travelling Rose we aim for compliance, i.e. walking together for the Travelling Rose, and for attitudinal change, i.e. walking together in non-Travelling Rose situations. The pedometer of the Travelling Rose can monitor the first effect, but the latter effect is far more difficult to measure. Even more interesting might be the design principles that lead to specific attitudinal transfer effects. For instance, would it be enough to use the Travelling Rose only once for an attitudinal change, or do we have to add new components to the Travelling Rose to motivate people for attitudinal effects. We expect the latter, and the Warden or the social communal working environment might play a pivotal role here. The value of the Traveling Rose is achieved when the game is embedded as an ongoing game in neighbourhood activities. It then enables older people to maintain their physical and social health, their active contribution to the neighbourhood, and to respond effectively to the physical, psychological and social challenges of older age, adding quality to their lives (Bowling et al., 2003).

REFERENCES

Adams, S. et al. (2005). The effect social desirability and social approval on self-reports of physical activity. *American Journal of Epidemiology*, *161*(4), 389–398. doi:10.1093/aje/kwi054 PMID:15692083

Baumeister, R. F., & Leary, M. R. (1995). The need to belong: Desire for interpersonal attachments as a fundamental human motivation. *Psychological Bulletin*, *117*(3), 497–529. doi:10.1037/0033-2909.117.3.497 PMID:7777651

Bowling, A., Gabriel, Z., Dykes, J., Dowding, L. M., Fleissig, A., Banister, D., & Sutton, S. (2003). Let's ask them: A national survey of definitions of quality of life and its enhancement among people aged 65 and over. *International Journal of Aging & Human Development*, *56*(4), 269–306. doi:10.2190/BF8G-5J8L-YTRF-6404 PMID:14738211

Busch, M. C. M., & Schrijvers, C. T. (2010). *Effecten van leefstijlinterventies gericht op lagere sociaaleconomische groepen*. RIVM.

Carstensen, L. L. (1993). Motivation for social contact across the life span. In JacobJ. (Ed.), *Nebraska Symposium on Motivation: Developmental perspectives on motivation* (pp. 40, 209-254). Lincoln, NE: University of Nebraska Press.

Diehl, M., Coyle, N., & Labouvie-Vief, G. (1996). Age and sex differences in strategies of coping and defense across the life span. *Psychology and Aging*, *11*(1), 127–139. doi:10.1037/0882-7974.11.1.127 PMID:8726378

Droomers, M. et al. (2001). Educational level and decreases in leisure time physical activity: Predictors from the longitudinal GLOBE study. *Journal of Epidemiology and Community Health*, *55*(8), 562–568. doi:10.1136/jech.55.8.562 PMID:11449013

Giles-Corti, B., & Donovan, R. J. (2002). Socioeconomic status differences in recreational physical activity levels and real and perceived access to a supportive physical environment. *Preventive Medicine*, *35*(6), 601–611. doi:10.1006/pmed.2002.1115 PMID:12460528

Godin, G., & Kok, G. (1996). The theory of planned behavior: A review of its applications to health-related behaviors. *American Journal of Health Promotion*, *11*(2), 87–98. doi:10.4278/0890-1171-11.2.87 PMID:10163601

Green, M., Brock, T., & Kaufman, G. (2004). Understanding media enjoyment: The role of transportation into narrative worlds. *Communication Theory*, *4*(4), 311–327. doi:10.1111/j.1468-2885.2004.tb00317.x

Hagger, M., Chatzisarantis, N., & Biddle, S. (2002). A meta-analytic review of the theories of reasoned action and planned behavior in physical activity: Predictive validity and the contribution of additional variables. *Journal of Sport & Exercise Psychology*, *24*, 3–32.

Johnson, R., & Meadows, R. (2010). Dog-walking: Motivation for adherence to a walking program. *Clinical Nursing Research*, *19*(4), 387–402. doi:10.1177/1054773810373122 PMID:20651066

Kato, P., Cole, S., Bradlyn, A., & Pollock, B. (2008). A video game improves behavioral outcomes in adolescents and young adults with cancer: A randomized trial. *Pediatrics, 122*(2), 305–317. doi:10.1542/peds.2007-3134 PMID:18676516

Lee, I., Shiroma, E., Lobelo, F., Puska, P., Blair, S., & Katzmarzy, P. (2012). Effect of physical inactivity on major non-communicable diseases worldwide: An analysis of burden of disease and life expectancy. *Lancet, 380*(9838), 219–229. doi:10.1016/S0140-6736(12)61031-9 PMID:22818936

Lin, J., Mamykina, L., Lindtner, S., Delajoux, G., & Strub, H. (2006). Fish'n'Stepts: Encouraging physical activity with an interactive computer game. *Lecture Notes in Computer Science, 4206*, 261–278. doi:10.1007/11853565_16

Mulder, I., & Kort, J. (2008). Mixed emotions, mixed methods: the role of emergent technologies to study user experience in context. In S. Hesse-Biber, & P. Leavy (Eds.), *Handbook of Emergent Methods in Social Research* (pp. 601–612). New York: Guilford Publications.

Oinas-Kukkonen, H. (2013). A foundation for the study of behavior change support systems. *Personal and Ubiquitous Computing, 17*(6), 1223–1235. doi:10.1007/s00779-012-0591-5

Riediger, M., Schmiedek, F., Wagner, G. G., & Lindenberger, U. (2009). Seeking pleasure and seeking pain. *Psychological Science, 20*(12), 1529–1535. doi:10.1111/j.1467-9280.2009.02473.x PMID:19891749

Rogers, J., Hart, L., & Boltz, R. (1993). The role of pet dogs in casual conversations of elderly adults. *The Journal of Social Psychology, 133*(3), 265–277. doi:10.1080/00224545.1993.9712145 PMID:8412041

Ryan, R. M., & Deci, E. L. (2000). Self-determination theory and the facilitation intrinsic motivation, social development, and well-being. *The American Psychologist, 55*(1), 68–78. doi:10.1037/0003-066X.55.1.68 PMID:11392867

Seeman, T. E. (1996). Social ties and health: The benefits of social integration. *Annals of Epidemiology, 6*(5), 442–451. doi:10.1016/S1047-2797(96)00095-6 PMID:8915476

Staiano, A., & Calvert, S. (2011). Exergames for physical education courses: Physical, social, and cognitive benefits. *Child Development Perspectives, 5*(2), 93–98. doi:10.1111/j.1750-8606.2011.00162.x PMID:22563349

Stralen, M., deVries, H., Mudde, A., Bolman, C., & Lechner, L. (2009). Determinants of initiation and maintenance of physical activity among older adults: A literature review. *Health Psychology Review*, *3*(2), 147–207. doi:10.1080/17437190903229462

Tomaka, J., Thompson, S., & Palacios, R. (2006). The relation of social isolation, loneliness, and social support to disease outcomes among the elderly. *Journal of Aging and Health*, *18*(3), 359–384. doi:10.1177/0898264305280993 PMID:16648391

VanderWijst, H. (2011). Al wandelend praat je anders. *Lopende Zaken, 57.*

Victor, C., Scambler, S., Bond, J., & Bowling, A. (2000). Being alone in later life: loneliness, social isolation and living alone. *Reviews in Clinical Gerontology*, *10*(4), 407–417. doi:10.1017/S0959259800104101

Visch, V., Vegt, N., & Anderiesen, H. & vanderKooij, K. (2013). Persuasive Game Design: A model and its definitions. In *Proceedings of CHI.* Paris: ACM.

Wendel-Vos, W., Droomers, M., Kremers, S., Brug, J., & vanLenthe, F. (2007). Potential environmental determinants of physical activity in adults: A systematic review. *Obesity Reviews*, *8*(5), 425–440. doi:10.1111/j.1467-789X.2007.00370.x PMID:17716300

KEY TERMS AND DEFINITIONS

Game World Experience: An game world is typically experienced by occurrence of the symptoms of being engaging, enjoyable, safe, free and by providing direct feedback.

Gamification: Design of game-elements applied on real-world attributes to create a user experienced game-world.

Motivation: To be moved to do something (Ryan & Deci, 2000).

Persuasive Game Design: Game design aiming to move the user from a real world experience towards a game world experience in order to change the user's behavior in the real world.

Real World Experience: The regular world experience of daily life.

SES (Socioeconomic Status): Socioeconomic status is commonly conceptualized as the social standing or class of an individual or group. It is often measured as a combination of education, income and occupation. Examinations of socioeconomic status often reveal inequities in access to resources, plus issues related to privilege, power and control (definition by American Psychological Association).

Transfer: Effect of user experienced game world on forming, altering, or reinforcing user-compliance, -behavior, or –attitude, in the real world.

User Centered Design: Involvement of users in the design process. Varying from user studies preceding the design, to user evaluation/participation during the design, to user evaluation after the design.

Chapter 7
Power Explorer:
Is Indoctrination Right?

Paschalina Skamnioti
University of Applied Sciences of Bremen, Germany

EXECUTIVE SUMMARY

Persuasive games are often designed for social, political, or environmental purposes to promote particular values and behaviours. In this chapter, the author analyzes the manner in which values are conveyed to the player through the persuasive game Power Explorer, a mobile game for reducing household energy consumption. For the analysis the author takes into account the diverse approaches of Values Education and the criticism they have received. In particular, a) the author illustrates the similarities of the values education designed in Power Explorer with the traditional approach of Character education, and b) the author raises ethical issues regarding the design, the objectives, and the effects of games for change using persuasive technologies[1], in terms of both the individual and the society.

VALUES EDUCATION IN GAMES FOR CHANGE

Games for change (G4C) are designed with the objective to have a social impact regarding political, social and environmental issues. Building awareness upon real-world problems such as poverty, violence, climate change, or human rights is a sensitive task. All these problems are, in fact, rooted in our society's values, which consequently originate from our personal ethical values, as active citizens, motivating and guiding our behaviours. The question is how do we learn and 'change' values intentionally?

DOI: 10.4018/978-1-4666-6206-3.ch007

Copyright ©2014, IGI Global. Copying or distributing in print or electronic forms without written permission of IGI Global is prohibited.

Values Education is a field that examines *how* the moral person (i.e. the player) should be educated. It also touches upon the layers of the *what* (i.e., which values should be promoted?), and the *why* (i.e., what are the end-state goals?). Generally it is defined as "the conscious attempt to help others acquire the knowledge, skills, attitudes, and values that contribute to more personally satisfying and socially constructive lives" (Kirschenbaum, 1995, p.14). However, what is considered 'good' for the individual or the society is not always so clear. The conception of how the moral person should act and think in modern society can be complex and diverse. Throughout the history of Values Education there is substantial literature suggesting, analyzing and criticizing various approaches for educating values; reflecting at the same time changes in the society and its perceptions of education and morality.

It is significant, therefore, when we talk about designing games with socio-political and environmental focus to take into consideration the theories and practices of Values Education. We need to examine the particular ways that values can be reflected through such games, as well as to recognize and be able to deal with some ethical concerns: Who decides which values G4C aim to promote? Which are the educational objectives for the moral agent? Are the players given the opportunity to develop moral reasoning or to discover their own values? What exactly is the change envisioned by G4C, and to what extent are the problems addressed?

For answering in the above questions, I have developed the theoretical Model VEGA (Values Education in Games Analysis). VEGA is the result of a thorough study of the main approaches of Values Education and their connection to the design of G4C. In this chapter, I employ VEGA for analysing the persuasive game Power Explorer. Power Explorer, designed by Interactive Institute in 2009, is a pervasive mobile game for promoting energy awareness. It aims at changing the attitudes of players by measuring their household energy use in real time.

The chapter is organized as follows: First, I present the objectives of the game, the gameplay, and the results of the game trial. Second, I introduce the VEGA Model, clarifying the three main approaches of Values Education according to their basic differential characteristics. Third, I proceed to the analysis of the game. Primarily, I illustrate that the design of Power Explorer corresponds to the approach of Character Education; an indoctrinative approach targeting to instill the 'right' values to the learners by focusing on their behaviour. Given that Character Education has received much criticism, I continue by putting forth significant ethical issues that should be taken into account when using this approach of educating values.

My perspective is that of a game researcher, with some experience in designing games with environmental objectives[2]. VEGA Model is developed during my Postgraduate Studies. In the current analysis, although I appreciate the work that has been done in Power Explorer, I highlight significant inadequacies of the game, under a values educational scope. My intention is, on one hand, to find explanations and ideas for solutions, provided by the field of Values Education. On the other hand, is to offer the ground for a deeper understanding of the ethical and educational aspects of persuasive technologies in G4C. I expect that this analysis can serve as well as an inspirational source for a more effective and ethically complete design of G4C.

CASE DESCRIPTION

Power Explorer

Power Explorer is a pervasive action-oriented multiplayer game designed to develop the skills for energy-efficient consumption at home and a positive attitude towards conserving electricity. The target group of the game is teenagers, and their families. The game focuses on measuring the household energy consumption of the players in real time with sensors that provide feedback when the players switch on and off devices. (Bäng, Svahn, & Gustafsson, 2009; Gustafsson, Bång & Svahn, 2009b)

A main concern for designing Power Explorer was to achieve a persistent and long-term effect in the players' conserving energy attitudes, as it was previously shown that a similar persuasive game, Power Agent (Bang, Gustafsson, & Katzeff, 2007; Gustafsson, Katzeff, & Bang, 2009a), had resulted in extreme but short-term measures, lasting only during the game. Indeed, Power Agent's trial showed that although energy consumption has been reduced remarkably during the game, it "basically returned to normal within a few days after the game trial had ended" (Gustafsson, et al. 2009b). As a way to improve this, the design of Power Explorer provided: a) a sensor system with instant feedback of the players' energy use, rather than every 24 hours in Power Agent, and b) a more casual game style, with small investments by players, than the committed style of Power Agent.

Along with this prototype, many efforts have been done lately for designing environmental games with similar persuasive technologies and design, as EnergyLife (Gamberini, Corradi, Zamboni, Perotti, Gadenazzi, Mandressi, Tusa, Spagnolli, Björkskog, Salo, & Aman, 2011) and LEY (Madeira, Silva, Santos, Teixeira, Romão, Dias, & Correia, 2011).

The Gameplay

In Power Explorer each player has an avatar, a monster blob, and tries to keep it happy and healthy in its garden environment. The goal of the game is to keep the CO2 cloud of your garden smaller than your opponents' clouds. This is controlled by the energy consumed in the player's household in real-time. A player can also choose to have a duel against another player, by manipulating household electric devices.

More precisely, the game provides four different environments with different modes of interaction:

- **The Habitat:** Every time the player is turning on some electric device, some weed is growing up around the monster blob, with a size corresponding to the amount of electricity being consumed *(Figure.1a)*. The monster blob then eats the weed and puffs out small clouds of CO2 in response, while its face looks unhappy. Conversely, when the player is turning off the devices and the energy consumption decreases, there are flowers appearing in the garden. The flowers have a positive effect on the avatar. In the sky there is also a medium-sized grey cloud, from the beginning of the game, which gets smaller with time. However, the CO2 emitted by the monster blob makes the cloud grow. Therefore, in the case of excessive increase of the household energy use, the cloud of CO2 could fill the entire screen, making the monster blob visibly sick.
- **The Pile:** The pile ranks the avatars in accordance with the size of their respective CO2 clouds *(Figure.1b)*. The avatar with the smaller cloud is the player that consumes the least, and as the current winner is on the top of the pile.
- **Duel in the Rainforest:** This duel is focusing on energy and more specifically on learning how much energy is consumed by different devices, when in continuous use *(Figure.1c)*. The two players are competing on a racetrack in a rainforest, while there are some obstacles appearing on their way, causing bad health to the avatars. Some of the obstacles are moving periodically and cannot be avoided by pressing the jump-button, but only by adjusting the speed of the avatar. The speed can be controlled by manipulating quickly the energy consumption; turning electric devices on makes the avatar run faster and vice-versa. Yet, increasing the speed causes a flood in the forest, which could make the avatars drowning. Therefore, the task is to maintain enough speed to overcome the obstacle, but without fatally flooding the forest.
- **Duel in the North Pole:** The player is trying to knock the opponent off the icecap into the water by throwing various objects *(Figure.1d)*. There are light objects such as snowballs, fish and seals, or heavy ones like polar bears and blue whales. The heavier the object is, the more brutal the impact to the opponent. In order to get a polar bear or a blue whale, the player has to turn on a major electricity-consuming device for some seconds, while for a snowball a lamp is sufficient. Seals slide on the ice when thrown and the opponent then has to jump to avoid them. As in the rainforest-duel using extra energy to win the battle is a double edged sword: the players combined consumption rate affects the strength of the sun, which means that the more consumption the less consistent the icecap is.

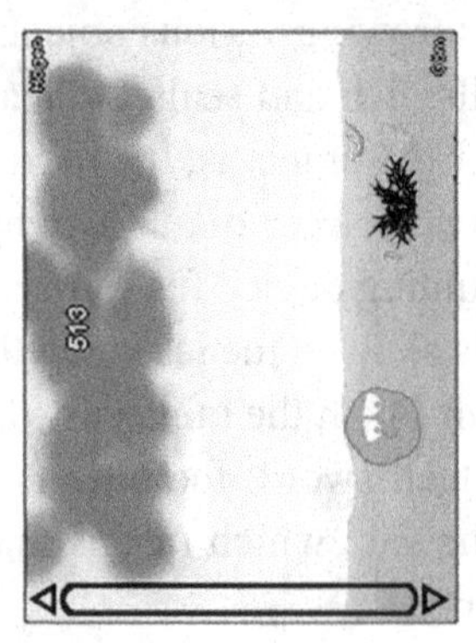

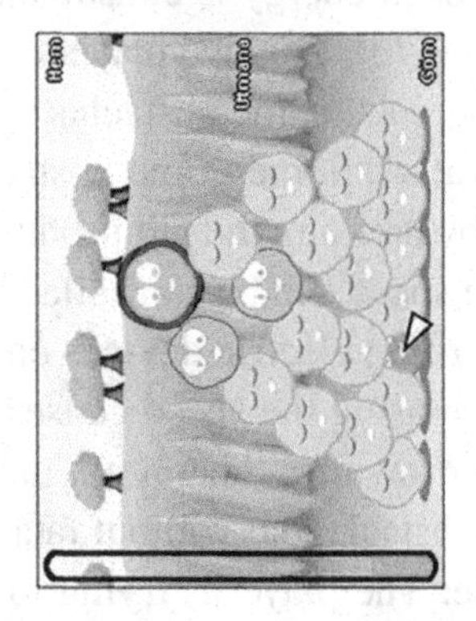

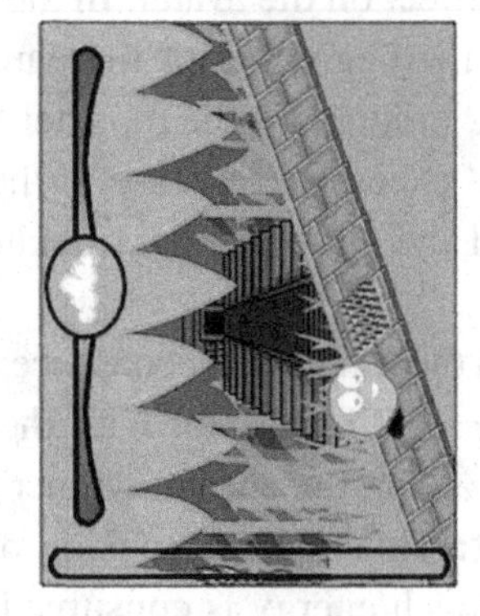

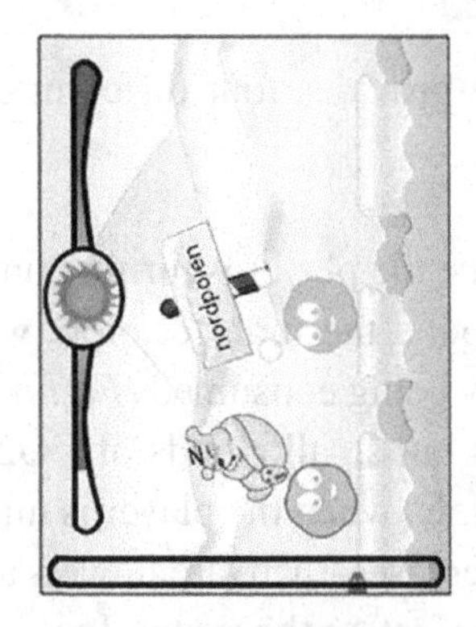

Figure 1. a) The habitat. b) The pile. c) The rainforest duel. d) The north pole duel.

When a player wins a duel, the avatar wears the golden scarf of victory in the pile, while also earns one hour long of immunity against weed. Therefore, his garden gets easier clean from the thistles and the extra CO2 emitted during the duels.

The design of Power Explorer does not provide any informative content, tips, or other reflective messages within the game. The game designers declare that they left intentionally the interpretation of the game events to the players, and their discussions with their family and peers outside the game.

The Game Trial

A test group of 15 players aged of 12-14 years old played Power Explorer for 7 days. The results have been evaluated by comparing the attitudes of the players to a control group of 20 households that did not play the game. The focus of the study was on the post-game effects of the players. Part of the players' interviews and experiences from this game trial are published in scientific papers, as well as the data from the electric measures of the households retrieved during and after the game. The participants were also subjected to a questionnaire before and after the game. Here I include some of the observations in order to get a deeper understanding of the whole project:

By testing this game prototype on teenagers the game designers found out that the energy consumption during and after the game period decreased in the player's household. During the game trial, only three of the 15 participants didn't show any decrease in the use of electricity. As for the 10 weeks following the game trial, the consumption was "on average 14% lower in the player group compared to the control group". "Significance of difference between the two groups however decreased shortly after the game trial ends, indicating a greater overlap between the two groups." "Findings also show a statistically significant positive change in the players' attitude towards saving energy", while the reference group "became slightly more negative". "..participants also saw themselves as more prone to promote energy conservation to their surroundings, indicating a change in self perception." "..the knowledge regarding different appliances was also widely shared and discussed among players in story like fashion". However, "the increase in explicit knowledge in regard to power rating on appliances is marginal" and "the players were actually worse at determining the amount energy used for different tasks after the game." Moreover, "..the game did not appear to have any positive effect on the players' attitude towards the environment in general. In fact, results rather indicated a more negative attitude in this regard". (Gustafsson et al., 2009b)

Players mentioned that what they did differently was "turning off lamps, TVs and computers", generally "things that were unnecessary or that you didn't use". According to the researchers, these are indications of tolerable "non-extreme measures", which could potentially remain after the game. Only in a few cases the players talked about energy saving measures that would "directly infringe on comfort levels", or as they also say "done for the game" (Gustafsson et al., 2009b, p. 5). The above results of this game trial are also compared to the results from the prior game Power Agent, in the pursuit of more sustained behaviour changes by the players. Power Explorer has managed to achieve an improvement in this aspect.

THEORETICAL MODEL

Values Education Game Analysis (VEGA)

Character education, *Moral development* and *Values clarification* are the main approaches to Values Education [5-6]. Each of these approaches is based on different philosophical positions, objectives, and practices. Additionally, they have all received a significantly constructive criticism.

The primary idea for creating the VEGA Model was that all this knowledge from the field of Values Education could be useful for analyzing and designing games, performing values education as well. VEGA Model is now developed, and presented here, with the following affordances:

1. **Identifying the values education approach of a game:** Based on the main differential characteristics of the three main approaches for Values Education, VEGA helps to identify the particular approach followed by a game. In this process, the five questions to be examined are:
    a. **Who decides upon what values introduced through the game:** Are these values considered as core values of the society, conceived as consensual, or is it about values to be questioned by the players? This can be answered by taking a close look at the objectives of the game. Or is it a free-value game, letting the players the freedom to decide upon their own values?
    b. **Right/wrong:** Are these clearly presumed by the game design or do the players search for them through controversies with their consciousness and personal pursuit? For positioning on this, it is necessary to examine the "procedural rhetorics" of the game (Bogost, 2007, p. 3), the arguments deriving from particular game processes.

c. **Role of the game designer as values educator:** Does the game perceive the players as passive receivers of a specific message, or does it provoke them to reflect upon issues and take moral decisions? Is there a judgment for their decisions within the game?

d. **End-state for the individual:** This refers to the objective of the game regarding the player as moral agent. What is the desirable effect of the game experience for the player?

e. **End-state for the society:** This refers to the objective of the game regarding the society. How is society depicted by the game, what would be the ideal social change supported by the game? However, it is not necessary for every game of change to have an end-state goal for the society, since many games are developed with the objective of the player's personal development.

2. **Strategies suggested by each approach:** This, in fact, is the first academic attempt to identify the game strategies for performing and designing Values Education, based on the practices and the theoretical recommendations of each approach, including their criticism. The classification of these design strategies is according to the different approaches, and through case studies I deduct new conclusions to be integrated in the Model. This work can be advanced in the future, by creating a catalogue of design patterns especially for the case of G4C. Unfortunately, in this chapter, I refer only slightly to this part of the Model, due to my emphasis on the next point.

3. ***Ethical matters* that should be taken into account when using each approach:** Ethics open the door for a deeper criticism of the Values Education performed by games. These matters address the *deficiency of indoctrinating*, *moral relativism*, the *consistency of the means and the ends*, and how the *wider view of the problem* is reflected in the game. Finally, another important issue examined is the *commitment* of the players to the promoted values, after the game. These issues are important in my analysis, for offering explanations of significant inadequacies of Power Explorer, and potential solutions provided by the field of Values Education.

Character Education Approach

Character is derived from a Greek word c*harasso (χαράσσω)*, which means, "to mark" or "to engrave". According to Lapsley and Narvaez (2006), "it points to something deeply rooted in personality, to its organizing principle that integrates behaviour, attitudes and values." (Lapsley & Narvaez, 2006, p. 2) There are numerous attempts in defining character, mainly through its relation to behaviour and values. Denoting the connection between these concepts, Character Education aims "to shape directly and systematically the behaviour of young people by influencing explicitly the non-relativistic values believed to bring about that behaviour" (Lockwood, 1997, p. 5).

Character Education is not a new movement. With a history dating back to the beginning of the 20th century, it has been the most dominant approach to Values Education for many decades, always "focusing on the inculcation of desirable habits" (Althof & Berkowitz, 2006, p. 497). Even as a term, it is sometimes mistakenly used as synonym of values education in general. In Europe and US, numerous state governments, organizations, and boards of education still advocate its implementation.

In the early 1990s however, the field of Character Education changed and more methods have been integrated for teaching values. Notwithstanding the new models for implementation aroused in the New Character Education movement, indoctrination remains the most dominant method; the role of the educators is to inculcate specific values and desirable behaviours to the moral agents. The aim of the New Character Education is more explicitly described in the following sentence: "The task of Character Education therefore is to help children develop an increasingly refined understanding of the core values, a deeper commitment to living according to those values and a stronger tendency to behave in accordance with those values" (Lickona, 1996, p. 96).

The Character educators claim that they promote core or universal ethical values to the agents, as most of them see values as socially and culturally accepted standards or rules of behaviour, "starting with the myriad values we all share" (Etzioni, 1998, p. 447). As it is evident, right and wrong values are completely defined and the objective for the students is to incorporate these values into their own value system and the right behaviours into their lives. As for the society, the advocates of Character Education believe that by achieving the character education of the individuals, the morality of the society will be also restored.

Character education programs are most commonly using the following practices: clear expectations of the moral agents, with a special focus on controlling their behaviour, rewards and punishments, exemplars, involvement of the family, peers and community, and praising specific values for the day/week.

Moral Development Approach

In the case of the Moral Development approach, the aim is to develop the critical thinking and moral reasoning skills of the players, by exposing them in dilemma-situations and alternative viewpoints. According to Kohlberg (1973), the role of the educator is to move the student through the above stages of moral development and to make them able to function as rational actors and to deal autonomously with ethical issues. The approach of Moral Development is based on the structure of judging morally and not upon specific beliefs, emphasizing the processes of thinking and not its content.

The only values been recognized as universal and directing the whole approach are justice and democracy. Gilligan's research (1977) includes also care. In this developmental process, the educator plays a deliberate role that involves valuing, criticizing, and evaluating and is considered to be partly indoctrinative, as it has the direction of justice.

Right and wrong in this approach depend on the reasoning of each individual, which depends on which moral stage the agent belongs to at the current state. Progressively, right could be; obeying the authority and avoiding punishment; fulfilling immediate needs; what pleases others and gets approved by them; honouring and sustaining the law by doing one's duty; the general rights and obligations of individuals for a functioning society beyond the minimum standards set by law; or at the sixth stage of moral development, right is based on individual decisions of conscience made in accordance with universal principles of justice.

The main strategies proposed by the Moral Development theory are: moral dilemmas, role-playing, fostering empathy, conflict resolution, and choosing endings in unfinished stories.

Values Clarification Approach

The departure point of Values clarification is that both individual and societies are confronting a wide range of conflicting values (Simon, Howe, & Kirschenbaum, 1972), on the one hand performing great acts of charity and construction and on the other hand perpetrating moral atrocities and environmental destruction (Kirschenbaum, Harmin, Howe, Simon, 1977). Values Clarification is aiming to resolve this value confusion, not by identifying and transmitting the right values, but by helping the moral agents clarify and obtain their own values. Values are viewed as relative, personal, and situational. Individuals should be capable then, to prize and cherish more of their choices, beliefs and activities, to experience a stronger self-concept and to give a greater meaning in their lives.

Consequently, right is what the moral agent is valuing, as soon as he is reflective upon his values. Therefore, Values Clarification is an approach that is based upon the process of discovering one's own values, and not upon the content, which is considered to be values-free. As for the role of the educator, the rejection of the traditional method of indoctrination is a *raison d'être* for Values Clarification movement. The educators are suggested to be absolutely neutral, and non-judgmental, accepting all the different viewpoints and values, without intervening in the discussions by arguing or valuing.

However, to make moral value decisions a matter of personal taste is to make it possible for any and every ethical system to be justified (this is what relativism supports). This means that Values Clarification does not have a predetermined idea of what would be the end-state goal of society. Sometimes even fundamental societal values are treated in the same way as any other value.

As Values Clarification process is focusing on assisting the moral agents to discover their own values, the strategies suggested are mostly activities of evaluating and identifying the most desired pursuits in life.

The above conclusions are summarized in the following table (Table 1), indicating according to the VEGA Model the main differential characteristics of the main approaches to Values Education.

ANALYSIS OF THE CASE

Character Education in Power Explorer

In this first part of my analysis, I examine Power Explorer regarding to the above differential characteristics of the diverse philosophies for Values Education. My aim is to demonstrate that the values education designed in Power Explorer is attributed to Character Education.

Table 1. VEGA model: identifying the values education approach adopted by a game for change

	Character Education	Moral Development	Values Clarification
Whose Values	Values considered universal	Player's values to be questioned	Player's values to be clarified
Right/Wrong	Clearly defined Right/ wrong presumed	Depending on the stage of moral development Universal justice	Only the player's own values No 'right/wrong'
Role of the Game Designer	Indoctrinative Focus on behaviour	Partly indoctrinative Direction of justice	Absolutely neutral Non indoctrinative Non judgmental
End-State for the Individual	Good character Right behaviour	Moral judgment through reasoning Autonomous critical thinking	Clarified, purposeful, committed
End-State for the Society	Preserving the right values, changing the individuals holding the wrong ones	Global justice, care, democracy	Not under value confusion, with members acting upon their clarified values

Which are the values addressed by the game, are they considered 'universal'?

The game targets on changing the attitudes of teenagers towards reducing energy use at home. Domestic energy conservation is a central goal both of environmental policy and environmental education, along with other practices serving the more general concept of sustainability. Presumably, the behaviour promoted by Power Explorer is clearly addressing universal values of our society; "Sustainable living must be the new pattern for all levels: individuals, communities, nations and the world. To adopt the new pattern will require a significant change in attitudes and practices of many people." (IUCN/UNEP/WWF, 1991, p. 5) Furthermore, the long-term commitment of the players in this sustainable behaviour of energy conservation is also an indisputable objective. The education for sustainability gives also emphasis to the importance of long term improvement than only immediate environmental actions (Tilbury, 1995).

The game, in order to make the expected sustainable behaviour more explicitly clarified, engages the players in learning specifically about the impact of different electric appliances. This focus on the behaviour and the clear directions for achieving it are also characteristics of Character Education.

Are right and wrong values clearly defined and presumed in the game?

The investigation at this point needs to be conducted by taking a look at the procedural rhetorics of the game, which are actually based upon this premise: consuming less energy is the good behaviour to be adopted, while keeping the electric appliances on, when not in use, is considered negligent and has to be avoided. The duels, on the other hand, challenge the players to learn about the concept of continuous electricity consumption - by leaving "the right combination of consuming appliances on, (though not too much)", as well as the concept of power used by an appliance - turning quickly the right appliances on/off (Bäng et al., 2009, p. 5). The winning strategies in each case indicate directly the right and wrong behaviours to be adopted.

Is the educational role of the game designer indoctrinative?

Saving energy at home is a ready-made truth directly conveyed to them through the game. The players are motivated to learn the ways to conserve electricity at home, and not to reflect upon this environmental issue, which would affect the game. In case of any opposition from the part of a player, the only option given by the game design is not to comply to the expected attitude, and as a consequence to get a position in the bottom of the pile. There is no other means of interaction with the game, even if the players can discuss upon different perspectives on this matter, outside the game. This means that the flow of the knowledge is only from top to down, and apparently, that the role of the game designers as educators is indoctrinative.

What is the end-state goal for the individual?

The end-state goal for the individual moral agent is adopting the right behaviour of saving energy. At the same time much attention has been given to clarify to the players the energy and power consumed by each appliance. One of the challenges as well, is to make the players commit to this behaviour persistently, after the game. These objectives correspond exactly to the ones advocated by the Character Education approach; to "understand the core values, adopt or commit to them, and then act upon them in their own lives" (Lickona, 1993).

What is the end-state goal for the society?

The long range goal of Power Explorer is contributing to the global energy problem by changing the energy usage behaviours of the players. It is premised that small individual efforts at home might be crucial for reducing the emissions of CO_2. (Bäng et al., 2009; Gustafsson et al., 2009b) The game is also seen as a challenge towards sustainable living, a social transformation that is also expected by the people adopting energy saving habits. Hence, the end-state visions of the game are in fact dealt by changing the individuals towards the right behaviours.

Since the design of Power Explorer fulfils all of the characteristics of Character Education, then this game educates values to the players following this approach. What does this signify for the players, for the game design or for the aims of the game? This is analysed in the following part.

Ethical Issues around Character Education

Character education raises a number of critical questions that its advocates have not satisfactorily addressed. These issues are clarified below, in order to contribute to the analysis of Power Explorer.

Indoctrination

A major area of criticism surrounding character education involves the fact that it is an indoctrinative approach. Indoctrination is commonly associated with such terms as inculcating, moralizing, preaching, and instilling. What actually means is that the educator sets particular values and behaviours to the moral agents, without allowing them "to subject the idea to objective reasoning" (Huss & Patterson, 1993, p.239). Several theorists have rejected this educational method, from many different perspectives.

Hindering moral reasoning and critical thinking: Primarily, the matter mostly recognized as deficient in indoctrination is that it hinders moral reasoning and critical thinking. The moral agents learn the right behaviour, without appealing to any reason for internalising this behaviour, they are committed to the what, but they do not understand the why (Raywid, 1980). Character education, "must recognize the possibility of people's doing the right thing for the wrong reason, or for no reason at all." (Pritchard, 1988, p. 475)

Examining the design of Power Explorer, it is evident that the game does not provide any reason to the teenagers for adopting the promoted habit of conserving energy at home. The cloud of CO2 has some connotation to the real problem, but no actual connection e.g. to the real impact of the greenhouse gas emissions or the consequences of the global warming. The urgent need for humans to consume less non-renewable energy not only is not emphasized, but it is not reflected at all.

As a result, the players do not acquire any understanding on why to save energy, and to continue saving, as it appears in the interviews of the players after the game trial. Sara, for example, on the question of what she thought about the subject of saving energy, she answered that she didn't have an opinion. Willem, the most engaged player, had no arguments for supporting the behaviour of consuming less energy and seemed not convinced by its significance: "Well… you don't have to use it unnecessary. I don't feel that it is that important but… You can't just use… nothing. You have to use some". Adam, on the other hand, although he clearly felt that not using too much electricity was important, when he talks about his reasons he states: "..it is good to save electricity because it costs very much and so… it is not good". (Gustafsson et al., 2009b, p.6)

Limiting moral autonomy and imagination: Constraining the learners to think in a prescribed way, limits their moral autonomy, as well as their imagination and their ability to think of alternatives (Tan, 1990). It is also associated to a future disability of the moral agents to adjust and find new solutions in potential problematic situations in the future (Paske, 1986).

In the specific case of the game, no alternatives are presented to the players (i.e. renewable energy sources) neither there is any teaser to challenge them to discover new saving energy techniques in their home (i.e. better isolation). The players are focused just on consuming less than the others, while experimenting with the devices that they already use. In the previous game (Power Agent), the players improvised by using i.e. candles for illumination (Gustafsson et al., 2009a). However, this was considered by the game designers as an "extreme" strategy that would not be continued after the game, and therefore was replaced by more "casual" activities (Gustafsson et al., 2009b).

In my view, challenging the imagination of the players to seek for alternative and even radical solutions, could only add creativity, knowledge and fun to the game experience, even if these strategies last only during the game. Especially regarding the energy problem, it would not be surprising to see inventions of new devices using renewable sources that could be used for a lifetime and even promoted further, shared and improved. The fact that the players are not committed to saving energy after the game should be more associated to the lack of reasoning, than with this feature.

Danger of educating ill values: As Character education is targeting mainly in promoting specific values chosen by the educators, it is very important to take a close look on the particular values transmitted to the agents when this approach is used. Indeed, Character Education could end up dangerous, if the promoted values support an ill system. It is a fact that Character Education tradition has actually "characterized the German school system during much of his history prior to the Nazi takeover, and certainly during the Nazi era itself." (Primack, 1986, p. 12)

In Power Explorer, this danger does not exist. As I have illustrated already, conserving less energy is considered a commonly accepted behaviour, serving the indisputable goal of sustainability.

Contradiction with democracy: The argument here is that imposing to the moral agents a "bag of rules that he or she may not understand or accept and may very likely resist", undercuts the very democratic process itself; "The most deadly pedagogical sin is moral imperialism" (Beach, 1992, p. 31). The uncritical acceptance of behaviours and values as "dogmas" has been also considered dangerous for the continuation of democracy, which needs active critical thinkers rather than "blindfolded adherents" (Kilpatrick, 1972, p. 52).

The game of Power Explorer does not provide any means for discussion or decision-making among the peers that could affect the gameplay. As a democratic feature, I can only imagine the common targets and responsibilities of the family members in saving energy. This of course does not make the game non-democratic, but perhaps it could inspire for more democratic affordances of the persuasive design for change, especially when addressing values as sustainability.

Universal Values vs. Relativism

Ethical Relativism is a philosophical theory, claiming that "there is not a single objectively true morality but only many different moralities".(Harman, 2012, p. 1) The relativists held that values differ from society to society, and therefore we are not justified in making cross-cultural judgments. As this theory has been generalized to persons, this means that each ones' values have to be respected, and that there are no ethical standards for judging. Consequently, any universally accepted value is trivial, as there are no absolute truths. This theory is fundamental for the approach of Values Clarification, opposing the absolutism of Character Education.

Moral relativism is also associated with the idea of tolerance (Benedict, 1934, p. 37), as the individuals are expected to be tolerant of diverse moral practices, "as the one virtue he must accept" (Harrison, 1976, p. 131). However, much criticism of relativism lays on this understanding. The reason is that tolerance may lead to "moral neutrality and inaction" in situations as for example genocides, slavery, rapes etc. in which basic human rights are violated (Hatch, 1997, p. 372). Some other theorists find that the basis of the theory is the *enculturation* (Renteln, 1988, p. 62), which means that our "evaluations are relative to the cultural background out of which they arise" (Renteln, 1988, p. 59). Our family, our peers and our social environment provide us with different ethical traditions that form the ground of our morality.

Character Education does not support the theory of relativism, since believes in universal common values. The values addressed by the game, as I have demonstrated, are also considered universal. Then how the theory of relativism could contribute to my analysis? What I need to do here is a more in depth investigation of the values addressed by the game, with the new prism of relativism. Starting with the value of sustainability, I believe that it can be definitely considered as indisputable. Does the same apply to the behaviour of not wasting energy? If we think about the consequences, burning less fuel, as a way to tackle the problem of global warming, affects not only all the humans but all the fauna and flora on earth. So far, the game is not relativistic.

However, why do we use the term 'wasting' energy? Do we use energy only for covering our basic needs, or for empowering a global industrial market, manufacturing continuously new products far from our needs? Is it a universal value to spend a non-renewable energy source, in order to feed our consumerist needs? The answer is no, since first of all 2–3 billion people worldwide currently lack access to modern forms of energy. The players and their families, on the contrary, belong to this part of the global society that consumes energy to a large extent, and the teenagers have been raised with this comfort. The behaviour of consuming energy has been internalised during their socialization.

The question here is: does the game do any moral judgment on this matter, or does it keep a 'tolerant' position? The game not only takes the players' way of living as granted, without any intention to change it, but by adopting a casual game style it gets even more adjusted to their daily habits. This leads to the conclusion that the game is relativistic on this matter. An additional remark is that the players compete to each other for reducing less energy, without any other, objective standards that they need to reach.

Moreover, the social environment of the teenagers is probably accounted for the connection that they do, as it seems from the interviews, between the energy consumption and the economical cost. In my short visit at the official website of the Swedish Energy Agency, in the section of sustainability, I found a brochure informing about saving domestic energy, which has the general title: "Save Energy - and make your housekeeping money last longer." (Swedish Energy Agency, 2009) This money-related perception of sustainability is also not addressed by the game, and therefore remains.

Exploring the values of the social environment of the players, in order to identify why they might have this wrong behaviour that needs to change, is significant for designing in line with the suggestions of Values Education. Kohn, based upon evidences of social psychology, states that "much of how we act and who we are reflects the situations in which we find ourselves" (Kohn, 1997, p. 5). If we need the players to change an attribute which characterizes their surroundings, then we need first to acknowledge it and then to find ways to address it in the game design. Otherwise, the easy way for them is always to go back to what they are accustomed to, and what all the others do. The importance of "accounting for culture in design" is already recognised in the field of designing serious persuasive games (Khaled, Barr, Biddle, Fischer, & Noble, 2009, p. 7).

Wider View of the Problem

Kohn (1997) accuses the advocates of Character Education for having a deficient view of social problems, as they usually ignore the political and economic realities; "Never mind staggering levels of unemployment in the inner cities or a system in which more and more of the nation's wealth is concentrated in fewer and fewer hands", the character educators just focus on imposing changes to individuals (Kohn, 1997, p. 430). Perry Glanzer (1998) talks as well about "structural injustices" that should be considered when pursuing values education, as well as about the position of the people within those structures (Glanzer, 1998, p. 435).

In the case of Power Explorer, the problem is the energy waste, a global problem with well-known causes. The majority of energy consumption does not occur in the domestic level (11% of the total global energy use), but mainly by the sectors of manufacturing, construction, mining and transportation (Eurostat, 2011). Without underestimating the worth of any individual effort for saving energy, it must be acknowledged that the game, fails on giving a representative picture of the causes of the problem. The players learn that the more efficient they are in using domestic energy, the closer they get to the global environmental sustainability. Consequently, they may feel then content with their environmental consciousness, just by turning off some appliances when not in use, and by keeping their avatar happier than the neighbours' one.

Moreover, the game design does not refer at all to the global consequences of the problem, which could help also the players giving a real meaning to their individual efforts for tackling it. By presenting to them that climate change, for example, is a human generated problem due to the greenhouse gases, they might feel a greater responsibility for changing their domestic behaviours. Additionally, showing that some of the predicted consequences of climate change are already present in our world, and that they are not only potential risks to be avoided by making our avatar jumping, gives a more real dimension of the effects.

What is also ignored by the game is the significant role of politics in the global energy problem. It is well-known that sustainable consumption, apart from the consumers, requires action by the industry and the governments. Energy efficiency for example, particularly by households, depends to a large extent on the infrastructure provided. The role of the state is also crucial in the implementation of environmental legislation in the industry structure and processes, as also in the public transport. State energy policies also determine the choice of the energy sources: Is it about maximizing energy efficiency and minimize energy use? Is it about developing new sources of clean energy? Or the focus is simply on finding the cheapest energy whatever the ecological costs? All these are political decisions that directly effect on the household energy consumption in any given country, and citizens should be therefore well informed.

These important aspects are missing in Power Explorer. This is probably also the reason that the general environmental concerns of the players after the game were decreased. It is possible that the players felt that they completed their duties to the environment by having played the game, and therefore they do not need to care anymore. It appears that the game design would be more efficient under a values educational scope, by providing to the players the real and larger context of the problem (causes, consequences, and structural aspects). In my view, this would stress more the need for taking action towards domestic energy conservation, and could potentially even lead them to alternative actions towards global energy conservation, as for example consuming less industrial products in general, acting locally for environmental causes, or being a more active and environmentally concerned citizen.

Consistency of Means and Ends

Sometimes, although we might have an ultimate goal to achieve in our lives, we might find ourselves using means different or even conflicting to this end-state goal. Rokeach (1973) has distinguished two fundamental types of values, according to their purpose: *terminal values*, which are preferred end states corresponding to the needs of human beings, and *instrumental values*, preferred ways of behaving as means for attaining terminal values (Rokeach, 1973). The problem that Kohn (1997) has noticed often in character education is that the supported behaviours might be "potentially controversial" to the terminal desirable values, or that there is "a superficial consensus" that might dissolve when we take a closer look.

The practice, for example, of setting individual players or groups opposing each other in a quest for triumph, cannot be claimed that promotes 'cooperation'. Even if the groups work together, "cooperation becomes the means and victory is the end" (Kohn, 1997, p. 433). Such inconsistency is encountered often in games. Especially regarding persuasive techniques, it is common that they "might produce desirable ends, from the perspective of the creator or sponsor of a persuasive technology. But they do not necessarily produce desirable means." (Bogost, 2008, p. 16)

In Power Explorer, the terminal values are sustainability and energy conservation. Although, the teenagers collaborate with their family members, the game is competitive in the sense that they compete against each other. Competing is not at all compatible with the concept of a sustainable planet. Evidently, there is an inconsistency of means and ends; while collaboration is the end, the means is competition. This competitiveness, according to Kohn, could lead the moral agents "to accept competition as natural and desirable, and to see themselves more as discrete individuals than as members of a community" (Kohn, 1997, p. 433). Moreover, the particular inconsistency of the game design makes more difficult to acknowledge if the players save energy because they are aware of the significance of this action, or because they just want to defeat the other opponents.

Moreover, although the main objective of the game is to consume less energy, during the duels the players are asked to turn on devices, wasting energy. The more energy-consuming is the device, the faster your avatar runs, and the heavier object you can throw to your opponent. The only constrain in the energy wasting is the risk to overflow the rainforest. Moreover, winning a duel gives to the player a one hour long immunity against weed, enough for the cloud and the garden to recover. This "conflicting design" is recognized by the designers (Gustafsson et al., 2009b, p. 7), and according to their statements, "seems to have contributed positively to the dynamics of the game". However, it is certainly one more inconsistency of the game, from the values educational perspective.

Commitment

One of the main challenges of Power Explorer is to achieve persistent behaviour changes, not only during the game, but also after the game ends. The commitment of the people to their values after the educational process, and ideally for a life time,

Figure 2. The player is throwing a polar bear to the opponent, result of increasing much and fast his consumption

is a goal discussed by almost all the approaches of Values Education. Hence, it is important to take a look on their philosophies regarding this issue.

The approach emphasizing less the matter of commitment to our ethical values, is Moral Development. Kohlberg believes in the opportunity for a continuous moral growth through the development of cognitive skills such as critical thinking and moral decision-making. When the agent reaches a new stage of moral development, cannot reason as in the previous stage. This higher thinking is what works as a 'safety valve' for the commitment of the moral agents. However, individuals do not always behave according to what they think it is right. This gap between moral judgment and action is one of the deepest criticisms of Kohlberg's theory (Blasi, 1983).

In Values Clarification approach on the other hand, commitment plays a different role. The moral agents after choosing their own values free (between alternatives and after thoughtful consideration of the consequences), they should be also willing to publicly affirm them, and to act upon these values repeatedly, in some pattern of life (Raths, Harmin, & Simon, 1966). These activities constitute the fundamental criteria for defining a 'value', which means that behaving according to a value is considered essential. Yet, the problem that the advocates of values clarification fail to address is that of determining what behaviours are consistent with what values (Lockwood, 1975).

In Character Education, these behaviours are clearly defined. The major concern and one of the main principles of the movement is to help the moral agents to "know the good, love the good and do the good" (Ryan, 1993, p. 16): *Knowing the good* in this approach, can be achieved through explanation and exemplars. *Doing the good* can be achieved through opportunities for moral action and controls. *Loving the good* is what actually connects the other two, what motivates the individuals while knowing the right, to act according to the right and to continue doing that. This so-called "moral emotion" (Ryan, 2002) in Character Education, is what "serves as the bridge between judgment and action" (Lickona, 1996)

During the game trial of Power Explorer, the players get to know the good, by learning explicitly how to save energy, and they do the good (they do save energy). The problem is that when the game finishes, although they still know that saving energy is good, they are not as much motivated to continue doing it. The question then is: What happens with loving the good? Do the players feel emotionally committed to conserving energy?

The answer is: yes, but only through the fun of the game. They players like to compete and as stated in the interviews, they like "to see how much electricity you use!" (Gustafsson et al., 2009b, p.6) Despite this, however, the game design does not attempt to evoke any other feelings to the players, related to the behaviour of conserving energy (i.e. appreciation for the environment, or the feeling of responsibility). They might only feel some empathy for their virtual monster blob being suffocated. As a result, Power Explorer achieves to make the players loving the good game, but not necessarily loving the good impact of saving energy. As the game ends, there is no moral emotion to drag them into action.

The game designers seem to acknowledge this too; "As the motivation to change behaviours comes from winning the game and not from the benefits from adopting the behaviour there is always a risk that the participants will fall back into their old behaviour when the game ends." (Gustafsson et al., 2009b, p. 2) However, their attempts to improve the commitment of the players (real time feedback and casual game style), rather focus on making the game better, than on making the players emotionally committed to the game values and objectives.

CONCLUSION

Education of values occurs in any game for social and environmental awareness. VEGA Model contributes to the analysis and the design of G4C with significant conclusions from the field of Values Education. In this chapter, I used VEGA for analysing Power Explorer, a persuasive game aiming at conserving energy at home while learning about the electric appliances.

According to my analysis, the game corresponds to all the characteristics of the Character Education approach: a) the values promoted by the game are commonly accepted, universal values, with a focus on the behaviour, b) the right/wrong behaviours are predefined for the players, c) the role of the game designers is indoctrinative, as the game imposes the specific values/behaviour to the players, d) the end-state objective for the individual is to adopt this behaviour and to commit to it after the game, e) the end-state goal for the society is sustainability, but actually depends on correcting the behaviours of the individuals.

As a next step, I clarify the main pitfalls of Character Education, stressed by the critiques of the movement. These issues can be seen as well as guidelines for ethical game design when using this indoctrinative approach. Analysing Power Explorer in relation to these issues offers explanations for the inadequacies of the game, and potential ideas for improvement.

More specifically, the game should develop *moral reasoning* to the players for internalizing the specific values/behaviour, and provide the *real extensions of the problem - consequences, other causes, political structures*, in order to give a meaning to their individual efforts for tackling it. These seem to be the reasons that, in the interviews, the players are not aware of the energy problem, and their general environmental concern has decreased. Furthermore, the game does not challenge the *imagination* of the players (i.e. seeking for alternative solutions, or other behaviours towards sustainability), in order to prepare them to confront possible future situations (*moral autonomy*). Especially since the game addresses sustainability, it could also have a more *democratic character* (decision-making). It is highlighted for the designers to take also into account *the values of the players' social environment*, which might justify why the players are holding the wrong behaviours. Power Explorer is relativistic, as it does not address consumerism. The design is also *not consistent between the ends (objectives) and the means* used for achieving them; Competing for sustainability and wasting energy for learning how to conserve. Moreover, the game design, according to the theory of Character Education, had to invest on fostering *moral emotion* to the players for the promoted values/behaviours in order achieve the commitment of the players after the game ends. Otherwise, the players save energy only during the game, just because they like to play.

The above conclusions provided by the VEGA Model, offer the ground for further reflection and discussion about the ethical affordances of the G4C using persuasive technologies. Consequently, the following critical questions rise: Do G4C based on persuasive technologies follow by definition the Character Education approach for educating values to the players? How the game-designers following this approach deal with the ethical issues stressed above? I prefer not to give answers here to the above questions. I will only refer to Ian Bogost, who in his book 'Persuasive Games', differentiates persuasive games from persuasive technology and the concept of 'captology' introduced by B. J. Fogg (Fogg, 2003). He states:

... all of Fogg's techniques use technology to alter actions or beliefs without engaging users in a discourse about the behavior itself or the logics that would recommend such actions and beliefs." "..not fundamentally concerned with altering the user's fundamental conception of how real-world processes work. Rather it is primarily intended to craft new technological constraints that impose conceptual or behavioural change in users." "..the phrase (captology) itself conjures the sense of capture, of arrest and incarceration by an authority. A better name for Fogg's work would perhaps be manipulative technology". "..Fogg is perhaps unaware of the ideology he himself inhabits: one in which existing power structures always device ethical and desirable goals. (Bogost, 2007, p. 59, 61,62)

Consequently, the most important question is: Do games for change attempt to make socially emancipated, and moral autonomous thinkers, or to correct the behaviours of the players in the name of indoctrination and obedience?

RECOMMENDATIONS

VEGA Model provides a Values Education prism that sets up many creative spaces of ideas for designing games for change. In this chapter, I introduced alternative philosophical perspectives for Values Education, in which the objectives for the individuals and the society, as well as the role of the game designers differ significantly from the Character Education approach.

A first idea, for dealing with the weaknesses of Character Education, is by integrating effective practices from the approaches of Moral Development and Values Clarification, as i.e. moral dilemmas, decision making, role playing, empathy, considering the impacts of an action, or alternative actions. The games could include these new features in their gamestory, while keeping the persuasive technologies. Social media and web-content could also be helpful towards this direction. The design of these strategies could be supported as well by multidisciplinary counselling (i.e. political science, social/environmental education).

A second idea is related to the five foundational characteristics of Character Education. What if the designers could 'play' each time with one or two of them, while keeping the others stable? In this way, they would maintain persuasion to some extent, while at the same time creating different ethical experiences. This does not mean that the whole approach would change. The element of controlling the behaviours of the players, for example, could remain, as behaviour is the real enactment of values, and control is a foundational quality of games. I do not personally disagree also with the promotion of some values as universal. However, what if although there could be an end-state universal value, persuasive technologies were used for expressing different or opposing perspectives? What if right and wrong are not predefined, but the game lets the players find them? Or if persuasion was applied both for the right and the wrong behaviour? What if two different right values/behaviours were promoted, but at some point the players had to prioritize them? In few words, what if the route leading to the predefined values could belong to a larger extent to the explorative player? The role of the game designers then would be to enrich it with ethical adventures.

In general, I think that persuasive design for change should become more trustful to the human nature and its cognitive and emotional skills. It is on the hands of the game designers, and not on the technologies, to let more freedom to the players to interact with the game, and reflect on their own values as well as on the values promoted by the game. Personally, I would love to see and participate in such efforts, changing what we think, and not only what we do.

ACKNOWLEDGMENT

Prof. Barbara Grüter, Adrià Alcoverro, all the Skammioti family, and Vassili Bagiati.

REFERENCES

Althof, W., & Berkowitz, M. W. (2006). Moral education and character education: Their relationship and roles in citizenship education. *Journal of Moral Education*, *35*(4), 495–518. doi:10.1080/03057240601012204

Bång, M., Gustafsson, A., & Katzeff, C. (2007). Promoting new patterns in household energy consumption with pervasive learning games. In *Persuasive Technology* (pp. 55–63). Springer. doi:10.1007/978-3-540-77006-0_7

Bäng, M., Svahn, M., & Gustafsson, A. (2009). Persuasive design of a mobile energy conservation game with direct feedback and social cues. In *Proceedings of DiGRA 2009 Breaking New Ground: Innovation in Games, Play, Practice and Theory*. DiGRA.

Beach, W. (1992). *Ethical education in American public schools*. Washington, DC: National Education Association.

Benedict, R. (1934). *Patterns of culture* (Vol. 8). Boston: Houghton Mifflin Harcourt.

Blasi, A. (1983). Moral cognition and moral action: A theoretical perspective. *Developmental Review*, *3*(2), 178–210. doi:10.1016/0273-2297(83)90029-1

Bogost, I. (2007). *Persuasive games: The expressive power of videogames*. Cambridge, MA: The MIT Press.

Bogost, I. (2008). Fine processing. In *Persuasive technology* (pp. 13–22). Springer. doi:10.1007/978-3-540-68504-3_2

Etzioni, A. (1998). How not to discuss character education. *Phi Delta Kappa International*, *79*(6), 446–448.

Eurostat. (2011). *Energy, transport and environment indicators*. Luxemburg: Publications Office of the European Union.

Fogg, B. J. (2003). *Persuasive technology: Using computer to change what we think and do*. San Francisco: Morgan Kaufmann Publishers.

Gamberini, L., Corradi, N., Zamboni, L., Perotti, M., Gadenazzi, C., & Mandressi, S. et al. (2011). Saving is fun: Designing a persuasive game for power conservation. In *Proceedings of the 8th International Conference of Advances in Computer Entertainment Technology, ACE'11*. ACM. doi:10.1145/2071423.2071443

Gilligan, C. (1977). In a different voice: Women's conceptions of self and of morality. *Harvard Educational Review*, *47*(4), 481–517.

Glanzer, P. L. (1998). The character to seek justice: Showing fairness to diverse visions of character education. *Phi Delta Kappan*, *79*(6), 434–436.

Gustafsson, A., Bång, M., & Svahn, M. (2009b). Power Explorer – A casual game style for encouraging long term behavior change among teenagers. In *Proceedings of the International Conference on Advances in Computer Enterntainment Technology* (pp. 182-189). ACM. doi:10.1145/1690388.1690419

Gustafsson, A., Katzeff, C., & Bång, M. (2009a). Evaluation of a pervasive game for domestic energy engagement among teenagers. *Computers in Entertainment*, *7*(4), 54. doi:10.1145/1658866.1658873

Harman, G. (2012). Moral relativism explained. Unpublished, written for a volume edited by Bastian Reichard.

Harrison, G. (1976). Relativism and tolerance. *Ethics*, *86*(2), 122–135. doi:10.1086/291986

Hatch, E. (1997). The good side of relativism. *Journal of Anthropological Research*, *53*(3), 371–381.

Huss, H. F., & Patterson, D. M. (1993). Ethics in accounting: Values education without indoctrination. *Journal of Business Ethics*, *12*(3), 235–243. doi:10.1007/BF01686451

IUCN/UNEP/WWF. (1991). *Caring for the earth: A strategy for sustainable living*. London: Earthscan.

Khaled, R., Barr, P., Biddle, R., Fischer, R., & Noble, J. (2009). Game design strategies for collectivist persuasion. *Proceedings of the 2009 ACM SIGGRAPH Symposium on Video Games* (pp. 31-38). ACM. doi:10.1145/1581073.1581078

Kilpatrick, W. H. (1972). Indoctrination and respect for persons. In *Concepts of indoctrination: Philosophical essays* (pp. 37–42). Routledge.

Kirschenbaum, H. (1995). *One hundred ways to enhance values and morality in school and youth settings*. Needham Heights, MA: Allyn and Bacon.

Kirschenbaum, H., Harmin, M., Howe, L., & Simon, S. (1977). In defense of values clarification. *Phi Delta Kappa International*, *58*(10), 743–746.

Kohlberg, L. (1973). Stages and aging in moral development-some speculations. *The Gerontologist*, *13*(4), 497–502. doi:10.1093/geront/13.4.497 PMID:4789527

Kohn, A. (1997). How not to teach values: A critical look at character education. *Phi Delta Kappan*, *78*, 428–439.

Lapsley, D. K., & Narvaez, D. (2006). Handbook of child psychology: Vol. 4. *Character education* (6th ed., pp. 248–296). Wiley.

Lickona, T. (1993). The return of character education. *Educational Leadership*, *51*(3), 6–11.

Lickona, T. (1996). Eleven principles of effective character education. *Journal of Moral Education*, *25*(1), 93–100. doi:10.1080/0305724960250110

Lockwood, A. (1975). A critical view of values clarification. *Teachers College Record*, *77*(1), 35–50.

Lockwood, A. (1997). *Character education: Controversy and consensus*. Thousand Oaks, CA: Corwin Press.

Madeira, R. N., Silva, A., Santos, C., Teixeira, B., Romão, T., Dias, E., & Correia, N. (2011). LEY! Persuasive pervasive gaming on domestic energy consumption-awareness. In *Proceedings of the 8th International Conference on Advances in Computer Entertainment Technology*. ACM. doi:10.1145/2071423.2071512

Paske, G. H. (1986). The failure of indoctrination: A response to Wynne. *Educational Leadership, 43*(4), 11–12.

Primack, R. (1986). No substitute for critical thinking: A response to Wynne. *Educational Leadership, 43*(4), 12–13.

Pritchard, I. (1988). Character education: Research prospects and problems. *American Journal of Education, 96*(4), 469–495. doi:10.1086/443904

Raths, L. E., Harmin, M., & Simon, S. B. (1966). *Values and teaching: Working with values in the classroom*. Columbus, OH: Charles E. Merrill.

Raywid, M. A. (1980). The discovery and rejection of indoctrination. *Educational Theory, 30*(1), 1–10. doi:10.1111/j.1741-5446.1980.tb00902.x

Renteln, A. D. (1988). Relativism and the search for human rights. *American Anthropologist, 90*(1), 56–72. doi:10.1525/aa.1988.90.1.02a00040

Rokeach, M. (1973). *The nature of human values*. New York: Free Press.

Ryan, K. (1993). Mining the values in the curriculum. *Education, 51*(3), 16–18.

Ryan, K. (2002). Six E's of character education. *Issues in Ethics, 13*(1).

Simon, S. B., Howe, L., & Kirschenbaum, H. (1972). *Values clarification: A handbook of practical strategies for teachers and students*. New York: Hart Publishing.

Swedish Energy Agency. (2009). *Save energy and make your housekeeping money last longer*. Retrieved from https://www.energimyndigheten.se/Global/Offentlig%20sektor/Energiarbete%20i%20kommun,%20l%C3%A4n%20och%20region/Kommunala%20energi-%20och%20klimatr%C3%A5dgivare/Broschyr%20andra%20spr%C3%A5k/Energispartips_en_lu.pdf

Tan, C. (1990). *Indoctrination, imagination and moral education*. Paper presented at the 2nd International Conference on Imagination and Education. Vancouver, Canada.

Tilbury, D. (1995). Environmental education for sustainability: Defining the new focus of environmental education in the 1990s. *Environmental Education Research, 1*(2), 195–212. doi:10.1080/1350462950010206

KEY TERMS AND DEFINITIONS

Character Education: An approach for values education, mostly flourished at the end of the nineteenth century and through the first four decades of the twentieth century, essentially interrupted by World War II. Focusing on changing the behavior of the moral agents towards excellence, it supports the inculcation of particular values, considered as the 'right' ones.

Ethical Game Design: The design of games as moral objects, taking into consideration the ethical values imprinted in their story or their mechanics, and the morality of the play-experiences.

Games for Change: Games for social, political and environmental awareness, or for self-awareness. They can be also referred as serious games, or social impact games.

Sustainability: A value related to our responsibility to care for the nature (biodiversity, interspecies equity, balanced interdependence between man and earth), as well as to care for each other (social justice, human needs and rights, equity and participation).

Values Education: The scientific field studying the educational ways to convey values to the students/moral agents. The aim is the personal development along with the achievement of a socially constructive life. This field has existed as long as humans have thought about how to raise each subsequent generation, and there have been different approaches developed for educating values.

VEGA Model (Values Education Game Analysis): A Model for analyzing how values are conveyed through games for change, providing also guidelines for designing games for educating values in ethical ways. The Model is based on the diverse theories of values education and derived from both theoretical and empirical studies. It is primarily conceived and created by the author of this Book Chapter.

ENDNOTES

1. For scientific accuracy, I prefer not to use the term 'persuasive games', but games using 'persuasive technologies' based on the concept of 'captology' introduced by B. J. Fogg. Captology is the acronym of 'Computers As Persuasive Technologies'. This distinction is explained more explicitly in the conclusions.
2. (1) "Aegean 2037" is a serious flash-game developed for my MscThesis at the Aegean University of Greece, in Mytilini, in 2007. The design is based on real scientific data regarding the environmental and social effects of climate change in the Mediterannean region, and the fictional –utopian and dystopian- ways of the citizens for tackling them. Supervisors: Myrivili Eleni, Pavlogeorgatos

Gerasimos, Kawa Abraham. (2) "Artemis Calling" is an environmental location-based game for children aged 10-12 years old, that took place in Buergerpark in Bremen, March 2009. The children meet spirits of the forest and learn more about the Gas-Monster, its threats, and ways to save energy for helping the forest.

Chapter 8
Rhetoric of Game:
Utilizing the Game of Tavistock Method on Organizational Politics Training

Ben Tran
Alliant International University, USA

EXECUTIVE SUMMARY

The Tavistock method, commonly known as group relations, was originated from the work of British psychoanalyst Wilfred Ruprecht Bion. The Tavistock method's basic premise is that an aggregate cluster of persons becomes a group when interaction between members occurs. Within a group, there is organizational politics, and there are two features of organizational politic that should be considered when investigating its relationships with employee attitudes and behaviors. First, perceptions are more important than reality. Second, organizational politics may be interpreted as either beneficial or detrimental to an individual's well-being. Thus, organizational politics perceptions may result in differing responses to organizational policies and practices depending on whether politics are viewed as an opportunity or as a threat. How well one survives within an organization is correlated with how well one navigates these organizational politics. The Tavistock method is utilized as a game to assess and train individuals on organizational politics.

BACKGROUND

The connotation of game, more often than not, is a negative one. The term game often implies deceits and questionable ethics. In the field of business, to practitioners more so than to researchers, business is often perceived as a game, and like a game, it possesses its own special rules for playing. Similarly, lies of omission,

DOI: 10.4018/978-1-4666-6206-3.ch008

Copyright ©2014, IGI Global. Copying or distributing in print or electronic forms without written permission of IGI Global is prohibited.

overstatements, puffery and bluffs are morally acceptable within business because it, like a game, has a special ethic which permits these normally immoral practices (Shapiro & Carr, 2012). Although critics of this reasoning have used deontological and utilitarian arguments (Bowie, 1993) to show that deceit in business is just as immoral as it is in any other realm of human practice, little attention has been paid to the fact that the argument is one of analogy.

A (computer) game, according to Tran (2014a), can embody more than one rhetoric, but play as a form of conflict and contest (power), a means of expressing an identity and belonging to a group (identity), as imagination and creativity (the imaginary), and a means of relaxation and escape (the self), are the most common forms experienced in computer games. Regardless of the embedded play rhetoric, one of the primary concerns of industrial and organizational practitioners is the utilization of game as a method on training and assessing gamers (potential and current employees) in an organizational setting. Understanding of such psychological gamer is important not just in studying video game players but also for understanding behaviors and characteristics of individuals who are non-players of video game [non-player characteristics (NPCs)]. As an industrial and organizational practitioner and researcher, it appears that currently there is still a gap in literature concerning both the understanding of these psychological factors and the utilization of these psychological factors of the gamer in assessing and training gamers in organizational settings.

The gap in literature concerning the understanding of these psychological factors and the utilization of these psychological factors in relation to assessing and training employees on organizational politics are due to three reasons. First, in the field of psychology, the American Psychological Association (APA) has 56 divisions, but none of which addresses this topic and area of study. Second, this topic and area of study is not in the area of business (management or human resources), because both the research (academic) and the practice (practitioners) of business is not clinically-based. Third, more often than not, industrial and organizational (I/O) practitioners [for the business arena (also known as industrial and organizational psychologists for the non-business arena)] are the group of individuals who will, more likely than not, study and utilize this area in assessment. However, it does not mean that it is a common practice for I/O practitioners to adopt this methodology and I/O researchers to select this route of research.

The understanding of the psychological gamer is important not just in studying video game players but also for understanding behaviors and characteristics of individuals who are non-players of video game. Currently there is a gap in literature concerning the utilization of the gamer in selecting and training potential and current employees in organizational settings. The benefits and competitive advantage of utilizing the game of Tavistock Method in assessing and training gamers are: 1) identifying qualified gamers who are a good fit to fulfill the organizational needs, and 2) assisting the organization to achieve and maintain competitive advantage over its competitors. Organizations are strongly encouraged to utilize the game of Tavistock method to assess gamers on their other characteristics (O) and to train gamers on organizational politics. The O factor is part of the knowledge, skills, and abilities factors, commonly known as KSAOs. The emphasis will be on the game of Tavistock Method and on utilizing the game of Tavistock method to train gamers on organizational politics.

GAME THEORY

Game theory is the formal study of conflict and cooperation, and the game theoretic concepts apply whenever the actions of several agents are interdependent (Varoufakis, 2001). These agents may be individuals, groups, firms, or any combination of these. The concepts of game theory provide a language to formulate structure, analyze, and understand strategic scenarios. In other words, a gamer refers to an interactive situation involving two or more players making strategic decisions. Game theory is a branch of applied mathematics concerning optimal or purposeful behavior in different types of situations involving strategy and rational decision.

The Game in Game Theory

The object of study in a game theory is the *game* which is a formal model of an interactive situation. The game in game theory typically involves several *players*. A *player* is defined as a rational agent in which a rational agent is not necessarily a person as a rational agent could be an institution or a firm, where rationality consists of "complete knowledge of one's interests and flawless calculation of what actions will best serve those interests" (Dixit, Reiley, & Skeath, 2009, p. 30). An individual decision or choice of a player is defined as a mover and a series of moves of a given player is a strategy. A unique combination of players' strategies will result in a game outcome.

A game with only one player is usually called a *decision problem*. The formal definition defines the players, their preferences, their information, their strategic action available to them, and how these influence the outcome. Games can be described formally at various levels of detail. A *coalitional* (or cooperative) game is a high-level description, specifying only what payoffs each potential group, or coalition, can obtain by the cooperation of its members. As such, the result of a game for a given player is defined as the payoff (Straffin, 1996, p. 3). Cooperative game theory investigates such coalitional games with respect to the relative amounts of power held by various players, or how a successful coalition should divide its proceeds. This is most naturally applied to situations arising in political science or international relations, where concepts like power are most important.

The Gamers in Business Games

In the field of business, according to Tran (2014b), for practitioners more so than researchers, business is often perceived as a game, and like a game, it possesses its own special rules for playing. While players normally do not tolerate deceit, in the game of poker, bluffing is not only acceptable but is expected. Similarly, lies of omission, overstatements, puffery and bluffs are morally acceptable within business because, like a game, it has a special ethic which permits these normally immoral practices (Carr, 1968). Although critics of this reasoning have used deontological and utilitarian arguments (Bowie, 1993) to show that deceit in business is just as immoral as it is in any other realm of human practice, little attention has been paid to the fact that the argument is one of analogy. This oversight is unfortunate, given the strong intuitive appeal Carr's argument has to both business persons and to commerce students (Koehn, 1997). As such, Koehn (1997) identified nine traits in gamers and game playing: 1) A game is played to win, 2) In games, losers suffer few consequences, 3) A game is constituted by certain rules, 4) The rules of the games are fixed, 5) The rules of the games are accepted by all who played the game, 6) Players act intermittently, 7) In games the scope for bluffing is quite narrow and well-understood, 8) Players in a game risk only what is theirs to risk, and 9) In a game, it is clear to whom any gain belongs.

But is business managers are allowed to engage in such maneuvers, there is no reason why, qua gamesters, they cannot execute these strategies in such a way as to maximize their own personal gains. The firm may have been driven into the ground under this business management. But if corporate ethics are those of the game, the manager is a good player who has played exceptionally well, availing himself of every legal opportunity to haul in the biggest pot. The problem, of course, is that firms will find it close to impossible to exist if management ceases to consider the larger corporate interest but ponders only how to increase its own income and wealth.

SETTING THE STAGE: THE TAVISTOCK METHOD

The Tavistock method (Shapiro & Carr, 2012), commonly known as group relations (Banet, Jr. & Hayden, 1977), was originated from the work of British psychoanalyst Wilfred Ruprecht Bion. In the late 1940s, Bion conducted a series of small study of groups at the Centre for Applied Social Research in London's Tavistock Institute of Relations. Bion's previous experience with military leadership training and the rehabilitation of psychiatric patients convinced him of the importance of considering not only the individual in treatment, but also the group of which the individual is a member. Schooled in the psychoanalytic tradition of Melanie Klein, Bion employed her innovative methods of direct, confrontive intervention while working with the study groups and reported his experiences in a series of articles for the journal Human Relations (1959). Later published in book form as Experiences in Groups (1961), this seminal work stimulated further experimentation at Tavistock and other locations with Bion's novel approach of taking a group and viewing it as a collective entity. With that said, other original contributions to the Tavistock method are of A. Kenneth Rice (1966), Eric Trist (Trist & Sofer, 1959), Eric Miller (1976, 1989; Miller & Rice 1967) and other pioneers, who developed these conferences as learning laboratories for the study of leadership, authority and organizational life (Banet, Jr. & Hayden, 1977; Shapiro & Carr, 2012).

Basic Premises

An aggregate cluster of persons becomes a group when interaction between members occurs, when members' awareness of their common relationship develops, and when a common group task emerges. Various forces can operate to produce a group—an external threat, collective regressive behavior, or attempts to satisfy needs for security, safety, dependency, and affection. Other, more deliberate forces that result in the birth of a group are the conscious choices of individuals to band together to perform a task. Essential to the Tavistock approach is the belief that when an aggregate becomes a group, the group behaves as a system—an entity or organism that is in some respects greater than the sum of its part—and that the primary task of the group is survival. Although this primary preoccupation and latent motivating force for all group members. This emphasis on survival provides the framework for the exploration of group behavior and all the overt and covert manifestations of the primary task (Banet, Jr. & Hayden, 1977; Shapiro & Carr, 2012).

Appreciating the group as a whole, according to Banet, Jr. and Hayden (1977) and Shapiro and Carr (2012), requires a perceptual shift on the part of the observer or consultant, a luring of individual separateness and a readiness to see the collective interactions by group members. As Bion (1961) noted, we may observe individual gears, springs, and levers and only guess at the proper function, but when the pieces of machinery are combined, they become a clock, performing a function as a whole, a function impossible for individual parts to achieve. When individuals become members of a group, behavior changes and a collective identity emerges: a task group, an athletic team, a lynch mob, a utopian community, an organization—all become a new Gestalt in which the group is focal and the individual members become the background.

Basic Assumptions

On the basic assumption level of functioning, behavior is "as if" behavior: the group behaves as if a certain assumption is true, valid, and real and as if certain behaviors are vital to the group's survival. As Bion (1976) has recently pointed out, both words—basic and assumption—are important to understanding the term. Basic refers to the survival motivation of the group, where assumption underscores the fact that the survival motivation is based, not on the fact or reality, but on the collective projections of the group. Bion identified three distinct types of basic assumptions: dependency, flight/flight, and pairing. Turquet (1974) has added a fourth—oneness. Bion's theory is the cornerstone of the Tavistock method: it serves as a framework for the group-as-a-whole approach. Extensions of the theory to work groups and psychotherapy situations are provided by many authors (Banet, Jr. & Hayden, 1977):

- **Basic Assumption of Dependency:** The essential aim of this level of group functioning is to attain security and protection from one individual—either the designated leader or a member who assumes that role.
- **Basic Assumption of Fight/Flight:** In this mode of functioning, the group perceives its survival dependent on either fighting or fleeing from the task. In flight functioning, leadership is usually bestowed on an individual who minimizes the importance of the task and facilitates the group movement away from the here-and-now.
- **Basic Assumption of Pairing:** Pairing phenomena include bonding between two individuals who express warmth and affection leading to intimacy and closeness. The pair involved need not be a man and a woman. Such a pair or pairs often provide mutual intellectual support to the extent that other members become inactive. When the group assumes this mode of functioning, it perceives that its survival is contingent on reproduction.
- **Basic Assumption of Oneness:** Described by Turquet (1974, p. 357), this level of functioning occurs "when members seek to join in a powerful union with an omnipotent force, unobtainably high, to surrender self for passive participation, and thereby to feel existence, well-being, and wholeness."

Bion's Theory

According to Banet, Jr. and Hayden (1977), groups, like dreams, have a manifest, overt aspect and a latent, covert aspect. The manifest aspect is the work group, a level of functioning at which members consciously pursue an agreed-upon objective and deliberately work toward the completion of a task. Although group members always have hidden agendas—parts of themselves that they consciously or unconsciously plan not to share with the group—they rely on internal and external controls to prevent these hidden agendas from emerging and interfering with the announced group task. However, groups do not always function rationally or productively, nor are individual members necessarily aware of the kinds of internal and external controls they reply on to maintain the boundary between their announced intentions and their hidden agendas. The combined hidden agendas of group members constitute the latent aspect of group life, the basic assumption group. In contract to the rational, civilized, task-oriented work group, the basic assumption group is comprised of unconscious wishes, fears, defenses, fantasies, impulses, and projections. A tension always exists between the work and the basic assumption group, a tension usually balanced by various behavioral and psychological structures, including individual defense systems, ground rules, expectations, and group norms (Banet, Jr. & Hayden 1977).

The Conference

Bolstered by Bion's theory, the conference design also showed the influence of Kurt Lewin and the experimental ideas of the National Training Laboratories in the United States. Under the guidance of A. Kenneth Rice, chairman of Tavistock's Centre for Applied Social Research and a member of one of Bion's early study groups in 1947-1948, the design emphasis shifted from the roles individuals assume in work groups to the dynamics of leadership and authority relations in groups. Rice's views, which echoed Bion's earlier, touchstone hypothesis that individuals cannot be understood—or indeed, changed—outside the context of the groups in which they live, shaped the contours of the group relations conference as a teaching modality (Shapiro & Carr, 2012). According to Shapiro and Carr (2012), under Rice's influence, experimental group work during the 1960s in England became synonymous with the group relations method. In contrast, experimental groups in the United States during the same period were becoming quite diverse, moving away from the group dynamics focus of the early T-groups and on to personal growth and the study of interpersonal dynamics.

Rice directed all the Tavistock-Leicester conferences from 1962 to 1968. In 1865, Rice led the first group relations conference in the United States at Mount Holyoke College. This event, co-sponsored by the Washington School of Psychiatry and the Yale University Department of Psychiatry, was supported by Margaret Rioch, Morris Parloff, and F. C. Redlich, who became instrumental in the development of the Tavistock method in the United States. Currently, training in group relations is provided by the Tavistock Institute in England and by the A. K. Rice Institute and its affiliated centers in the United States. As seems fitting, no single person can be regarded as the founder of the group relations method, but the founding group would have to be include Bion and Rice in the United States.

The Tavistock approach, according to Shapiro and Carr (2012), developed by A. K. Rice and his colleagues, is founded not on teaching but on learning. Underlying this approach to learning is a psychoanalytically informed stance. But this does not imply that the task itself lies somewhere in the field of group psychoanalysis. There are three major psychoanalytic concepts that are extensively used in the conference. The basic notion is that of *unconscious functioning*: we are all moved about in life by much that is internal to us, but out of our awareness, both as individuals and in groups.

Our unconscious functioning becomes evident through *transference* and *counter-transference* and the use of *projective identification*. *Transference* refers to the ways our internalized images of others derived from our childhood experiences push us toward recreating familiar relationships in ways that can obscure the complexity of the people in our lives. *Counter-transference* refers to our unconsciously derived reactions to being seen as someone we do not feel we are. *Projective identification* (Klein, 1946; Shapiro & Carr, 1991; Zinner & Shapiro, 1972) refers to the way we unconsciously attempt to coerce others through covert actions to become the people we need them to be for our own unconscious and neurotic reasons. With that said, Shapiro and Carr (2012, p. 72-73) noticed three levels of group learning. The first level is the heightened recognition that individuals behave irrationally in the face of authority. The second level is the ability to recognize group functioning and see the ways in which conscious efforts toward collaborative work can be hampered by irrational thinking on the part of group members. The third level of learning involves a shift toward new ways of thinking. The participant discovers "a capacity to doubt the validity of perceptions which seem unquestionably true" (Palmer, 1979, p. 142). This requires developing a capacity for both involvement and detachment, similar to what Harry Stack Sullivan (1953) called "participant-observation."

The Components of Learning

According to Shapiro and Carr (2012), A. K. Rice and his colleagues at the Tavistock Institute in London held the first such conference in 1957, focusing on studying authority and the dynamics of institutional life. As might be expected, because the issues explored are persistent aspects of human life—for example, authority, responsibility, relationships, and relatedness—the design of the conference has not changed greatly. While subsequent conference directors have shaped conference structures to respond to changes in society, it is still organized around four basic components: holding and containment, a series of specific group contexts, shared group dynamics, and a determined focus on the group and the developing institution. The four components are combined in a skeletal conference design and interact to create opportunities for members to join, engage in, and study institutional dynamics (Shapiro & Carr, 2012):

1. **Holding and Containment:** The child psychiatrist, Donald Winnicott (1960) first described the "holding environment". One task of the staff is to be dependable so that members can feel secure and confident enough to cope with the anxiety, aggression, confusion, and new learning evoked by the lack of familiar guide-posts.
2. **Series of Specific Group Contexts:** Within the series of specific group contexts, by joining the conference, members have authorized the staff to work at this learning task in this particular way:
    a. **The Individual Member:** However deeply he or she becomes immersed in the life and dynamic of the conference, the member remains an individual and as such is responsible for his or her reactions to fellow members, staff, and other groups. In particular events, the individual will have the opportunity to receive and give delegated authority. This generates its own internal and group dynamic;
    b. **Specific Events:** Each event, such as large or small study groups, intergroup activities, or application groups, is a subsystem of the conference with its own specific task which evokes characteristic dynamics;
    c. **The Moment within the Event:** Whatever is happening in the here-and-now of the group process is an entry point for new learning; and
    d. **The Conference as a Whole:** Given the focus on what is happening in the moment, any member's role in the conference institution as a whole will be difficult to grasp. The staff with the director, however, will be constantly working on grasping the whole conference, regularly offering their developing views to the members.
3. **Shared Group Dynamics:** During the conference the member is located somewhere on a spectrum from using others to being used by them. Within the framework of events, individuals have the opportunity to work at the task of the conference—to study authority and leadership—with little explicit help from the staff except for group interventions. Over time, the group begins to share

unconscious assumptions (Bion, 1961). Though, as the conference institution takes shape, these shared unconscious assumptions inevitably become more refined and complex, initially they are of three basic types[1]:

a. **Dependency:** Dependency is in which the group becomes passive and looks, usually, to its staff consultant for rescue from confusion;
b. **Pairing:** Pairing is a variant of dependency, when the group turns to a couple, who are seen as an idealized hope for producing a solution for the group's problems; and
c. **Fight/Flight:** Fight/flight is quite different and more volatile assumption, in which the group acts as if fighting with the task or fleeing from the work are the only alternatives.

4. **Determined Focus on the Group and the Developing Institute:** This approach makes the group unequivocally the focus of attention and interpretation. This is not done, as with a therapy group, as a means to assists the individual to develop greater self-awareness and understanding. The group itself, and the group alone, is the focus of study. From this perspective, the *group* is defined as any collection of individuals linked by a shared task (Shapiro & Carr, 1991, 2006).

THE POLITICS OF ORGANIZATIONS

Ferris, Adams, Kolodinsky, Hochwarter, and Ammeter's (2002) review presents a number of key attributes of the various definitions of politics over roughly the last two decades. These definitions of politics in organizations have been traditionally framed in the negative. For example, "individual or group behavior that is informal, ostensibly parochial, typically divisive, and above all in a technical sense, illegitimate—sanctioned neither by formal authority, accepted ideology, nor certified expertise (Mintzberg, 1983, p. 172) or "actions by individuals which are directed toward the goal of furthering their own self-interests without regard for the well-being of others or their organizations (Kacmar & Baron, 1999, p. 4). These definitions, and others discussed by Ferris et al. (2002) share the following key attributes: politics is informal, discretionary, promotes self-interest on the part of the actor, potentially threatens others, is indifferent to or may run counter to organizational goals, and is based on social influence. This is because, according to Goodman, Evans, and Carson (2011), generally, it is expected that ambiguity in the work environment contributes to organizational politics (Ferris, Russ, & Fandt, 1989), and organizational politics correlates with organizational learning.

Although organizational learning recently has been acknowledged as a fundamentally political process (Blackler, 2000; Burgoyne & Jackson, 1997; Coopey &

Burgoyne, 2000; Fox, 2000), there has been relatively little theory development systematically connecting organizational politics and organizational learning. There are at least three critical reasons to connect politics—the dynamics of power in organizations—to organizational learning. First, scholars interested in organizational learning have called for the development of research and theory that are cumulative and integrative (Crossan & Guatto, 1996; Huber, 1991), but power and politics have remained largely ignored. Second, Lawrence, Mauws, and Dyck (2005) believe that any theory of organizational learning without an understanding of its political dynamics will always be incomplete. Third, bringing power and politics into research on organizational learning should provide a more effective foundation for understanding why some organizations are better able to learn and why some of the available useful innovation are embraced by organizations.

In examining the politics of organizational learning, Lawrence et al. (2005) take as their point of departure the model of organizational learning developed by Crossan, Lane, and White (1999). Crossan et al.'s model provides a rich, coherent framework that specific four general processes through which organizational learning occurs. Three characteristics of the model stand out as particularly important in the development of a general model of organizational learning: (1) it is multilevel, bringing together individual, group, and organizational levels of analysis; (2) it is dynamics, bridging the levels with specific mechanisms; and (3) it clearly articulates four processes—intuiting, interpreting, integrating and institutionalizing the "4Is— that allow learning to feed forward to the organizational level and feed back to the individual. Nevertheless, Crossan et al.'s model typifies the tendency in studies of organizational learning to focus on learning as a principally social psychological phenomenon, with less attention paid to its political dynamics.

The 4I's Model of Organizational Learning

Crossan et al. (1999) argue that organizational learning is a multilevel process that begins with individual learning, which leads to group learning, which then leads to organizational learning. These levels, they argue, are connected by bidirectional processes that involve both the creation and application of knowledge. More specifically, they describe four processes that connect individual learning to organizational learning: intuiting, interpreting, integrating, and institutionalizing. According to Lawrence et al. (2005), Crossan et al. (1999, p. 525) define *intuiting* as "the preconscious recognition of the pattern and/or possibilities inherent in a personal stream of experience". *Interpreting*, is "the explaining, through words and/or actions, of an insight or idea to one's self and to others" (Crossan et al., 1999, p. 525). Integrating is the first level that occurs at the group level. It is "the process of developing shared understanding among individuals and of taking coordinated action through mutual adjustment" (Crossan et al., 1999, p. 525). Institutionalizing is learning that has occurred among individuals and groups is embedded into organizations through "systems, structures, procedures, and strategy" (Crossan et al., 1999, p. 525).

Power, Politics, and Organizational Learning

Although power is often still defined narrowly in management theory and research (Fiol, 2001; Shen & Cannella, 2002), in a significant stream of research, scholars have recognized the wide variety of forms power can take in organizations (Clegg, 1989; Covaleski, Dirsmith, Heian, & Sajay, 1998; Hardy & Clegg, 1996; Lawrence, Winn, & Jennings, 2001). For the purpose of this chapter, the definition of power is adapted from earlier work that emphasizes two distinct modes in which it operates—systemic and episodic (Clegg, 1989; Foucault, 1977; Giddens, 1984; Hardy & Clegg, 1996; Lawrence et al. 2001)—and argue that both modes have distinct implications for organizational learning. Episodic power refers to discrete, strategic political acts initiated by self-interested actors. This mode of power has been the traditional focus of organizational research and theory, with its emphasis on examining which actors in organizations are most able to influence organizational decision making (Pfeffer, 1981). In contrast, systemic forms of power work through the routine, ongoing practices of organization. Rather than being held by autonomous actors, systemic forms of power are diffused throughout the social systems that constitute organizations (Clegg, 1989; Scott, 2001). Examples of systemic forms of power include socialization and accreditation processes (Covaleski et al., 1998; Shaiken, 1984).

ORGANIZATIONAL POLITICS (OP)

Ferris, Russ, and Fandt (1989) note two features of organizational politic that should be considered when investigating its relationships with employee attitudes and behaviors. First, perceptions of organizational politics are more important than reality (Ferris et al., 1989). As Lewin (1936) noted many years ago, people respond to their perceptions of a situation, which may be different from the situation itself. Second, organizational politics may be interpreted as either beneficial or detrimental to an individual's well-being (Ferris et al., 1989). Organizational policies and practices that are viewed as highly political can create situation of potential gain as well as potential loss (Ferris et al., 1996). Thus, organizational politics perceptions may result in differing responses to organizational policies and practices depending on whether politics are viewed as an opportunity or as a threat: 1) politics as a threat (Kacmar & Baron, 1999); 2) Politics as an opportunity (Murphy & Cleveland, 1995;

Longenecker, Gioia, & Sims, 1987); 3) Understanding (Ferris et al., 1989; Kacmar, Bozeman, Carlson, & Anthony, 1999; McGrath, 1976; Sutton & Kahn, 1986); and 4) Control (Ferris et al., 1996; Ferris et al., 1989; Sutton & Kahn, 1986).

Organizational Politics: The Importance of Organizational Politics

The importance of organizational politics (OP), according to Vigoda (2000), lies in its potential consequences and effect on work outcomes. The concept of organizational politics, according to Ullah, Jafri, and Dost (2011), started getting interest of academicians and practitioners in the last decade. The body of literature on organizational politics is expanding but still the research remains distorted with respect to theory and research methodologies adopted. Despite a lot of empirical data conceptual vagueness still exist. Organizational politics is proved to be fact of life (Vigoda-Gadot, 2001).

Organizational politics has proved to be an important part of both public and private sector organizations, therefore researchers argue for the need for further investigation of the issue (Drory & Romm, 1990; DuBrin, 1988; Mayes & Allen, 1977; Mintzberg, 1983; Parker, Dipboye, & Jackson, 1995; Pfeffer, 1981, 1992). The literature on organizational politics was systematically reviewed to discover how academicians and practitioners have defined and examined organizational politics in existing literature. Theoretical arguments suggest that politics often interferes with normal organizational processes and damages productivity and performance on individual and organizational levels. Some studies found a negative relationship of OP to job attitudes or stress-related responses (Drory, 1993; Ferris, Frink, Bhawuk, & Zhou, 1996; Ferris, Frink, Galang, Zhou, Kacmar, & Howard, 1996). More recent works suggested that politics enhances withdrawal behaviors and turnover intentions (Bozeman, Perrewe, Kacmar, Jochwarter, Brymer, 1996; Cropanzano, Howes, Grandey, & Toth, 1997), but others found no such relationship (Parker, Dipboye, & Jackson, 1995). All these studies overlooked the relationship between OP and other possible work outcomes, such as direct negligent behavior and actual job performance.

Few studies have examined issues related to OP in the public sector. At first glance several studies appear to have done so, but in fact they were conducted mainly at universities (Christianse, Villanova, & Mikulay, 1997; Ferris, Frink, Bhawuk, & Zhou, 1996; Ferris, Frink, Galang, Zhou, Kacmar, & Howard, 1996; Welsh & Slusher, 1986). Or they used mixed samples of private and semipublic agencies like hospitals and governments-owned industries (Drory, 1993; Ferris & Kacmar, 1992; Kumar & Ghadially, 1989). Also, most studies of OP refer to the US private sector (Bozeman et al., 1996; Cropanzano et al., 1997; Hochwarter, witt, & Kacmar,

1997; Wayne, Liden, Graf, & Ferris, 1997). With the exception of that of Parker et al. (1995) no study has examined the effect of perceived organizational politics on work outcomes among public sector employees who serve citizens.

Organizational Politics: Defining Organizational Politics

The disagreement of researchers over the definition of politics in organizations has remained an issue in the field of management. This lack of consensus is also an indication of ongoing debate over the issue of organizational politics (Vigoda-Gadot & Drory, 2006). The interchangeable use of identical terms like: political skill, influence tactics, impression management, political behaviors, and political maneuvering have made research in the field of organizational politics fragmented. Therefore, some conceptual ambiguities are necessary to be removed before starting report on literature reviewed.

Organizational politics (OP), according to Ullah, Jafri, and Dost (2011), is defined as the existence of multiple interests and incompatible goals, beyond the goals of organization, and the influence techniques used to defend them. OP, on the other hand, according to Vigoda (2000), is usually defined as behavior strategically designed to maximize self-interests (Ferris, Russ, & Fandt, 1989) and therefore contradicts the collective organizational goals or the interests of other individuals. For the purpose of this chapter, OP is defined as behavior strategically designed to maximize self-interests (Drory, 1993; Ferris & Kacmar, 1992) regardless of the existence of multi interests and compatible of goals, and the influence techniques used to defend these self-interests. OP is dependent is political behavior (PB), where PB means observable behaviors of the individual who is pursuing one's goals.

These political PBs, according to Ullah, Jafri, and Dost (2011), are nominated in the field of management as political strategies, influence tactics, political tactics, and political maneuvering. Some PBS are given specific names like: impression management, blandishment, and referencing. With that said, based on the concept of game theory, and the definition of games, as well as the various types of games available, the following is a case study of a game.

TAVISTOCK METHOD: CASE STUDY OF A GAME

One California State University campus, among twenty-three, decided to offer a new program in 2005. The program is funded by a grant from the U.S. Department of Education to help students with disabilities succeed at the university. The focus of the program is on retention whereas the focus of the department that houses the program is on legal accommodations. While the program is housed within the department the program and the department do not occupy the same facility, does not share a common goal (retention vs. accommodation), but do share the same counselors and director.

The original grant for the program was written to be based on two individuals who were to be the director of the department and the program coordinator for the program. The program (Figure 1) was to have one full time Program Coordinator (PC), one three-fourth time Assistive Technology Instructor/Counselor (ATIC), and one half-time Wellness Counselor (WC) (Figure 2). The Director of the Department (DD) will oversee and manage both the department and the program while counselors of the department will dedicate approximately one-third of their time to the program (Figure 3). However, prior to the approval of the grant, the DD left the department and accepted another project within the university, then only to resign to pursue her doctorate. On the other hand, the "to be" PC, due to politics, left the university.

Upon approval of the grant, the department did not have a director, and the program did not have a PC. A search for a new director was underway, recruited, and selected. The PC was anything but recruited and selected. The PC was the office manager for the department who transitioned oneself into the role of the PC. The PC has been with the university for more than nine years without a promotion but has been unwillingly transferred among departments. The PC has requested for promotions in the past with no success, applied for other openings in the past without success, and has been rejected for requests of transfer to other departments by individuals who have worked with the individual in the past or have heard of the individual.

Upon approval of the grant and the opening of the program, the program is to have one full time PC, one three-fourth time ATIC, and one half-time WC. The individual came out of retirement, joined the university as the DD couple of month prior to the ATIC joining the university, and will oversee and manage both the department and the program while counselors of the department will dedicate approximately one-third of their time to the program. The DD has no prior experience in an educational (or higher educational) environment before becoming the DD. While the ideological paradigm is practical, the philosophical paradigm is struggling. The department focuses on legal accommodations, the program focuses on retention, and the students unrealistic expectations/sense of entitlements are on overdrive.

Students frequently compare and contrast the department and the program services and attempt to utilize the relationship to their advantages due to the lack of communication between the department and the program. Since time was of the essence, the PC, who was the office manager for the department at the time, requested a transfer and was approved. The Program opens its door in 2005 with four staff members—the PC, the ATIC, a WC, and an intern. Due to a lack of boundaries and

Figure 1. The program's office structure

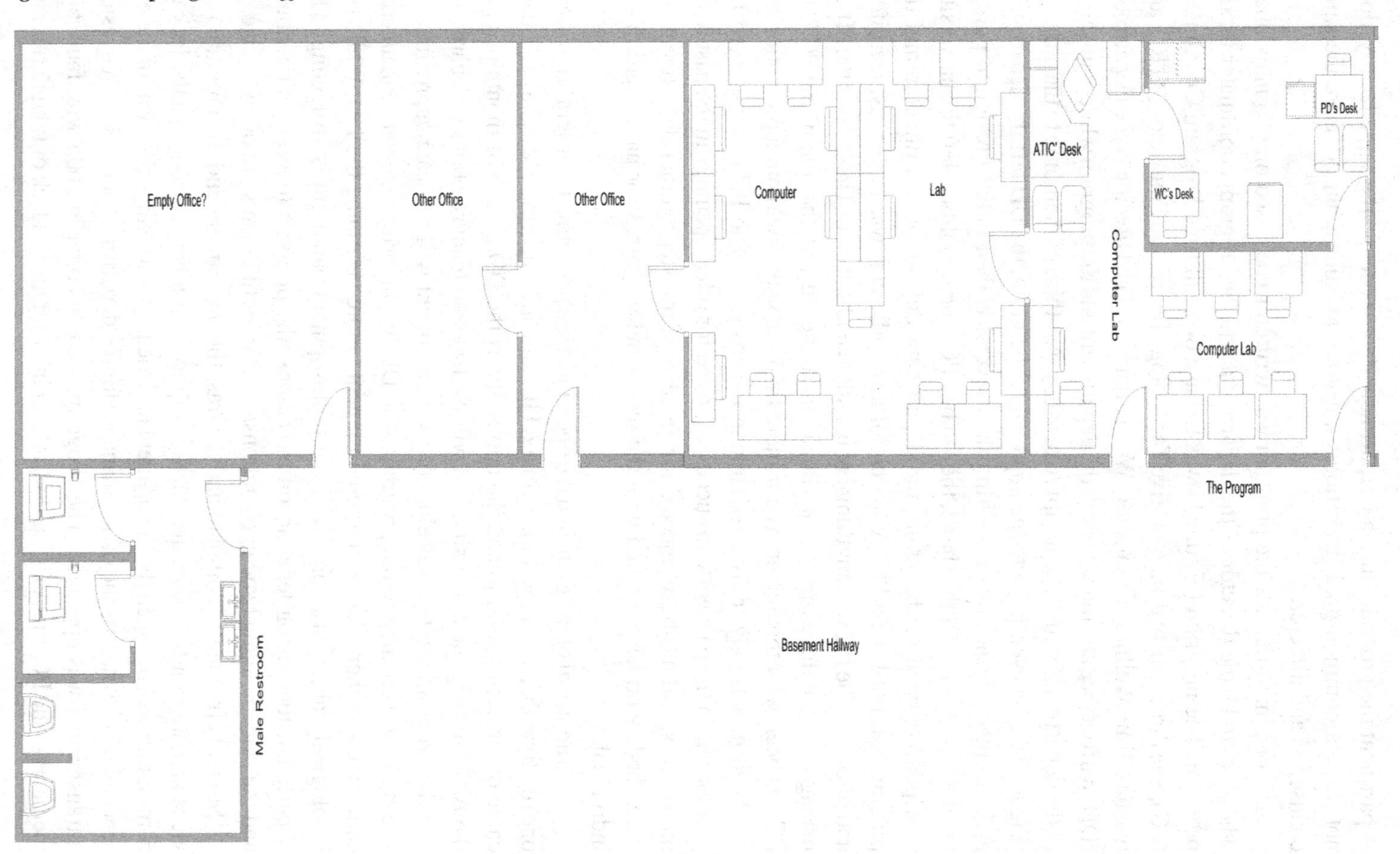

Figure 2. The program

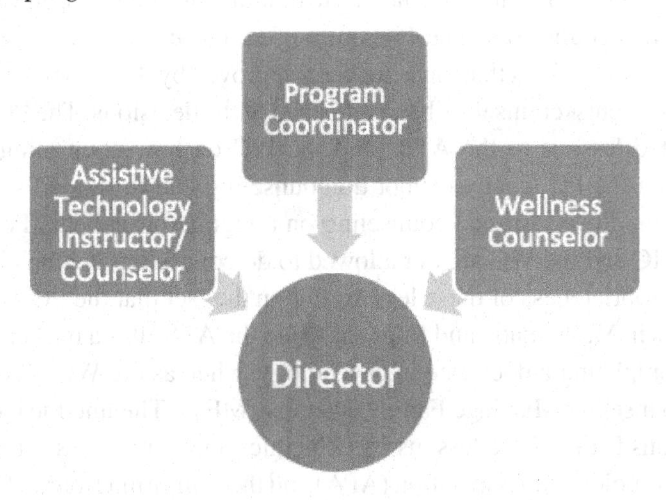

Figure 3. The official structure

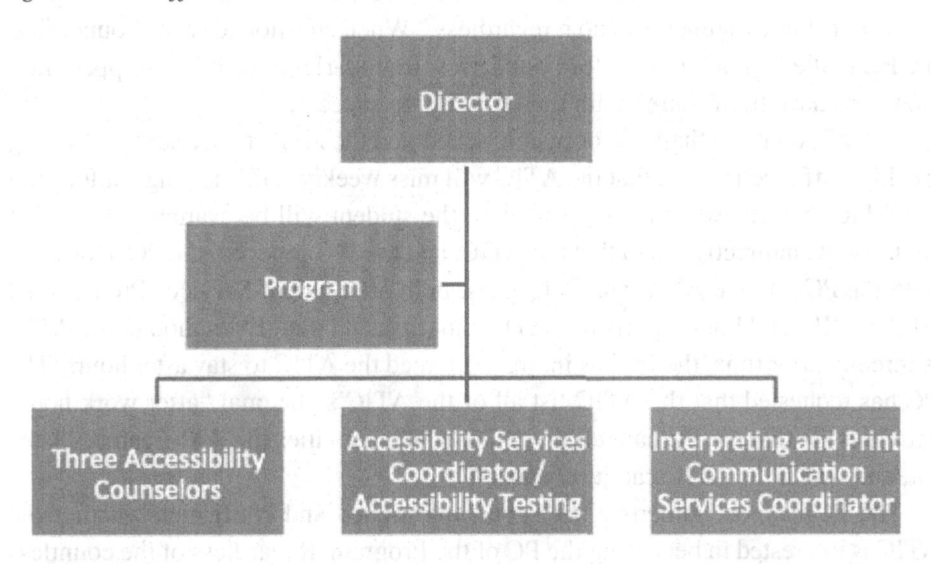

ethics of the PC, both the ATIC and the WC left the program in less than a year, followed by the intern. The intern turned down a paid internship with the Program and accepted another pro bono internship.

Towards the end of the first year since the Program opened its door, the Program recruited, and selected a new ATIC and a new WC. However, the lack of boundaries and ethics of the PC persists, and the PC indirectly runs the Program as if it is its own department, and "self-elected" oneself as the Director. The PC of the Program quite often indirectly, in a subliminal manner, disrespects boundaries, ethics, and exercises one's status on behalf of students to the students' benefits, and undermines the department's counselors. The PC, quite often put in actions on behalf of the students, accommodations that have not been approved by the students' counselors, nor were the students' counselors been informed of the decisions. The PC frequently gives uninvited lessons to the ATIC and the WC on how counseling should be enacted. While the PC legally cannot do counseling due to the PC's educational background, the PC performed counseling on a regular basis behind closed door, while the ATIC and the WC are not allowed to do counseling behind closed door.

The inappropriateness of these lessons fall on the fact that the PC's educational background is in Mathematic and English, while the ATIC has a master in psychology and is completing a doctorate in psychology, whereas the WC is completing a Master in Counseling/Marriage Family Therapy (MFT). The unethical component of these lessons is that these lessons violate ethics and boundaries set forth by the American Psychological Association (APA) and the California Board of Behavioral Sciences (BBS). For example, one of the many lessons is that, "A counselor does not advise students on what to do, a counselor tells students what to do, but a good counselor 'forces' students to do it regardless." When questioned about boundaries, the PC replied, "I do not care for boundaries. Boundaries have never stopped me." Concerns have been shared with the DD with no success.

The PC, on more than one occasion, scheduled the ATIC to see students during weekly staff meetings, so that the ATIC will miss weekly staff meetings, informing the ATIC that this was the only time that the student will be available. With that said, the PC indirectly asked that the ATIC, and the WC report to the PC. However, both the PC and the ATIC are colleagues, both are Student Services Professional III (SSP III), and both reports to the DD, not to each other. Even though the ATIC is three-quarter time, the PC has indirectly forced the ATIC to stay after hours. The PC has requested that the ATIC list all of the ATIC's personal "after work hour" activities, so that the PC can tell the ATIC which activities the ATIC can be "late" for, and which the ATIC can just "cancel."

The PC has, on numerous occasions, questioned and confirmed whether the ATIC is interested in becoming the PC of the Program. Regardless of the countless confirmations that there is no such interest from the ATIC, the PC remains skeptical, for the ATIC is completing one's doctorate, and the PC was unsuccessful with two different doctoral programs. The PC, on numerous occasions, expressed sorrows and regrets, that one has missed out on the opportunity to pursue one's doctorate due to age constraint. However, the PC has admitted to the fact that the PC truly believes that a business degree is the most fraud degree that the American universities have offered in history to date, for a business degree is the most useless degree. According to the PC, "what are you going to do with a business degree? You go to school for four years, come out with a business degree, and just learn how to balance a check book? It is so unfortunate that America values a business degree so much."

Towards the end of the second year, the Program hired an external consultant, a retired U.S. Department of Education grant evaluator, to evaluate the progress of the program. Based on the consultant's evaluation, the result indicated that the program lacks boundaries, and thus ethics came into question. If such written feedback is to be presented to the President of the University, the University has thirty days to fix it, or the (grant) program will be eliminated. These were some of the same concerns that the ATIC shared with the DD with no success.

During the same year both the ATIC and the WC received their annual evaluations. Based on six criterions, the ATI received six-out-of six on four criterions, and five-out-of six on two criterions. The Program decided to hire the ATIC and have the ATIC continue service with the Program. The ATIC was also invited and served on the Counselor, Media Production Manager, and Wellness/MFT search committees. On the same note, the Program recruited two interns, both lasted less than two months, and both left due to a lack of ethics and boundaries.

After the departure of the two interns, the second WC left, and the Program hired a new intern, the fourth intern. Without a WC, the PC fought with the grant, and Human Resources (HR), so that the PC can teach a Wellness course. The Wellness course can only be taught by a WC or an individual with a graduate degree in psychology. The PC made it clear that the PC will not allow the ATIC, who has a master in psychology, to teach the course, but the PC will. Within the first week, the PC gave the intern a counseling lesson, a lesson that violates the ethics and boundaries set forth by the APA and the BBS. The intern expressed concern with the ATIC on the lack of ethics and boundaries forced upon by the PC's lesson. The ATIC acknowledged the concern, confirmed that such concerns have been shared with the DD with no success, and encouraged the intern to meet with the DD.

Approximately four weeks after being hired by the Program after the annual evaluation, three weeks after the consultant's evaluation and feedback on the Program, and two weeks after the intern's concern about the PC's counseling lesson, the ATIC was let go "on the spot." The DD informed the ATIC that, "your service is no longer needed here. We revoked the offer on hiring you. You must leave now." One year later, at the end of the DD's second year, the DD too has left the university. A couple of years thereafter, the PC was no longer the PC for the Program, but was placed in the Department of Academic Advising and Career Education as an Academic Counselor for the university.

SOLUTIONS AND CHALLENGES

Given the research and knowledge on politics and organizational politics, academia still lacks courses in organizational politics, and a formal course of study in organizational politics, while degree fields as such leadership development, strategic leadership, and organizational development are available. Organizational politics, in the workplace in the U.S., is slowly making its mark and is no longer considered as taboo as it used to be. More and more workplaces are starting to tackle organizational politics through the usage of the Tavistock method, such that, according to Tran (2014a), the Tavistock approach has been capitalized by corporate America in various formats under the reputation of reality television. With that, to many non-academicians, the apprentice, giving you the business, mystery diners, and restaurant stakeout are all simplified interpretations of the Tavistock approach. These simplified interpretations of the Tavistock approach are derived for two reasons: (a) reality made exciting for viewers based solely on ratings and for those who do not have a stake on its end result, and (b) a softer version of a scientific approach to assessing, analyzing, and resolving organizational dilemmas at hand. Among the various Tavistock approach formats, The Apprentice has been adopted internationally, and the PC game version of The Apprentice was released on March 2005 by Legacy Interactive.

Take for example, the Tavistock method provides the environmental setting within the game, where an aggregate cluster of persons becomes a group when interaction between members occurs, when members' awareness of their common relationship develops, and when a common group tasks emerges. These common, and sometime uncommon group tasks emerges and challenges both the individual and the group, allowing businesses to assess the individual's KSAOs in handling these tasks. Paramount components within the KSAOs may vary among businesses, for the right decisions, to some businesses are not desired, but possessing the necessary KSAOs to deriving the desired decisions are. Utilizing the Tavistock method, businesses can utilize the enneagram method to elicit, and assess the necessary desired KSAOs.

In the business arena game, rhetorically has a negative connotation. Game, when used in the business arena, often conjures questionable practices and ethical dilemmas. As such, in the business arena, the term game is not being used as it is in the gaming industry, or in the research and academic (psychology) arena. In the business arena, under the usage and purpose of the gaming industry and the research and academic (psychology) arena, game is known as "training" (or testing) methodology where firms utilize these methodological resources to assess their current and potential employees. Historically, businesses have been using these training methodologies as assessments, but have not been giving these methodological assessments their appropriate credits. For example, many businesses utilize this training for new employees' orientations and continuing education/certification (i.e. watch a video and successfully complete an online test) to demonstrate that the required knowledge, skills, and abilities (KSAs) have been acquired.

Furthermore, government and state jobs in the U.S. often assesses interested candidates by utilizing these personality tests, but often via paper-and-pencil due to budget constraint in implementing assessments through games. This constraint yields a different side of validity, because when situations are presented in words on paper, decisions made due to reactions based on words will definitely be different compared to decisions made due to reactions based on simulation in a game. It is the simulation within a game that captures gamers, such that it is the simulation of life-like situations via the Tavistock method within a game that elicit true personality, demeanors, behaviors, knowledge, skills, abilities, and other characteristics (KSAOs). With that said, provided are questions regarding the case study to be considered, and reflect upon so that other programs can be better governed:

1. What could the DD done when the situation was first brought to the DD's attention?
2. Was the DD's leadership effective and how would the DD's leadership be categorized as?
3. What was the relationship between the DD and the PC?
4. What types of organizational political power did the PC possess within and within the department?
5. What types of organizational political power did the PC possess with and within the university?
6. Was the ego and the self-esteem of the PC a result of the PC's organizational political power with the university and within the department?
7. What could the ATI or the WC done differently as new members of the program?
8. How could this program be more effectively governed so that organizational political power from any one individuals could not be exerted?
9. If such organizational political power is to be exerted, what are the strategies that can be implemented to address the situation, so that the situation could be handled?

REFERENCES

Banet, A. Jr, & Hayden, C. (1977). A tavistock primer. In J. E. Jones, & J. W. Pfeiffer (Eds.), *The 1977 annual handbook for group facilitators* (6th ed.). Pfeiffer & Company.

Bion, A. C. (1961). *Experiences groups*. New York: Basic Books. doi:10.4324/9780203359075

Bion, A. C. (1970). *Attention and interpretation: Scientific approach to insight in psychoanalysis and groups*. New York: Basic Books.

Bion, W. R. (1977). Attention and interpretation: Container and contained. In *Seven servants: Four works*. New York: Jason Aronson.

Blackler, F., & McDonald, S. (2000). Power, mastery and organizational learning. *Journal of Management Studies*, *37*(6), 833–851. doi:10.1111/1467-6486.00206

Bowie, N. E. (1993). Does it pay to bluff in business? In T. L. Beauchamp, & N. E. Bowie (Eds.), *Ethical theory and business* (pp. 443–448). Englewood Cliffs, NJ: Prentice Hall, Inc.

Bozeman, D. P., Perrewe, P. L., Kacmar, K. M., Hochwarter, W. A., & Brymer, R. A. (1996). *An examination of reactions to perceptions of organizational politics*. Paper presented at the 1996 Southern Management Association Meetings. New Orleans, LA.

Burgoyne, J., & Jackson, B. (1997). The arena thesis: Management development as a pluralistic meeting point. In J. Burgoyne & M. Reynolds (Eds.), *Management learning: Integrating perspectives in theory and practice* (pp. 54-70). London: Sage.

Carr, A. Z. (1968). Is business bluffing ethical? *Harvard Business Review*.

Christiansen, N., Villanova, P., & Mikulay, S. (1997). Political influence compatibility: Fitting the person to the climate. *Journal of Organizational Behavior*, *18*(6), 709–730. doi:10.1002/(SICI)1099-1379(199711)18:6<709::AID-JOB811>3.0.CO;2-4

Clegg, S. (1989). *Frameworks of power*. London: Sage. doi:10.4135/9781446279267

Coopey, J., & Burgoyne, J. (2000). Politics and organizational learning. *Journal of Management Studies*, *37*(6), 869–885. doi:10.1111/1467-6486.00208

Covaleski, M. A., Dirsmith, M. W., Heian, J. B., & Sajay, S. (1998). The calculated and the avowed: Techniques of discipline and struggles over identity in the big six public accounting firms. *Administrative Science Quarterly*, *43*(2), 293–327. doi:10.2307/2393854

Cropanzano, R., Howes, J. C., Grandey, A. A., & Toth, P. (1997). The relationship of organizational politics and support to work behaviors, attitudes, and stress. *Journal of Organizational Behavior, 18*(2), 159–180. doi:10.1002/(SICI)1099-1379(199703)18:2<159::AID-JOB795>3.0.CO;2-D

Crossan, M., & Guatto, T. (1996). Organizational learning research profile. *Journal of Organizational Change Management, 9*(1), 107–112. doi:10.1108/09534819610107358

Crossan, M., Lane, H., & White, R. (1999). An organizational learning framework: From intuition to institution. *Academy of Management Review, 24*, 522–537.

Dixit, A. K., Skeath, S., & Reiley, D. (2009). *Games of strategy* (3rd ed.). New York: W. W. Norton & Company.

Drory, A. (1993). Perceived political climate and job attitudes. *Organization Studies, 14*(1), 59–71. doi:10.1177/017084069301400105

Drory, A., & Romm, T. (1990). The definition of organizational politics: A review. *Human Relations, 43*(11), 1133–1154. doi:10.1177/001872679004301106

Dubrin, A. J. (1988). *Human relations: A job oriented approach*. Englewood Cliffs, NJ: Prentice-Hall.

Ferris, G. R., Adams, G., Kolodinsky, R. W., Hochwarter, W. A., & Ammeter, A. P. (2002). Perceptions of organizational politics: Theory and research directions. In F. Dansereau, & F. J. Yammarino (Eds.), *Research in multi-level issues* (Vol. 1). Oxford, UK: Elsevier Science/JAI Press. doi:10.1016/S1475-9144(02)01034-2

Ferris, G. R., Frink, D. D., Bhawuk, D. P. S., & Zhou, J. (1996). Reactions of diverse groups to politics in the workplace. *Journal of Management, 22*(1), 23–44. doi:10.1177/014920639602200102

Ferris, G. R., Frink, D. D., Galang, M. C., Zhou, J., Kacmar, K. M., & Howard, J. L. (1996). Perceptions of organizational politics: Prediction, stress-related implications, and outcome. *Human Relations, 49*(2), 233–266. doi:10.1177/001872679604900206

Ferris, G. R., & Kacmar, K. M. (1992). Perceptions of organizational politics. *Journal of Management, 18*(1), 93–116. doi:10.1177/014920639201800107

Ferris, G. R., Russ, G. S., & Fandt, P. M. (1989). Politics in organizations. In R. A. Giacalone, & P. Rosenfield (Eds.), *Impression management in the organization* (pp. 143–170). Hillsdale, NJ: Erlbaum.

Fiol, C. M. (2001). All for one and one for all? The development and transfer of power across organizational levels. *Academy of Management Review*, *26*, 224–242.

Foucault, M. (1977). *Discipline and punish: The birth of the prison*. New York: Vintage Books.

Fox, S. (2000). Communities of practice: Foucault and actor-network theory. *Journal of Management Studies*, *37*(6), 853–867. doi:10.1111/1467-6486.00207

Giddens, A. (1984). *The constitution of society: Outline of a theory of structuration*. Cambridge, MA: Polity Press.

Goodman, J. M., Evans, W. R., & Carson, C. M. (2011). Organizational politics and stress: Perceived accountability as a mechanism. *The Journal of Business Inquiry*, *10*(1), 66–80.

Hardy, C., & Clegg, S. R. (1996). Some dare call it power. In S. R. Clegg, C. Hardy, & W. R. Nord (Eds.), *Handbook of organization studies* (pp. 622–641). London: Sage.

Hochwarter, W. A., Witt, L. A., & Kacmar, K. M. (1997). *Perceptions of organizational politics as a moderator of the relationship between conscientiousness and job performance*. Paper presented at the Southern Management Association Meeting. Atlanta, GA.

Huber, G. (1991). Organizational learning: The contributing processes and the literatures. *Organization Science*, *2*(1), 88–115. doi:10.1287/orsc.2.1.88

Kacmar, K. M., & Baron, R. A. (1999). The state of the field, links to related processes, and an agenda for future research. In G. R. Ferris (Ed.), *Research in personnel and human resources management* (Vol. 17, pp. 1–39). Stamford, CT: JAI Press.

Kacmar, K. M., Bozeman, D. P., Carlson, D. S., & Anthony, W. P. (1999). An examination of the perceptions of organizational politics model: Replication and extension. *Human Relations*, *52*(3), 383–415. doi:10.1177/001872679905200305

Klein, M. (1946). Notes on some schizoid mechanisms. *The International Journal of Psycho-Analysis*, *27*, 99–110.

Koehn, D. (1997). Business and game-playing: The false analogy. *Journal of Business Ethics*, *16*(12/13), 1447–1452. doi:10.1023/A:1005724317399

Kumar, P., & Ghadially, R. (1989). Organizational politics and its effects on members of organizations. *Human Relations*, *42*(4), 305–314. doi:10.1177/001872678904200402

Lawrence, T. B., Mauws, M. K., Dyck, B., & Kleysen, R. F. (2005). The politics of organizational learning: Integrating power into the 4i framework. *Academy of Management Review*, *30*(1), 180–191. doi:10.5465/AMR.2005.15281451

Lawrence, T. B., Winn, M., & Jennings, P. D. (2001). The temporal dynamics of institutionalization. *Academy of Management Review*, *26*, 624–644.

Longenecker, C. O., Sims, H. P., & Gioia, D. A. (1987). Behind the mask: The politics of employee appraisal. *The Academy of Management Executive*, *1*(3), 183–193. doi:10.5465/AME.1987.4275731

Maslyn, J., Fedor, D., Farmer, S., & Bettenhausen, K. (2005). *Perceptions of positive and negative organizational politics: Roles of the frequency and distance of political behavior*. Paper presented at the 2005 Annual Meeting of the Southern Management Association. Charlotte, NC.

Mayes, B. T., & Allen, R. W. (1977). Toward a definition of organizational politics. *Academy of Management Review*, *2*(4), 672–678.

McGrath, J. E. (1976). Stress and behavior in organizations. In M. D. Dunnette (Ed.), *Handbook of industrial and organizational psychology* (pp. 1351–1395). Chicago, IL: Rand McNally.

Miller, E. (1976). *Task and organization*. London: Wiley.

Miller, E. (1989). *The Leicester model: Experimential study of group and organizational processes*. London: Tavistock.

Miller, E. J., & Rice, A. K. (1967). *Systems of organization*. London: Tavistock Publications.

Mintzberg, H. (1983). *Power in and around organizations*. Englewood Cliffs, NJ: Prentice Hall.

Murphy, K. R., & Cleveland, J. N. (1995). *Understanding performance appraisal: Social, organizational, and goal-based perspectives*. Thousand Oaks, CA: Sage.

Palmer, B. (1979). Learning and the group experience. In W. G. Lawrence (Ed.), *Exploring individual and organizational boundaries* (pp. 169–192). London: John Wiley and Son.

Parish, M. (2007). Reflections on the administrator's role. *Organisational and Social Dynamics*, *7*(1), 61–72.

Parker, C. P., Dipboye, R. L., & Jackson, S. L. (1995). Perceptions of organizational politics: An investigation of antecedents and consequences. *Journal of Management*, *21*(5), 891–912. doi:10.1177/014920639502100505

Pfeffer, J. (1981). *Power in organizations*. Marshfield, MA: Pitman.

Pfeffer, J. (1992). Understanding power in organizations. *California Management Review*, *34*(2), 29–50. doi:10.2307/41166692

Rice, A. K. (1966). *Learning for leadership: Interpersonal and intergroup relations*. London: Tavistock Publications.

Shaiken, H. (1984). *Automation and labor in the computer age*. New York: Holt, Rinehart & Winston.

Shapiro, E. R., & Carr, A. W. (1991). *Lost in familiar places: Creating new connections between the individual and society*. New Haven, CT: Yale University Press.

Shapiro, E. R., & Carr, A. W. (2006). These people were some kind of solution: Can society in any sense be understood? *Organisational and Social Dynamics*, *6*(2), 241–257.

Shapiro, E. R., & Carr, A. W. (2012). An introduction to Tavistock-style group relations conference learning. *Organisational & Social Dynamics*, *12*(1), 70–80.

Shen, W., & Cannella, A. A. Jr. (2002). Power dynamics within top management and their impacts on CEO dismissal followed by inside succession. *Academy of Management Journal*, *45*(6), 1195–1206. doi:10.2307/3069434

Straffin, P. D. (1996). *Game theory and strategy*. Washington, DC: The Mathematical Association of America.

Sullivan, H. S. (1953). *The interpersonal theory of psychiatry*. New York: Norton.

Sutton, R., & Kahn, R. L. (1986). Prediction, understanding, and control as antidotes to organizational stress. In J. Lorsch (Ed.), *Handbook of organizational behavior*. Englewood Cliffs, NJ: Prentice-Hall.

Tran, B. (2014a). Rhetoric of play: Utilizing the gamer factor in selecting and training employees. In T. M. Connolly, T. H. Hainey, E. Boyle, G. Baxter, & P. Moreno-Ger (Eds.), *Psychology, pedagogy, and assessment in serious games* (pp. 175–203). Hershey, PA: Premier Reference Source/IGI Global.

Tran, B. (2014b). Game theory vs. Business ethics: The game of ethics. In B. Christiansen, & M. Basilgan (Eds.), *Economic behavior, game theory, and technology in emerging markets* (pp. 213–236). Hershey, PA: Premier Reference Source/IGI Global.

Trist, E., & Sofer, C. (1959). *Exploration in group relations*. Leicester, UK: Leicester University Press.

Turquet, P. M. (1974). Leadership: The individual and the group. In G. S. Gibbard, J. J. Hartman, & R. D. Mann (Eds.), *Analysis of groups*. San Francisco, CA: Jossey-Bass.

Ullah, S., Jafri, A. R., & Dost, M. K. B. (2011). A synthesis of literature on organizational politics. *Far East Journal of Psychology and Business*, *3*(3), 36–49.

Varoufakis, Y. (2001). General introduction: Game theory's quest for a single, unifying framework for the social sceinces. In Y. Varoufakis (Ed.), *Game theory: Critical concepts in the social sciences* (Vol. 1). London: Routledge.

Vigoda, E. (2000). Organizational politics, job attitudes, and work outcomes: Exploration and implications for the public sector. *Journal of Vocational Behavior*, *57*(3), 326–347. doi:10.1006/jvbe.1999.1742

Vigoda-Gadot, E. (2001). Stress-related aftermaths to workplace politics: The relationships among politics, job distress, and aggressive behavior in organizations. *Journal of Organizational Behavior*, *23*(5), 571–591. doi:10.1002/job.160

Vigoda-Gadot, E., & Drory, A. (2006). *Handbook of organizational politics*. Cheltenham, UK: Edward Elgar. doi:10.4337/9781847201874

Wayne, S. J., Liden, R. C., Graf, I. K., & Ferris, G. R. (1997). The role of upward influence tactics in human resource decision. *Personnel Psychology*, *50*(4), 979–1006. doi:10.1111/j.1744-6570.1997.tb01491.x

Welsh, M. A., & Slusher, E. A. (1986). Organizational design as a context of political activity. *Administrative Science Quarterly*, *31*(3), 389–402. doi:10.2307/2392829

Winnicott, D. W. (1960). *The maturational processes and the facilitating environment*. New York: International Universities Press.

Zinner, J., & Shapiro, E. R. (1972). Projective identification as a mode of perception and behavior in families of adolescents. *The International Journal of Psycho-Analysis*, *53*, 523–530.

ADDITIONAL READING

Allen, R. W., Madison, D. L., Porter, L. W., Renwick, P. A., & Mayes, B. T. (1979). Organizational politics-tactics and characteristics of its actors. *California Management Review*, *22*(1), 77–83. doi:10.2307/41164852

Andrews, M. C., & Kacmar, M. K. (2001). Discriminating among organizational politics, justice, and support. *Journal of Organizational Behavior*, *22*(4), 347–366. doi:10.1002/job.92

Aryee, S., Chen, Z. X., & Budhwar, P. S. (2004). Exchnage fairness and employee performance: An examination of the relationship between organizational politics and procedural justice. *Organizational Behavior and Human Decision Processes*, *94*(1), 1–14. doi:10.1016/j.obhdp.2004.03.002

Bacharach, S. B. (2005). *Get them on your side: Win support, convert skeptics, get results*. La Crosse, WI: Platinum Press.

Beugre, C. D., & Liverpool, P. R. (2006). Politics as determinant of fairness perceptions in organizations. In E. Vigoda-Gadot, & A. Drory (Eds.), *Handbook of organizational politics* (pp. 122–135). Cheltenham, UK: Edward Elgar. doi:10.4337/9781847201874.00015

Buchanan, D., & Badham, R. (1999). *Power, politics, and organizational change: Winning the turf game* (2nd ed.). Sage Publications, Ltd.

Buchanan, D. A. (1999). The logic of political action: An experiment experience with the epistemology of the particular. *British Journal of Management*, *10*(s1), 73–88. doi:10.1111/1467-8551.10.s1.7

Coopey, J., & Burgoyne, J. (2000). Politics and organizational learning. *Journal of Management Studies*, *37*(6), 869–885. doi:10.1111/1467-6486.00208

Coulter, M., & Robbins, S. P. (2006). *Management* (9th ed.). Prentice Hall.

Doldor, E. (2007). *Conceptualizing and investigating organizational politics: A systematic review of the literature*. M. S. thesis, Cranfield University, Cranfield, Bedfordshire, United Kingdom.

Dufour, G. (2010). *Managing your manager: How to get ahead with any type of boss*. McGraw-Hill.

Fairholm, G. W. (2009). *Organizational power politics: Tactics in organizational leadership* (2nd ed.). Praeger.

Farrell, D., & Petersen, J. C. (1982). Patterns of political behavior in organization. *Academy of Management Review, 7*(3), 403–412.

Ferris, G. R., & Treadway, D. C. (2011). *Politics in organizations: Theory and research considerations (SIOP Organizational Frontiers Series).* Routledge.

Gabarro, J. J., & Kotter, J. P. (2008). *Managing your boss.* Harvard Business School Press.

Gandz, J., & Murray, V. V. (1980). The experience of workplace politics. *Academy of Management Journal, 23*(2), 237–251. doi:10.2307/255429

Goldstein, M., Read, P., & Cashman, K. (2009). *Games at work: How to recognize and reduce office politics.* Jossey-Bass.

Goman, C. K. (2011). *The silent language of leaders: How body language can help—or hurt how you lead.* Jossey-Bass.

Harris, K. J., James, M., & Boonthanom, R. (2005). Perception of organization politics and cooperation as moderators of the relationship between job strain and intent to turnover. *Journal of Managerial Issues, 17*(1), 26–42.

Kakabadse, A., & Parker, C. (1984). *Power, politics, and organizations: A behavioural science view.* Chichester: Wiley & Sons.

Kurchner-Hawkins, R., & Miller, R. (2006). Organizational politics: Building positive political strategies in turbulent times. In E. Vigoda-Gadot, & A. Drory (Eds.), *Handbook of organizational politics* (pp. 351–358). Cheltenham, UK: Edward Elgar. doi:10.4337/9781847201874.00029

McIntyre, M. G. (2005). *Secrets to winning at office politics: How to achieve your goals an increase your influence at work.* Saint Martin's Griffin.

Morris, R. A. (2010). *Stop playing games! A project manager's guide to successfully navigating organizational politics.* RMC Publications.

Ralston, D. A., Giacalone, R. A., & Terpstra, R. H. (1994). Ethical perceptions of organizational politics: A comparative evaluation of American and hong kong managers. *Journal of Business Ethics, 13*(12), 989–999. doi:10.1007/BF00881669

Sowmya, K. R., & Panchanatham, N. (2011). Organizational politics—behavioral intentions of banking sector employees. *Journal of Commerce, 3*(1), 8–21.

Sussman, L., Adams, A. J., Kuzmits, F. E., & Rocho, L. E. (2002). Organizational politics: Tactics, channels, and hierarchical roles. *The Journal of Applied Psychology, 90*(5), 872–881.

Tulgan, B. (2010). *Its okay to manage your boss: The step-by-step program for making the best of your most important relationship at work*. Jossey-Bass.

Zanzi, A., & O'Neill, R. M. (2001). Sanctioned versus non-sanctioned political tactics. *Journal of Managerial Issues*, *13*(2), 245–262.

KEY TERMS AND DEFINITIONS

Bion's Theory: Groups, like dreams, have a manifest, overt aspect and a latent, covert aspect. The manifest aspect is the work group, a level of functioning at which members consciously pursue an agreed-upon objective and deliberately work toward the completion of a task.

Coalitional (or Cooperative) Game: Is a high-level description, specifying only what payoffs each potential group, or coalition, can obtain by the cooperation of its members.

Game Result: For a given player is defined as the payoff.

Game Theory: Is the formal study of conflict and cooperation, and the game theoretic concepts apply whenever the actions of several agents are interdependent.

Game: Which is a formal model of an interactive situation.

Individual Decision (or Choice of a Player): Is defined as a mover and a series of moves of a given player is a strategy.

One Player Game: Is usually called a decision problem. The formal definition defines the players, their preferences, their information, their strategic action available to them, and how these influence the outcome.

Player: Is defined as a rational agent in which a rational agent is not necessarily a person as a rational agent could be an institution or a firm.

The Tavistock Method (Commonly Known as Group Relations): Was originated from the work of British psychoanalyst Wilfred Ruprecht Bion.

The Tavistock: The basic assumption level of functioning of behavior is "as if" behavior: the group behaves as if a certain assumption is true, valid, and real and as if certain behaviors are vital to the group's survival.

ENDNOTES

1 Wilfred Bion first described these basic assumptions in *Experience in Groups*. There have been suggestions about other basic assumptions, but none has as yet carried complete conviction.

Chapter 9
Emergently–Persuasive Games: How Players of *SF0* Persuade Themselves

Neil Dansey
University of Portsmouth, UK

EXECUTIVE SUMMARY

This case identifies and exemplifies a potential subset of persuasive games, called emergently persuasive games. These are games that focus more on unspecified, player-led persuasion as opposed to persuasion based on specific, designer-led outcomes. The game described in this case is SF0, and its players have been observed to have become more outgoing, creative, and wise, despite only an incidental, general level of pre-designed persuasion being advertised. It is demonstrated that the ambiguous rules of the game allow the players to customise the gameplay based on their everyday needs, and therefore decide for themselves whether and how they want to be persuaded. These creative interpretations of the rules are actively encouraged, rather than being discouraged as they would be in other games. The ongoing player-discussion of conflicting interpretations facilitates a very effective constructivist environment for self-improvement and understanding. Data was analysed from 24 players of SF0, and a Grounded Theory was generated both to explain the general observations of the player data and to identify the diverse ways in which real-world benefit has arisen.

DOI: 10.4018/978-1-4666-6206-3.ch009

Copyright ©2014, IGI Global. Copying or distributing in print or electronic forms without written permission of IGI Global is prohibited.

BACKGROUND

The research described in this case encompassed the main study of my PhD thesis on the subject of real-world benefit through gameplay. The study was carried out in the UK at the University of Portsmouth's School of Creative Technologies, where I teach as a Senior Lecturer on the BSc (Hons) Computer Games Technology course. I am a member of the School's Advanced Games Research Group, which brings together the diverse research interests of the staff and students, and our members' recent outstanding successes include the creation of the games studio *thechineseroom*, whose research-informed game *Dear Esther* (2012) was released to widespread critical acclaim. Over the past 6 years I have been heavily involved in organising and running educational "Game Jam" events in conjunction with partners in the UK, France, Denmark, and the USA, as part of my ongoing goal to build links with games developers and other academic institutions, while leveraging the social, educational, and employability benefits of problem-based game development exercises.

SETTING THE STAGE

This project was originally intended to be an exploratory study on the use of *ambiguity* in experimental game design. Typically in games, ambiguity is something to be avoided; it is regarded as a nuisance, or a hallmark of badly-worded rules or poorly-considered systems design. Generally speaking, if the current game state or pertinent rules of play are ambiguous, players will need to compromise, reach an agreement, or consult an adjudicator if harmonious play is to continue.

However, there are a number of games in which ambiguity, and the resulting discussion on conflicting interpretations, is *encouraged* as a key feature of gameplay. This case concerns such a game: *SF0* (sometimes written *SFZero*) is a highly-ambiguous game played both on the internet (at www.sf0.org) and in the physical everyday world. 24 players of *SF0* were interviewed, with the intention being to investigate the nature of the gameplay and the appeal of such an ambiguous game. However, due to unexpected developments in the interviews, the focus of the study was drawn away from ambiguity towards the more noticeable issue of real-world benefits that the players were getting as a result of persuasiveness in *SF0*, hence the applicability of the case to this book.

CASE DESCRIPTION

The Game of *SF0*

The overall aim of *SF0* is difficult to determine with confidence due to the quirky, vague language used throughout the game's website. Nevertheless, it is possible to make out that the gameplay encourages players to challenge the ingrained behaviours and "inaccurate" interpretations they already have of their everyday lives and surroundings. This is mainly achieved via the completion of "tasks", which involve the player going out into their everyday environment and taking part in various activities in exchange for points. The challenge for the player, however, is that the instructions for the tasks tend to be written in an abstract, metaphorical, or poetic manner, avoiding too much detail on how the outcome might be achieved. For example, the task *Saint George* (n.d.) instructs players to "Slay a dragon. Celebrate in verse, song or painting". This means that before a player completes the task they need to consider different ways of representing the fairly abstract notion of "slaying a dragon". It might be by confronting an intimidating colleague at work, or more literally, by smashing up a statue of a dragon. The ambiguity in the task description drastically reduces the likelihood of two players' experiences being exactly the same, and this is one of the main ways that *SF0* creates complex gameplay through a relatively small set of formal rules. Table 1 lists more of these tasks, and illustrates the vague nature of the instructions given to the players.

Once a player chooses and completes a task they must submit evidence of the task completion, such as photographs, prose or audio recordings, to the *SF0* website in order to receive points and progress in the game. Accumulating points will unlock more significant tasks for the player to try, increasing the difficulty of the game accordingly.

The player receives a predetermined amount of mandatory points for completing a task and uploading evidence, but the evidence is also discussed with other *SF0* players, with bonus points being awarded if they are suitably impressed, perhaps by the amount of effort, creativity, tenacity, or courage on display. While it is generally

Table 1. A small sample from the hundreds of tasks in SF0

Task Name	Instructions
Things you can run through	"Find some."
The speed of time	"Accelerate or decelerate an aging process."
Keep marching on	"Destroy a piece of your past."
Leave clues	"Leave clues."

understood that there is no 'right' way to complete a task, providing an *interesting* or *impressive* interpretation of the instructions is key if a player wants to achieve a high score in the game, particularly amongst experienced players. It is not uncommon for a player to receive many more bonus points than mandatory points: in one notable case ("*Campaign Trail*", n.d.) a player received 577 bonus points for a 15-point task.

Figure 1 shows the *SF0* task interface, which is the primary way for the players to browse and select tasks. Each task is colour-coded to show its relevance to one or more in-game *groups*, similar to factions, that represent particular thematic interests. One such group, *Biome*, favours "an awareness of the organisms populating the City with special emphasis on their locations and identities". Therefore, tasks that relate to such themes are presented in green to indicate their applicability to players interested in the *Biome* play-style. Elsewhere, the *Chrononautic Exxon* group rejects the notion of past and future, and therefore the purple tasks pertaining to this group require the player to interact with time in unconventional ways. For example, the task *CTRL+Z* (n.d.), otherwise known as the Windows keyboard shortcut for 'undo', has the following instructions:

The law of entropy: a system tends to degenerate over time.

It is easy to shatter glass, but difficult to put it back together. It is easy to create toast, but almost impossible to turn it back into bread. It is easy to make mistakes. Destroy relationships. Regret choices.

But we know that time is flexible. Now undo.

Each player has a "grouposis" section on their *SF0* profile page that tracks the amount of interest they have in the tasks of each group. As they complete more and more tasks, their grouposis updates to reflect this, and they automatically become affiliated with the group that governs the majority of their task completions. Certain tasks, being heavily themed, require affiliation with a particular group before they can be attempted.

The game also places great emphasis on cooperation and collaborative tasking. Perhaps the best example of this is the task *Journey to the End of the Night* (n.d.), which is a signature task of *SF0* and a favourite for completion at group gatherings, to the point that the task has been re-issued many times in order for it to become a repeatable annual event and a highlight of the *SF0* calendar in different locations around the world. The task seems to fully embody the *SF0* ethos, namely the willingness to better explore local surroundings and let chance dictate the outcome:

Figure 1. The SF0 task interface (© 2013, Sean Mahan, Ian Kizu-Blair, and Sam Lavigne. Used with permission)

For one night, drop your relations, your work and leisure activities, and all your usual motives for movement and action, and let yourself be drawn by the attractions of the chase and the encounters you find there [...] Document your experience.

To aid the organisation of *Journey to the End of the Night* for large groups, the participating players are split into *runners* and *chasers*, with the runners being required to visit a number of predetermined checkpoints around the city without being caught by the chasers. When runners get caught they also become chasers, making the chase increase in intensity as it progresses. Regardless of role, players are encouraged to document interesting experiences as creatively as possible to create a large body of evidence for the *SF0* website. Again, it is up to the players to decide what they feel is interesting enough to warrant inclusion.

Evidence of all task completions, or "praxis", is publically available on the *SF0* website, and two further examples are presented here to illustrate the scope of the tasks. The first is from the task *Make it Soft* (n.d.), which simply requires the player to "make something soft". One player (Mahan, 2010) used his understanding of electronics and some components sourced online to make his own electric guitar effects unit. This was then embedded inside a teddy bear, with the eyes and belly button being replaced with control knobs, and the hands forming the input and output sockets. As a charming final touch, the LED power indicator was positioned under the bear's fur so its heart region glowed red when the device was switched on.

The second example of *SF0* praxis is from the task *Stayin' Alive* (n.d.), in which the players are required to "learn a new skill that will help you survive after the apocalypse". One player described a zombie-themed apocalypse in which he would probably not survive. His plan was to map out every spot in his house where the floorboards creaked, so he could learn to identify the exact location of intruders from the noise their footsteps made on the floor. Interestingly, while this project was clearly based on a hypothetical apocalypse, the lessons learned could quite conceivably be useful in a real burglary situation.

The Study

The initial aim of the study was to explore what it was like to play experimental games such as *SF0*, and to find out what the players thought of the tasks given the amount of interpretive effort they needed to put in to be able to play. 24 recently-active players were interviewed about the tasks in *SF0*, with 15 interviews being carried out via the *SF0* messaging interface. The remaining 9 players were members of a UK-based *SF0* team that were holding their annual weekend gathering in their home town to socialise and complete some collaborative tasks. I attended the weekend and interviewed the players face-to-face on an individual basis, and took

Figure 2. An effects unit made soft (© 2010, Sean Mahan. Used with permission.)

part in the tasks, having been a solo *SF0* player myself in the past. As expected, the main task for the weekend was *Journey to the End of the Night*, and was extremely valuable in getting first-hand experience of group tasking, particularly when being chased around an unfamiliar city by people I had only just met.

While the ambiguous nature of the game and its tasks was discussed to some extent by the players, there seemed to be a much more significant, unprompted and recurring theme throughout many of the interviews. It became apparent that in addition to being enjoyable to play, *SF0* was also helping the players to improve their everyday lives - the game was giving them the means and motive to take part in beneficial activities that they felt they could not, or would not, have done otherwise. It seemed that the players were being *persuaded* to try new things in order to better themselves.

This was not surprising at first: *SF0* does claim to promote a change of attitude in its players, namely by "sensitising them to new affordances" in their everyday environment (similar to the *ubiquitous games* described by McGonigal, 2006, p.46). Participant 6 described this process as a 'warping' of peoples' perceptions of their everyday surroundings. Therefore, while the preferred attitude is neither mandatory nor particularly specific, it would seem that an element of general persuasiveness is designed into the game in the traditional sense: the required outcome is determined in advance, and examples of good practice are broadcasted to the player throughout play in order to influence their behaviour.

However, it would seem that the design effort put into in the predetermined and relatively vague persuasiveness in *SF0* was fairly trivial compared to the effort made by the players when applying it specifically to themselves. The results of the study showed that many players voluntarily applied the game in a diverse range of personally-relevant ways that were not explicitly stated or obviously implied in the *SF0* game materials, and often to a remarkable extent given the lack of detail provided by the tasks.

For example, at many points throughout the annual gathering players mentioned problems with living in such a run-down and conservative town, where the hostile nature of the locals towards creative, playful or expressive types bordered on unsafe. They found that *SF0* could be used as a tool to diffuse this hostility by framing the 'weird' activities they normally enjoy in a context that the public could understand, namely a game. While taking part in *Journey to the End of the Night* I found the reaction of the local residents and the sense of awkwardness from running around in a public space to be quite noticeable at first. However, this quickly became insignificant as the game became more intense and progressed towards its climax at the last checkpoint - I barely registered the embarrassment of falling over amongst a crowd of people and chipping a bone in my elbow, despite the pain being very real. Participant 1 agreed that although they often got "weird looks" from people, it was easy to ignore when playing *SF0*.

Two more examples of significant real-world benefit being facilitated by *SF0* are presented here. The first is from Participant 14 (edited purely for brevity and clarity), an email respondent who used the game to maintain a good relationship with family in the wake of the side effects of heavy medication:

I am playing this game with my son. We do not plan on stopping play for any reason at this time. This game has become a really cool interaction between my 10 year old and me.

I was injured at work this year. I am having trouble walking - sitting - well, pretty much anything is beyond me now. This contributes a strange set of problems in life for my family... I cannot drive - I cannot walk around - I can't do anything really without lots of medication. Medication which can make me irritable & nasty...

*I am not a nasty person - I am a fun & silly person. I have learned to roll with the uber-silly side that the drugs let out instead of the nasty a****** they create. [SF0] also allows me to be able to interact with my child even while on heavy RX drugs. We can do things together - even when I cannot physically do much. These large* *quantities of drugs also affect my mind in another way. Depression. No one wants to hang out with some downer - even if it is your family. When I am creative - I am focused - lost in another world.*

SF0 allows my son to view me in a new light - hurt but still trying. Sad but still looking for joy. Lost but looking for the right direction. In pain but finding creative outlets in which to let it go. These are lessons that I hope he will never have to learn on his own, but if he does... at least he has a good reference point on how to deal with it.

The second example is from Participant 16 (again, edited purely for brevity and clarity), an email respondent who used the game to deal with difficult social issues and achieve a sense of belonging:

SF0 really helps me to feel as if I am a part of something.

[The game] gives you an opportunity to be a person you normally aren't. Heck, it's practically a requirement! I'm normally a laid back individual, and SF0 has really helped me to open up. It gives me an opportunity to share similar experiences with other people, even though it is online. I can relate to these players, I can laugh and comment about other experiences that players have. Even if I never meet these people, I carry around the knowledge that there are others who know a bit about me, and who can appreciate the tasks I submit.

I suppose I certainly could have [done this without the game], but I also know that before SF0, I wouldn't have. To be frankly, brutally honest [...] I am starved for social interaction of any kind whatsoever.

SF0 fills my quota for friends. I am rather ashamed of my lack of, well, other people. It's rather disheartening to go to clubs or bars alone. Just sitting there, while people have fun around you. I sit in bookstores and cafes, and read. How am I to get into parties without friends to give me an invite, or to even let me know when and where it is?

If I were to put all of my reasons in one sentence, it would be this: SF0 helps me to be free, it helps me to feel like I belong.

It is unclear whether particular subject matter and such a level of poignancy was intended by the creators of *SF0*, but it is clear from these examples that the players are using the game to help themselves in a variety of specific and deep ways that would be extremely difficult to control or predict given the relatively vague nature of *SF0*. The outcomes are surprisingly complex and significant, more so than the implied sum of their parts, which makes the process closely related to the systems concept of *emergence* (Johnson, 2001). Therefore, I propose the term *emergently-persuasive* to describe the subset of persuasive games such as *SF0* in which the game takes a more passive role in unpredictable, context-sensitive, player-instigated change. Coupled with the welcoming attitude of the game towards contradictions in viewpoint and the inevitable discussion that results between players, this represents a more constructivist approach to personal development than is normally expected from a persuasive game. Because so much information is left for the players to supply, a player is the main facilitator of their own behavioural change, only if they want to, and in a context that is relevant to their needs. Each task experience is based on the specific preferences of the player to whom it belongs, with the temptation to avoid conforming to the popular views of others.

As I gathered interview data from players of *SF0* it was analysed using a Grounded Theory methodology informed by the work of Glaser (1978, 1998; For full details see also Dansey, 2013). Grounded Theory is an emergence-based methodology where patterns in the low-level interview data lead to the development of medium-level "codes" that summarise recurring concepts, such as *enjoyment*, *organisation*, *friendship* and so on. The codes are then grouped into higher-level "categories" based on their relatedness - for example, the codes *friendship*, *discussion*, *sharing* and *collaboration* might be grouped into a category called *socialising*, or *cooperation*, based on the researcher's perception of best fit. Finally, one of the emerging categories is described as the "core category" that will form the backbone of the theory to explain the interview data, and the rest of the categories are described in regard to this core category. Throughout the process, the researcher appeals for more data based on the emerging theory, and the entire process is repeated a number of times along the way to include new data, to identify redundancies and to better express the theory.

The Grounded Theory generated as a result of the study summarises the many emergent benefits observed by the 24 players of *SF0*. The theory is presented in Figure 3. Generally speaking, *SF0* is helping its players to become more outgoing, creative and wise, by providing the means and/or motive for them to take part in fairly difficult activities that they felt they would not, or could not have done before. Such activities include playing unashamedly, self-exploration, demonstrating individuality and so on. Going back through the interview data from the 24 players, this theory can be observed at over 150 specific points, which demonstrates both that it is sufficiently grounded in the data and that the topic is of key significance to the participants.

Figure 3. A grounded theory of the emergently-persuasive properties of SF0

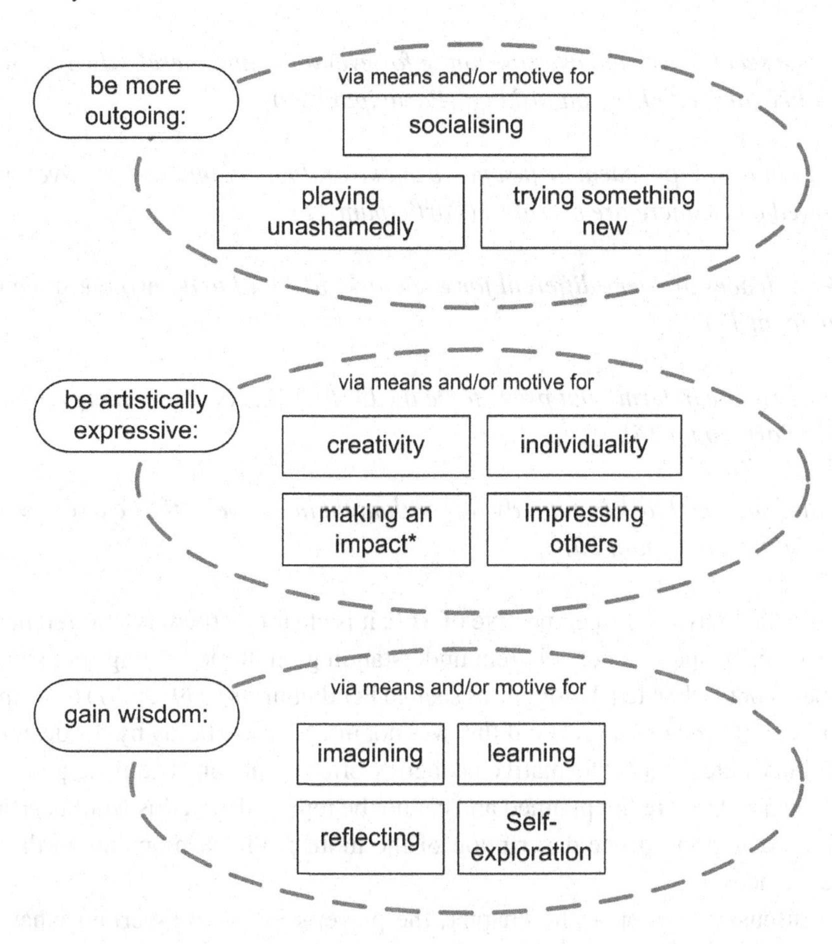

*There is currently no evidence for "the motive for making an impact".

Because of the complexity of the game, the potential range of human needs, and the welcoming of unique interpretations, the theory is certainly not exhaustive, but it can serve as a useful tool in arguing that a number of *specific* benefits exist as a result of player-led change rather than having been predicted or designed into the game at any detailed level, and therefore that the game is *emergently-persuasive*.

Learning Theories and Frameworks

The following quotes illustrate a typical player view of the scope of potential interpretations in *SF0*:

The game seems to try to have something for everyone, and there's always enough tasks I like for me to keep on tasking. (Participant 10)

I can try things, experiment or just make a joke or do it deathly serious, everything is allowed cause there are no rules. (Participant 11)

The experiences are very different for each task. It's hard to do two the same way. (Participant 12)

Most are simple in term - but prove to be the best. This is because interpretation is broad. (Participant 14)

Even the [simplest] task is actually rather thought intensive to figure out how to do it just right. (Participant 20)

From the individual's perspective of *SF0*, it is clear that there is a preference for the "personal responses not coherent understanding, multiple readings not singular arguments or teacher-led dialogue" as seen in Pendlebury's (1998, p.57) description of *performative pedagogy*, even if this was not intended explicitly by the designers. Pendlebury notes that performative pedagogy often results in "exquisite pedagogical moments" that are not planned and cannot be repeated, and this would certainly seem to be an appropriate description of the more poignant moments in the *SF0* task evidence.

As discussed earlier in the chapter, the player's job is to ascertain what *they* feel is appropriate for a task completion, and the flexibility of the game and the encouragement of alternate interpretations promotes a safe environment in which players can reflect with others on their level of success and try things out without the usual consequences of making mistakes. Indeed, according to Walker's (1999, p.81) description of "a learning community of self-determined individuals", in environments such as *SF0* not only is it inevitable that people make mistakes, but mistakes are *desirable* because they facilitate learning, and the player will learn over time to welcome errors as "opportunities for further learning". However, with *SF0*, it should be noted that because there is no 'correct' way to complete the tasks, the notion of 'mistakes' tends to be subjective and based on the player's analysis of their own performance compared to what they intended to achieve. Nevertheless, analysis is aided through discussion with the community, where assumptions, biases or avenues for further thought might be uncovered as the player compares their praxis with the experiences of other players. Here the player will almost certainly experience a level of schematic "disequilibrium" when they encounter interesting task submissions that do not fit with their own existing understanding of the potential of the game (Fox, n.d., p.11). Through critical dialogue with other members of the community the player remodels their own understanding to accommodate, but not necessarily agree with, the new information in a much more constructivist way than would be expected from other persuasive games. Furthermore, as the player's own praxis is viewed by other players in the future, it contributes to the ongoing reshaping of the community understanding of *SF0*. Gee (2007, p.17) agrees that this two-way process of allowing players to both "read" from and "write" to a domain (i.e. *SF0*), by constructing meaning within the domain as a result of their engagement (Gee, 2007, p.20), is incredibly valuable for learning. More specifically, therefore, it would seem that the "pedagogical emphasis on discussion, collaboration, negotiation and shared meanings" (Ernest, n.d., p.88) exhibited by *SF0* bears the hallmarks of a *socially*-constructivist approach, if indeed any particular approach was intended at all.

Such a level of community peer-review is important for player development, as would be expected to be the case for any type of reflective learning environment. This is explained by Hogan (1997, p.93), who concludes that:

Because our human efforts to understand are inescapably constrained by perspective, each of us will always understand incompletely, and in some degree differently... Our understanding can, through more disciplined, co-operative and sustained efforts in any field of study, become more complete even if it can never become absolute.

The cycle of reflection, comparison and self-improvement in *SF0* was observed by Participants 4 and 7, who noted that an experienced player can watch the play style of newer players changing over time as they interact more and more with the community, resulting in them becoming more adventurous and confident as they realise a wider potential. Indeed, throughout the study I got the impression that *SF0* is seen by the more experienced players as a lifestyle choice, or a way of thinking, as much as a game. This was confirmed by Participant 7 who mentioned that the game "withers away" after a while, implying that it leaves the kind of residual change in attitude that one would intend to happen with a persuasive game. In general, players of persuasive games might find that the required change in attitude is difficult to maintain, for example if the game involved persuading them to give up addictive substances, and Participant 24 demonstrated that this also applies to *SF0*, describing the feeling of being "saddled" because "it is both a burden and a purposeful rejiggering of natural talent, with an internalization of the values of *SF0*".

To enable such effective community learning, Packham (2008, p.15) advocates the need for a 'space' in which critical dialogue can happen as a result of (in the case of *SF0*) performative play. This space should enable political agency, opportunity and voice; it should allow the players to critically explore reality together, and should support an *epistemic community* in that it "shares and maintains resources for acquiring and developing knowledge" that is socially-constructed (Pendlebury, 1998, p.58). In terms of *SF0* the space is the game's website, with its player-created tasks, player-submitted praxis[1], and forum-style discussion boards. Furthermore, particularly evident in *SF0* are the *explicit* characteristics of epistemic communities discussed by Pendlebury, including methods for evaluating evidence (the scoring and discussion system), a language for description (the list of *SF0* "terms" (n.d.)), and "a critical mass of practitioners" (players) who share the values of the community.

A notable example of a task reflecting the community values of *SF0* is the task *Journey to the End of the Night 2* (n.d.), a sequel to the first version that encourages players to complete the task again but on a significantly higher scale, involving inter-city travel and over 1000 participants. Interestingly, the meagre 25 points available for completion of this task promotes the experienced attitude to playing *SF0* as if the score doesn't matter; as if the activities conducted in getting the score are more important than the score itself. This is similar to Dunne's (1995, p.152) discussion on the *internal and external goods of a practice*, where the 'practice' in this sense is the "coherent and invariably quite complex set of activities and tasks that has evolved cooperatively and cumulatively over time", bearing an uncanny resemblance to the game of *SF0*. The external goods are synonymous with the formal outcomes of the game such as the player's score and level. The internal goods, however, arise from the deep engagement with the practice, rather than just the surface-level engagement required to achieve the external goods. As noted by Dunne (1995, p.153), internal goods arise from the demonstration of honesty, humility, patience, courage and tenacity in a task, as well as inclusivity and the collaborative celebration of accomplishment and a job well done.

CURRENT CHALLENGES AND RECOMMENDATIONS

Emergently-persuasive games by definition are difficult to control. Following on from the discussion of the performative pedagogy approach to learning, Pendlebury (1998, p.57) notes that the interpretive freedom afforded by this approach means that it could be criticised for depending too much on luck or "happenstance" to make learning happen, implying that activities that rely on performative pedagogy might be too unpredictable to be of reliable use in a classroom environment, for example. As Participant 10 explained:

Sometimes you do a lot of planning, put a lot of thought into it, and get frustrated as it all fizzles. Sometimes all that planning turns into a brilliant time with you and some friends doing something strange [...] Sometimes you put yourself through all sorts of strange experiences without much planning at all.

A similar problem can be seen in the socially-constructivist approach that *SF0* seems to advocate - in particular, there is nothing stopping the construction of understanding from progressing towards non-helpful, or even actively unhelpful ends.

This tendency for unpredictable outcomes in *SF0* means that it is not likely to be suitable when an educational focus on a particular subject within particular time constraints is required. Instead, emergently-persuasive games such as *SF0* are better presented as a kind of passive service, where players can come and go as they please, and where there is neither pressure on the designers for the game to persuade successfully, nor on the players to be persuaded towards anything in particular. Thus, emergently-persuasive games are *games* first, and any real-world benefit is treated as a fortunate by-product of play.

If this is the case, however, it could then be argued that *all* games have the potential to be emergently-persuasive if interpreted in particular ways. One might find that playing *Puerto Rico* (2002) raises their awareness of human rights and questions their views on slavery, or that playing *Colin McRae Rally 2.0* (2000) makes them want to take up racing or buy a Subaru car. Indeed, other *persuasive* games might be *emergently*-persuasive in unexpected ways too. A game that advertises a brand of car might inadvertently persuade someone to purchase the soundtrack used for the in-game music, for example. Furthermore, a game that is designed to be persuasive might fail due to a lack of designer skill, or if players play the game in ways other than intended by the designers (see Gazzard's (2012) discussion on *appropriated play*).

Therefore, it is argued here that the study of persuasiveness in games should not be limited to games that have been *designed* to be persuasive - instead, focus might be better placed on games that have been *demonstrated* to be persuasive, regardless of designer intention. Unnecessary focus on predesigned persuasiveness could be limiting our understanding of the overall potential of persuasiveness and the scope of real-world benefit that can be obtained from *playing games in general*. Despite this, however, due to the unpredictability of player behaviour in games such as *SF0* it is understandable that pre-designed persuasive games are favourable in a research context; they are easier to discover and identify (being clearly labelled as such), and provide a dense, controlled and focused environment for data gathering, where efficient use can be made of any specialist hardware that is expensive or difficult to obtain. Nevertheless, it is important to consider the potential consequences that such an approach might imply.

REFERENCES

Campaign Trail. (n.d.). Retrieved April 28, 2014, from http://sf0.org/tasks/Campaign-Trail

Colin McRae Rally 2.0. (2000). Southam: Codemasters Software Company Limited. *CTRL+Z.* (n.d.). Retrieved April 28, 2014, from http://sf0.org/tasks/CTRL--Z

Dansey, N. (2013). *A grounded theory of emergent benefit in pervasive game experiences.* Retrieved from http://www.neildansey.co.uk

Dear Esther. (2012). Brighton, UK: thechineseroom.

Dunne, J. (1995). What's the good of education? In W. Carr (Ed.), *The RoutledgeFalmer reader in philosophy of education* (pp. 145–160). Abingdon, UK: Routledge.

Ernest, P. (n.d.). Varieties of constructivism: their metaphors, epistemologies and pedagogical implications. In R. Fox (Ed.), *Perspectives on constructivism* (pp. 73–92). Exeter, UK: University of Exeter School of Education.

Fox, R. (n.d.). Constructivist views of learning. In R. Fox (Ed.), *Perspectives on constructivism* (pp. 3–16). Exeter, UK: University of Exeter School of Education.

Gazzard, A. (2012). Re-coding the algorithm: Purposeful and appropriated play. In A. Lugmayr et al. (Eds.), *Media in the ubiquitous era: Ambient, social and gaming media.* Hershey, PA: Information Science Reference.

Gee, J. P. (2007). *What video games have to teach us about learning and literacy.* Basingstoke, UK: Palgrave MacMillan.

Glaser, B. (1978). *Theoretical sensitivity: Advances in the methodology of grounded theory.* Mill Valley, CA: Sociology Press.

Glaser, B. (1998). *Doing grounded theory: Issues and discussions.* Mill Valley, CA: Sociology Press.

Hogan, P. (1997). The politics of identity and the epiphanies of learning. In W. Carr (Ed.), *The RoutledgeFalmer reader in philosophy of education* (pp. 83–96). Abingdon, UK: Routledge.

Johnson, S. (2001). *Emergence: The connected lives of ants, brains, cities and software.* London: Penguin Books.

Journey to the End of the Night 2. (n.d.). Retrieved April 28, 2014, from http://sf0.org/tasks/Journey-To-The-End-Of-The-Night-2

Journey to the End of the Night. (n.d.). Retrieved April 28, 2014, from http://sf0.org/tasks/Journey-To-The-End-Of-The-Night

Mahan, S. (2010). *Make it soft.* Retrieved April 28, 2014, from http://sf0.org/swm/Make-it-Soft

Make it Soft. (n.d.). Retrieved April 28, 2014, from http://sf0.org/tasks/Make-it-Soft

McGonigal, J. (2006). *This might be a game: Ubiquitous play and performance at the turn of the twenty-first century* [PhD Thesis]. Retrieved from http://www.avantgame.com

Packham, C. (2008). *Active citizenship and community learning.* Exeter, UK: Learning Matters Ltd.

Pendlebury, S. (1998). Feminism, epistemology and education. In W. Carr (Ed.), *The RoutledgeFalmer reader in philosophy of education* (pp. 50–62). Abingdon, UK: Routledge.

Puerto Rico. (2002). Bernau am Chiemsee: Alea.

Saint George. (n.d.). Retrieved April 28, 2014, from http://sf0.org/tasks/Saint-George

Stayin' Alive. (n.d.). Retrieved April 28, 2014, from http://sf0.org/tasks/Stayin-Alive

Terms. (n.d.). Retrieved April 28, 2014, from http://sf0.org/terms

Walker, J. (1999). Self-determination as an educational aim. In W. Carr (Ed.), *The RoutledgeFalmer reader in philosophy of education* (pp. 74–82). Abingdon, UK: Routledge.

KEY TERMS AND DEFINITIONS

Affordances: Properties of an object or location that communicate possibilities for interaction.

Conflicting Interpretations (In Games): Different views on some aspect of the game, for example if there was a disagreement over the exact meaning of a particular rule, or whether or not a particular action was "legal".

Constructivism: The philosophical position that values knowledge constructed through direct experience, comparison, reflection and adaptation.

Context-Sensitive (Elements in Games): The ability to adapt and/or apply meaningfully to a range of different situations.

Creative Interpretation: The avoidance of an obvious or literal interpretation in favour of (for example) a more metaphorical, witty or subversive one.

Emergence: A phenomenon in which frequent but simple interactions between elements of a system produce unexpectedly complex high-level behaviour, for example the organisation of ant colonies.

Game Jam: A high-intensity game development competition in which teams try to design and build the best game in a relatively short space of time, typically 24-72 hours.

Grounded Theory: An inductive social science methodology that aims to generate conclusions guided primarily by data rather than by hypotheses based on established theories.

Real-World Benefit: The consequences of playing a game that are beneficial outside of the game, for example becoming healthier by playing tennis.

ENDNOTES

1 Interestingly, Packham (2008, p.14) also uses the term *praxis* to describe reflective and informed engagement with an activity, echoing the language used in *SF0* to describe the achievements of the players.

Chapter 10
Communicating the Obvious:
How *Agents Against Power Waste* Influenced the Attitudes of Players and their Families

Mattias Svahn
Stockholm School of Economics, Sweden

Annika Waern
University of Uppsala, Sweden

EXECUTIVE SUMMARY

This chapter describes the game design and study of Agents Against Power Waste (AAPW), a large-scale field experiment where a persuasive pervasive game was put to use to influence households' attitudes towards electricity consumption. This game is particularly interesting as, although it was only the children of the family who were playing, the whole family was affected and to some extent forced to take part in the play activity. The style of game design has been called "social expansion" (Montola, Stenros, & Waern, 2009; Montola, 2011).The chapter focuses on how this impacted the psychological process of persuasion in responding families and individuals.

DOI: 10.4018/978-1-4666-6206-3.ch010

Copyright ©2014, IGI Global. Copying or distributing in print or electronic forms without written permission of IGI Global is prohibited.

1. BACKGROUND

Energy consumption has a price in political and economic problems. In the OECD there has been a 160% rise in domestic electricity consumption during the 40 something years since the energy crisis in the seventies (IEA 2008 p II.43 in Gustafsson, 2010). Electricity is often produced by burning fossil fuels, with diminishing supplies of fossil fuels which causes political instability (IEA 2011). The rise in consumption of electricity also leads to global warming (Solomon, Qin, Z. Chen, & Avery, 2007) which is an ecological problem. Taking inspiration from the Twin River project (Robert H., 1978), (van Houwelingen & van Raaij, 1989) and (Abrahamse, Steg, Vlek, & Rothengatter, 2005), whom all three underline the value of human behaviour in relation to electricity, this thesis approaches electricity consumption by focusing on human behaviour in households. Human behaviour in households is approached by using the pervasive persuasive game Agents Against Power Waste (hereafter "AAPW") to shape human behaviour in households. AAPW is a game that lets real life electricity consumption be a parameter in the game.

AAPW was the apex of a series of games that applied pervasive game design to the task of visualizing electricity consumption and heightening awareness of it, *(Bang, Torstensson, & Katzeff, 2006; Bang, Gustafsson, & Katzeff, 2007; de Jong, Balksjö, & Katzeff, 2013; Gustafsson & Bang, 2008; Gustafsson, Bang, & Svahn, 2009; Gustafsson, 2010; Katzeff, 2010).*

The team for the field test of AAPW was led by staff from the Stockholm of Economics together with the Energy Board of Mälardalen, the company Mobile Interaction, and the Energy Design Studio of the Interactive Institute.

2. SETTING THE STAGE

It should not be difficult to persuade households to consume less electricity. If prices consistently rise over a longer period of time that ought according to any credo of the rational economic man lead to a decrease in consumption of said product. Still, consumption has remained more or less constant (The Swedish Energy Agency, 2012).

One issue may be that while there is a will to conserve electricity in the general population the knowledge of how to go about it is low. This was found by (van Houwelingen & van Raaij, 1989) A pre-study on Swedish households with teenage children (Torstensson, 2005) independently confirmed that. The situation seems ideal for a persuasive communications campaign. There are nevertheless difficulties.

2.1 Electricity is Invisible

Electricity is ubiquitous. Production and consumption of electricity while central to our society and ubiquitous in daily life is quite invisible. The means for controlling electricity tends to be concealed inside the walls of houses or under the designed exteriors of household products. Even visible parts such as electricity cords are deliberately tucked away for aesthetic reasons (Gustafsson, 2010).

Furthermore the cost of the electricity is not communicated when it is consumed and neither is the electricity bill specified at any level of detail. This can be compared to how prices of consumer goods must be advertised in a shop or how a grocery shop receipt is specified. Instead we get an aggregated bill monthly or even bi-monthly giving only the total electricity purchased with no info about the cost of each individual usage situation.

Furthermore a consumer is not interested in the fact that they are buying a batch of electrons. The consumer purchases what those electrons can do. We do not really consider the fact that what we are buying is electricity when the home is warm and the lamps are lit the smart phone is online or the coffee is hot (Torstensson, 2005). Electricity is a very highly commoditized product.

It is well known that the market for a commodity product will be characterized by shrinking margins and shrinking brand equity (Dolak, 2012). All in all, electricity has on a conceptual level sunk underneath an awareness threshold and become "backgrounded". A goal for a persuasive communications strategy for reducing electricity consumption must then be to create not just brand awareness for the brand of electricity bought but also product category awareness (Lange, 2003; Svahn & Lange, 2009). Consumers need to learn that they are buying electricity when going through daily life.

2.2 Second Difficulty: Social Consumption Issues

Many households consist of several individuals of differing ages and incomes. It is rare for teenagers to live on their own and support themselves financially. Still a home has only one contract with an electrical company as a shared resource for all. If the children of a family unit are teenagers they can be heavy users of electricity while not contributing towards the cost.

One strategy towards addressing these issues could be to deliberately foster social communication - world of mouth - as part of the persuasion strategy. For example Marcell et. al. reports on an experimental study where two groups of respondents got the same communications campaign on reducing general energy and greenhouse gas consumption but one of the groups also had a social word-of-mouth component. The campaign that had a social component proved to be more efficient (Marcell, Agyeman, & Rappaport, 2004). The question then arises; what media what design can cut through all these difficulties and achieve a change in attitudes?

3. CASE DESCRIPTION

The game ' AAPW' was designed with a deliberate eye towards the difficulties presented above. A key design decision was to work with foregrounding electricity consumption as part of the use of household appliances and commodities. The design gives electricity visibility as well as attaches a visible cost to it. The second design decision was that it would *force* social interaction and communication within the participating families. Although the game is introduced in schools as a school exercise it cannot be played without affecting the whole family. Hence the game foregrounds the way electricity is a shared resource in a family and triggers dialogue and discussion within families and fosters word-of-mouth communication and discussion. In this section we give a brief introduction to the game and how it is played.

3.1 Game Design

AAPW is an adventure game. The players act as secret agents with an overall mission to conserve household electricity. Each day, a player receives a new mission task. The first six missions centre on specific household areas and are lamps, activities in the kitchen, entertainment equipment, heating of the house, washing and cleaning and finally, showering and bathing. The seventh and final mission concerns the household as a whole, and asks the player to lower the total energy consumption of the household over an entire weekend.

The key to understanding this game is to realise that we are not talking about a virtual household but of the players' own home. At the back end the game server taps into the readings of the home electricity meter and can thus measure how well the players succeeded in each task. This is possible since almost all Swedish electricity meters are of the "smart meter" kind that has the capacity to report their status to an on-line server. The game was able to tap into this data by permission from both the electricity company the participating households. Hence to succeed in the game players must enthusiasm their whole household to join into the task of saving as much energy as possible.

AAPW is not a household against household game. Instead players collaborate with two other players (presumably from the same school) against a team of three other players from another town. This structure was chosen as a way of further motivating players to engage (by peer pressure from the rest of the team).

Figure 1. The game system from Gustafsson (2010) © Interactive Institute

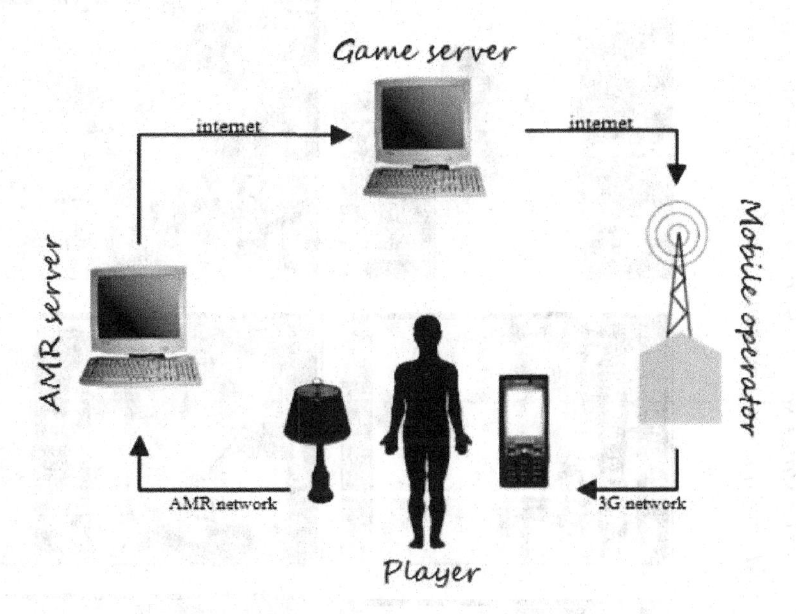

The player gets feedback in the form of a set of screens and bars which visualize the result in more detail, indicating the individual effort within the players′ team. There are also high-score lists showing the winning team for each of the missions and the overall game.

When the game has been completed a summarizing message is sent from the game. This message contains additional information on how much the achieved reduction in consumption corresponds to in money and carbon emission if it would be maintained over an entire year.

Finally 'AAPW' contains a component which is a more traditional computer game. In this game the player navigates a simple two-dimensional platform game to collect virtual 'batteries'. Each battery provides a tip on how this day's challenge can be solved. A clue could for example be "*Unplug wall sockets to prevent the DVD or the TV from using electricity when in standby mode*".

Implementation

AAPW was implemented for at the time standard java-enabled mobile phones. Since the electricity meters were connected to the game at the backend there was no need for custom equipment.

The smart meter systems typically make a data log entry on customers' electricity consumption on an hourly basis. Due to data transmission costs the data is only transmitted to the electricity company′s central server once every 24 hours. This

Figure 2. The AAPW interface © Interactive Institute Pictures 1 – 10 screen dumps from a mobile phone playing AAPW. From the upper left; (1) login screen, (2) login-splash screen, (3) presentation screen (4) agent profile/player profile (5) practice run in the form of a simple labyrinth game (6) a quiz (7) a mission presentation (8) a status screen (9) a navigation menu and (10) an end prognosis screen giving the player info on how much the player would save if the player continues living in the same way as during play.

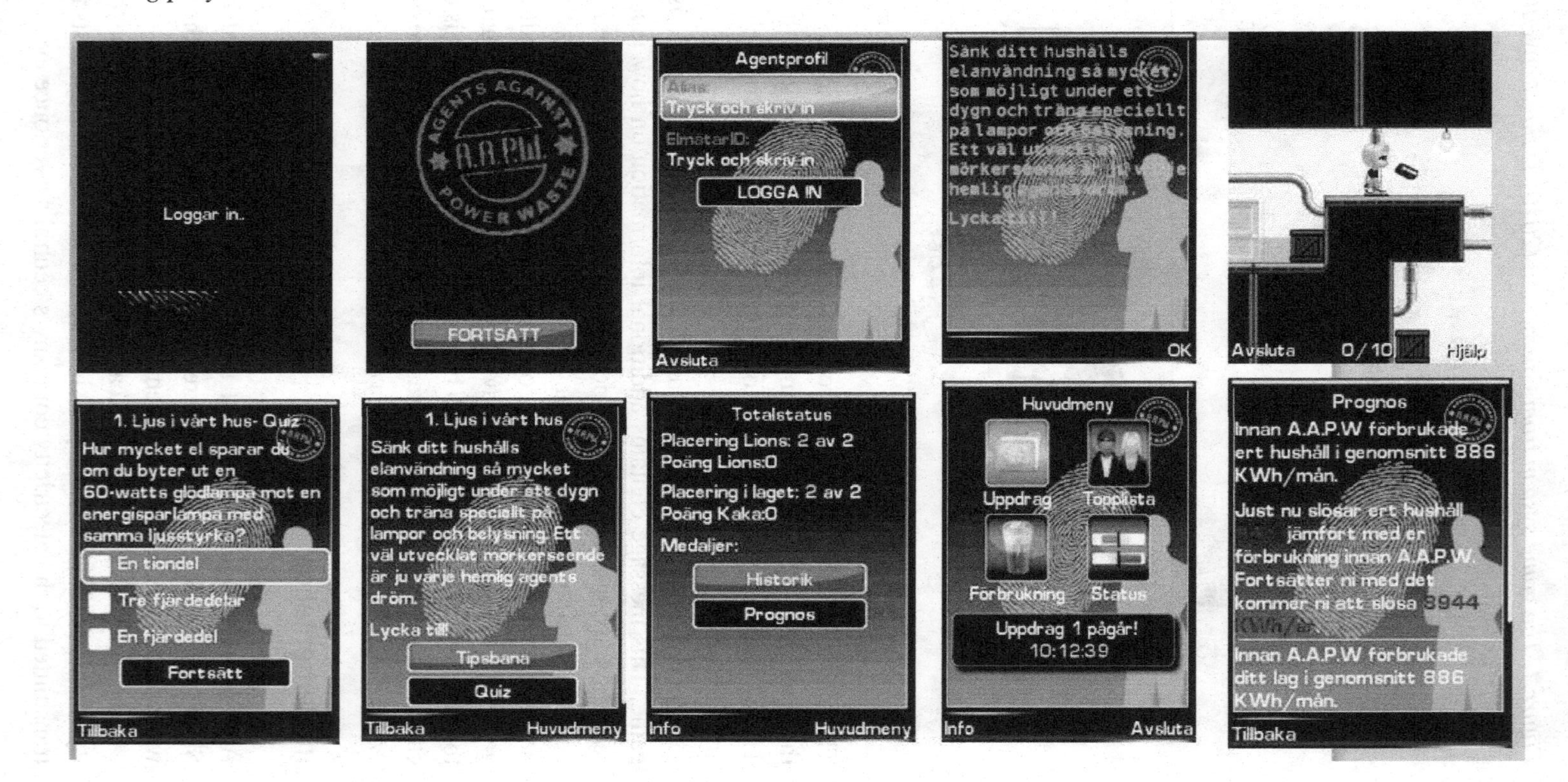

presents a challenge for gameplay design and motivated the design of AAPW as a slow turn-based role-playing game consisting of daily missions. The electricity consumption data was relayed to the game server during the early morning hours. The data was then processed on the game server and the players´ individual results calculated. The results were then communicated to the mobile phone and presented in the form of game graphics.

The overall research question is if this game design can manage to both "foreground" electricity and cut through the issue of social consumption.

4. FIELDWORK AND DATA COLLECTION OF AAPW

4.1 Field Test

AAPW was trialled in 2010 in the Swedish city *Eskilstuna.* The trial was performed by the aforementioned work team consisting of the company Mobile Interaction AB the Mälardalen Energy Board and the Energy Design Studio of the Interactive Institute, led by the Stockholm School of Economics. The electricity companies that had customers in the city provided the back end data on the players. The team recruited four school classes with pupils aged 13-15 years. The teens ran AAPW on their personal mobile phones. The teens and as already discussed also their families played the game for 28 consecutive days. All in all 127 households were involved in the project. The parents were not asked to take part in a game and they did not get any info about the project any more than being asked to sign a form that gave consent for the child to participate for their household electricity data to be used for "a project" and give the customer number of the household electricity contract.

4.2 Data Collection

Multiple sources of data were collected for the study. The participating pupils as well as their parents were asked to fill in questionnaires before and after the game was run. A control group was recruited who did not play the game but who filled in the same questionnaires at the same occasions in time. Finally the team collected log data from the game servers. Electricity meter data was successfully collected during the game to drive the game play but the team did not succeed in saving it for post-game analysis.

5. LEARNING THEORIES AND FRAMEWORKS

AAPW is designed to operationalize two areas of theory; game design theory in the form of pervasive-game design theory (Montola et al., 2009; Montola, 2011) and theories of social psychology, focusing on the Heuristic-Systematic Persuasion Model of (Chaiken, Lieberman, & Eagly, 1989; Chen & Chaiken, 1999).

5.1 Pervasive Games

AAPW is not a computer game although the game makes use of a mobile phone interface to communicate with its players. The game is played in the physical world and uses not only phones but several media forms to communicate the rules of the game and the results. It is possible to classify AAPW as a trans-media game (Evans, 2008), an alternate reality game (Szulborski, 2005), an adaptronic game (Reis & Correia, 2012), or a pervasive game (Montola, 2011). Each perspective brings out certain qualities of the design that may have a bearing on the analysis of the game as a vehicle for persuasion. In this article we focus on the analysis of AAPW as a pervasive game as the term has been discussed by (Montola et al., 2009). The chosen perspective highlights aspects of the design that we believe are crucial in understanding its effects.

According to (Montola et al., 2009), pervasive games are characterized by the way they expand into our everyday world. The argument relies on the recognition that ordinary games be they computerised or not are deliberately and clearly framed in time and space. We play in a set space be it an online server a football field or on a game board. We also play over a set time and the actions we take in the game have little or meaning outside the game and it is typically clear *who* plays a game. The content in games is fictive. 'Angry birds' is not about real birds and pigs and no pigs are harmed in the process. The popular alien hunter game "Halo" uses a close-to-reality rendering of a real-world island but the game is played in a virtual fictional counterpart. When a player plays a Second World War game which adequately reflects historical events, that player is not participating in the Second World War. The players are staging an imagined version of it.

Pervasive games expand out of these habitual demarcations of game versus reality. Spatial expansion means that the game is not physically limited to a controlled playspace. One effect of spatial expansion is that the affordances of the game become if not infinite then at least as richly variable and unpredictable as the real world into which they have moved (Davies, 2007). Advertising for tourism and educational issues for the history of a place match easily onto this design element (Benford et al., 2005.; Colvert, 2009; Raessens, 2007; Walz & Ballagas, 2007).

Temporal expansion occurs when a game interweaves the time of play with daily life so that there is no real "on" and "off" time of the game. The game may control when players should play and everyday actions may have a double meaning as both ordinary and game actions.

Finally social expansion is created when the game is intentionally designed to involve people who are not original players of the game. These people may or may not be aware they have become participant in a game. The role of these non-players may vary: they can become involved as resources for the players as co-players or adversaries or in some form of referee role (Montola et al., 2009).

AAPW is spatially, temporally and socially expanded. The way the game server is connected to the electricity meter of the home expands the playspace from the mobile phone screen to the entire household. It should be noted that this affords for an almost infinite number of ways that actions can influence the game; for example buying a new dishwasher could be an in-game action. AAPW uses a thin layer of fiction - the idea that the players are secret agents - but most of the game world is authentic our actual home. Furthermore though the game is played over a set time (some specific weeks), the actions in the game are normal everyday household chores - cooking dinner using the washing machine. These everyday actions get a secondary meaning from being part of the game but they are still ordinary actions. Finally as the activities of everyone in the household affect the game it is expands the notion of who is a 'player' - it is no longer just the person who holds the phone.

Although all of these design choices are interesting we the authors are particularly interested in the way AAPW manages to make *everyone* in the household a player. AAPW is socially expanded and the non-players are the other members of and visitors to the players' household. Since their electricity consumption will affect the game result these other people willingly or unwillingly become co-players in the player's team. In this chapter the children of the household can be called *primary players* and the parents can be called *secondary players*.

This type of social expansion has also been called integration of social context (Paavilainen, Korhonen, Saarenpää, & Holopainen, 2009). It should be noted that their involvement is independent of whether they actually are aware or unaware of the ongoing game. The intention with the design is to present a challenge to the player who cannot win unless he or she is able to enthusiasm siblings and parents to changing their electricity consumption behaviour in a way that fits the players' game objective.

The potential for educational and persuasive use of pervasive games has been discussed by (Palazzi, Roccetti, & Marfia, Dec 2010). Some previous examples of educational pervasive games include *Rexplorer*, an educational tourist game developed for Regensburg (Walz & Ballagas, 2007), and *Frequency 1550* an educational game for Amsterdam (Raéssens, 2007). Most examples have however focussed on the use of spatial expansion for physically grounding the learning experience in an authentic environment. There exists no previous examples of pervasive games that make deliberate use of social expansion to extend the persuasive effect outside the scope of the original players.

5.1.1 Unaware Participation

When games reach out to non-players they can do so in several different ways (Montola et al., 2009). Typically they do so through a sequence of levels of awareness. Initially non-players are *unaware* of the fact that a game is going on at all. It should be noted that they still can be a resource in the game for players. For example the college campus game "Killer" (Montola et al., 2009) uses a very simple form of social expansion; unaware non-players are in-game obstacles to be avoided. Non-players then go through a state of "ambiguous awareness" of the game during which non-players understand that something out of the ordinary is going on. The game" Cruel 2 B Kind" plays intensely on this state of awareness. In Cruel 2 B Kind players compete by doing good deeds e.g. handing out flowers to as many passers-by as possible. For the player the goal is to manage to hand out a flower to a particular other player - a person who has in her ruleset that she is allergic to flowers. However for the passer-by the person standing in a park handing out flowers is just slightly strange; albeit in a pleasant way. The final stage of awareness is when non-players understand that it is a game and is empowered to decide on their own involvement. This is possible even if they do not understand all details about the game. Non-players may for example decide to stay to watch a game event adopting a role as spectators even if they do not fully understand the rules of the game.

On a cursory glance it seems that expanded games offers up some ethical quandaries. It is important for the designer of an expanded game to provide bystanders with exits. Montola and Waern (Montola et al., 2009) stress the importance of offering non-players a way to *refuse* to play. Offering *aware* participants ways to refuse the game is normally rather easy; it is for example not emotionally stressful to ignore strange people running around in a street dressed up as Pac Man. The tricky part is making refusal a valid and real option for unaware or ambiguously aware participants. A particularly problematic form of non-player engagement occurs when somebody engages in a game while unaware of the fact that it is a game. This is particularly prone to happen when a game relies on blurring fiction and reality (Colvert, 2009; Waern & Denward, 2009). A person that believes that the game is "real" lacks the protective frame of considering themselves as playing and the experience can become quite upsetting in particular if the fictional storyline is scary.

In AAPW the household members of players can be considered as *aware* non-players recruited as co-players on the player team. However the most salient feature of social expansion in AAPW is that the rest of the household members never have made an active choice to be part of the game; the design of AAPW makes the rest of the household into players and players they become. There is no way a household member can refuse to be a secondary player. The household member can only choose to ignore the fact that play is ongoing.

6. LEARNING THEORY

6.1 Theoretical Framework: The Persuasive Processes

The notion that a game by being a designed structure of movement action challenges and responses creates a de facto effective learning environment can occasionally be found in game studies (Gee 2003). This structure supposedly common to all games are sometimes said to offer a player the capacity to try out alternative courses of action and afford players to experience the consequences of these alternatives before choosing the final course of action and that this trial-and-error process while advancing in the game is a learning process explainable by for example operant conditioning.

But for example (Düüs-Henriksen 2008; Egenfeldt-Nielsen, 2007) brings on a sobering perspective. They both point to the need for supervised reflection if a persuasive message is to be carried through into a changed mindset. (Linderoth, 2009), takes the sobering a bit further and points out that learning achieved through games can become illusory if common game design elements such as "power ups" are given to the player after an amount of invested time instead of after real achievement. What is necessary is to find a theory that has developed elements of reflection.

6.2 Heuristic-Systematic Experiencing and Games

The dual process modelling of attitude change is a paradigm that for some time has been influential in attitude research (Perloff, 2010), and does have well developed elements of reflection. The paradigm is made up of theory on the issue of how individuals processes and makes judgments about an experience due to variables interacting in between the individual and the experience. It is well researched; it is centred round the processing of experiences and has a focus on the relationship between variables in the individual and structures of the experience. Therefore it may be a suitable paradigm for applying to an attempt to explain AAPW.

One milestone in that paradigm is the Heuristic-Systematic Model of (Chaiken, 1987; Chen & Chaiken, 1999). It assumes that persuasion is achieved when the individual processes an experience according to two different modes that are distinctly different the *Systematic* and the *Heuristic* modes.

6.2.1 Systematic Processing

When we have a high need for confidence in our understanding of a situation humans process the perception of such a situation through logical and conscious thinking based on a desire to take in and understand the full spectrum of characteristics of that situation. Such processing is analytic and comprehensive. It is a judgment formed as a response to the actual content of the message. Systematic processing requires that the perceiver has cognitive ability capacity and the willingness to do so. This is what is called the Systematic Route. It runs from first perception of the situation through systematic processing to decision making about what to do and feel next (Chen & Chaiken, 1999).

6.2.2 Heuristic Processing

Heuristic processing is viewed as a processing mode offering a better cognitive economy than systematic processing. When processing heuristically humans focus on a subset of available information enabling the use of simple inferential rules schemata or cognitive heuristics to formulate their judgments and decisions (Chen & Chaiken, 1999). Such processing entails the activation and application of rules or heuristics learned and stored in memory. These judgments reflect easily processed judgment-relevant cues. The heuristic mode is constrained by principles of knowledge activation availability accessibility and applicability. That means that for heuristic processing to take place there must be heuristics available. Heuristic processing has been characterized as more exclusively theory driven than systematic processing and the mode-of-processing distinction was assumed to be not only more or less but of a different character (Chaiken et al. 1989). In this view heuristic processing is:

... more exclusively theory driven because recipients utilize minimal informational input in conjunction with simple (declarative or procedural) knowledge structures to determine message validity quickly and efficiently. (Chaiken et al., 1989)

In short; the heuristic mode of processing presumes that the individual at all has a personal theory about the object that is processed. That might be an issue in regard to pervasive games.

The Heuristic-Systematic Model assumes that a person's' main motivation in a persuasion setting is the desire to formulate valid or accurate attitudes and both modes of processing can drive attitudes (Chaiken et al., 1989)

6.2.3 The Sufficiency Continuum

The system of the Heuristic and Systematic judgements is not a bipolar binary situation. It is a fluid continuum described in the "Sufficiency Continuum"-construct of the Heuristic-Systematic Model. The Sufficiency Continuum construct states that in the human mind there is an ever ongoing tension between on the one hand the tendency towards cognitive economy pushing the mind towards Heuristic processing and on the other hand the wish to feel safe in ones´ perceptions and judgements of situations which pushes the mind towards Systematic processing. Perceivers strike a balance between minimizing cognitive effort on one hand and satisfying their motivational concerns on the other hand (Chaiken, Pomerantz, & Giner-Sorolla, 1995). Thus perceivers who are motivated to determine accurate judgments will exert as much cognitive effort as is necessary to reach a sufficient degree of confidence that their judgments will satisfy their accuracy goals. When low effort heuristic processing fails to confer sufficient judgmental confidence or cannot occur due to for example the absence of any judgment-relevant heuristic-cue information perceivers are likely to engage in systematic processing in order to close this gap (Chen & Chaiken, 1999).

The construct describes a continuum of judgemental confidence along which two critical points lie; one designating perceivers´ level of actual confidence; the other indicating the level of desired confidence. Perceivers will strive to have the two points meet. The difference between the two modes resonates well with the point of departure for AAPW. That consumption of electricity has fallen below an awareness threshold is an expression of an extreme point towards the heuristic end of the Sufficiency Continuum.

6.2.4 Processing Consequences

The HMS predict that attitudes acquired via the systematic route are expected to manifest greater temporal persistence be more predictive of behaviour and exhibit greater resistance to counter persuasion than attitudes acquired via the heuristic route. The greater resistance and persistence follow from the expectation that under the systematic route the issue-relevant attitude schema is accessed rehearsed and considered in the mind more often hence strengthening the interconnections among the components and rendering the schema more internally consistent accessible enduring and resistant than under the Heuristic route (Chen & Chaiken, 1999).The Heuristic-Systematic Model allows for systematic and heuristic processing to co-occur. It also holds that motivational variables have similar effects on systematic and heuristic processing. Thus for instance personal relevance is assumed to "influence not only the magnitude of systematic processing (but) also enhances the likelihood of heuristic processing because (it increases) the cognitive accessibility of relevant persuasion heuristics and/or increases the vigilance with which people search (the setting or their memories) for relevant heuristic cues" (Chaiken et al., 1989).

6.2.5 Validating the Heuristic-Systematic Model

When studying how people perceive persuasive communications and situations the dual-process model has found considerable use in analysing and predicting the effects of persuasive messages. A large number of studies have been made validating the dual process model on and through studies of advertising (Kruglanski & Stroebe, 2005) (Perloff, 2010). Still some critics have argued that the model is so explanatory and flexible that no study can falsify it (Perloff, 2010).

To this chapters´ authors´ knowledge no study has been made on dual process modelling of a pervasive persuasive game play situation whether it be play engendered by a persuasive game or play engendered by a pure entertainment game.

7. DATA ANALYSIS AND OPERATIONALIZATION OF THE THEORY

In this section we can go into if AAPW had an effect. That causal analysis of the *why* was done with Structural Equation Modelling, and is to be found in found in (Svahn, 2013). But even when analysing if there is any effect, we are met with methodological issues. The field of pervasive game design is very new, and the field of pervasive game design in the employ for advertising and learning is even newer still. Thus, there are no scales of measurement that have gone through iterations of mathematical and empirical validation. Hence, our results must be viewed as indicative rather than definitive.

It may be believed that the construct "attitudes to electricity" does not suffer from immaturity of scales and measurements. Attitude research is an old and mature field with many well validated scales and constructs; see for example (Albarracin, Johnson, & Zanna, 2005) for a thorough review. However the attitude constructs for AAPW are not so easy to pin down. They are supposed to concern attitudes towards electricity and nothing else. Previous research has discussed attitudes towards electricity disclosure, electricity transmission, electricity network reliability, or attitudes towards electricity generating options. Constructs for attitudes towards electricity as a neutral concept were difficult to find.

The AAPW research design therefore stood at a crossroads. One strategy could be to try and apply the operationalization of attitude constructs narrowly but stringently. The risk of this choice is that the data might not have captured the attitudes of the players but instead but something else instead - or perhaps nothing at all.

The route chosen was to design the data collection to cast a very wide net with many questionnaire items as well as other data points. This process captures a large amount of data, where the signal-to-noise ratio most certainly will become low. This is an explorative and inductive approach. The wide data capture method presents better chances to find attitude changes.

The data was purified based on theoretical models of attitude constructs and influences. Before going into results, we will therefore discuss the theoretical analysis underlying this purification process. A theoretical analysis of the context and purpose had identified three attitude constructs. The first was "*Feelings towards behaviours towards electricity consumptions*", which was considered a salient element due to the pervasive game structure, to drive behaviour in the real world. The second was "*Feelings of restrictions and empowerment, towards electricity consumption*" since AAPW was designed with the intent to inform and empower. To complement these, we included a third more generic construct "*General attitudes towards energy and the environment*". These three were all informed both the questions asked and the data gathered. The indicators were purified after data collection to represent coherence (within the constructs) and variability (between constructs). The final single attitude construct came to include seven variables, with all the original three theoretical attitude constructs represented. The items that in the end came to make up the electricity attitude measure were:

1. "I know very well that I can impact my households´ electricity consumption"
2. "My household cannot save electricity, we must use all the electricity we do right now" (Rescaled for ease of analysis)
3. "I usually urge my friends to conserve electricity"
4. "I can consider urging my friends to conserve electricity"
5. "I check how much electricity appliances consume"
6. "I turn out the lights when I leave a room"
7. "Saving electricity is cool"

The research question whether the games had an effect on the primary and/or secondary players is approached in a series of t-tests. These t-tests are meant to resolve the issue of whether or not there was an impact from pervasive-game design elements on the players and/or their parents. The analysis is approached as a series of statistical hypotheses.

7.1 Test One: Effects on the Attitudes of the Primary Player

H0: No effect on the attitudes on the primary player (the child) can be measured in the players of AAPW.

H1: Some effect on the attitudes on the primary player (the child) can be measured in the players of AAPW.

7.2 Test Two: Effects on the Attitudes of the Secondary Player

H0: No effect on the attitudes of the secondary player (the parent) can be measured in the players of AAPW.

H1: Some effect on the attitudes of the secondary player (the parent) can be measured in the players of AAPW.

7.3 Test Three: Conservational Effects on the Attitudes of the Primary Player

H0: No effect on the attitudes of the primary player (the child) in direction of conserving electricity can be measured.

H1: Some effect on the attitudes of the primary player (the child) in direction of conserving electricity can be measured.

7.4 Test Four: Conservational Effects on the Attitudes of the Secondary Player

H0: No effect on the attitudes of the secondary player (the parent) in direction of conserving electricity can be measured.

H1: Some effect on the attitudes of the secondary player (the parent) in direction of conserving electricity can be measured.

7.5 Test Five Wastefulness Effects on the Attitudes of the Primary Player

H0: No effect on the primary player (the child) in direction of being more wasteful with electricity can be measured.

H1: Some effect on the primary player (the child) in direction of being more wasteful with electricity can be measured.

7.6 Test Six Wastefulness Effects on the Attitudes of the Secondary Player

H0: No effect on the secondary player (the parent) in direction of being wasteful with electricity can be measured.

H1: Some effect on the secondary player (the parent) in direction of being wasteful with electricity can be measured.

These six tests together realize the research issue. T-tests are chosen because the data on the attitudes come from Likert scale based surveys and are non-multivariatelly normally distributed. The attitudes are interesting only as a construct, the individual measurements were never meant to represent anything else than aspects of a larger whole. They are not meant to carry meaning as singular entities. Therefore they were indexed into one construct before the effects measures were made.

The effects tests give the impression that AAPW did have an effect on the both primary and secondary players′ attitudes towards electricity consumption. Not a very strong one, but noticeable.

8. RESULTS

The overall research question was if this game design of AAWP could manage to both engage the backgrounding of electricity consumption and "foreground" it, and also cut through the issue of social consumption. The field test and the analysis show that there was a noticeable impact on both players and parents from playing AAPW. The design was meant to influence both players and parents to change their attitudes towards electricity to become more positive about saving electricity, and this effect could indeed be observed. From mere T-tests, we cannot conclude anything about why there was an effect. This has been further analysed in (Svahn, 2014), who shows that the element of social expansion can drive topical social interaction between primary and secondary player.

That the design did have an effect on secondary players, the parents, tells us that social expansion can drive thought. AAPW is a "game with a day job" (Gardner, 2001) and the design element of social expansion helped it do that job. It can be

Table 1. The tests

	Difference	Sig.	Null Hypothesis Confirmed/Rejected
5. The attitudes of the playing teens before, measured against the playing teens after.	+ 6%	0.00	Rejects the null hypothesis of test one Rejects the null hypothesis of test three Confirms the null hypothesis of test five
6. The playing parents before measured against the playing parents after.	+ 8%	0.00	Rejects the null hypothesis of test two Rejects the null hypothesis of test four Confirms the null hypothesis of test six
7. The playing parents after measured against control group parents after.	+ 7%	0.00	Also rejects the null hypothesis of test two Also rejects the null hypothesis of test four Also confirms the null hypothesis of test six
8. Playing teens after measured against control group teens after.	+ 6%	0.00	Also rejects the null hypothesis of test one Also rejects the null hypothesis of test three Also confirms the null hypothesis of test five

presumed that engaging with both the non-game world and the game world "overlappingly" and simultaneously was a state of affairs that that took cognitive effort, and hence drove both primary and secondary player to Systematic Thought in the terms of the Heuristic-Systematic Model.

9. CURRENT CHALLENGES

9.1 Practical Difficulties

The team ran into some practical problems with the logistical execution of the data collection. The team recruited the participants through contact with schools and that was a varied experience. The team initially relied on the teachers and the teens to distribute the surveys, and since the teens were underage also a consent form to be signed by the parents. This exchange of consent forms and surveys did not go smoothly. Some teachers were not enthusiastic and forgot to distribute the consent forms, which meant that some eager players could not play. Other teachers were more enthusiastic and took the opportunity to make the filling out of the surveys into a learning experience. This meant that all pupils were asked to fill out and discuss the survey in class; regardless of whether they had played the game or not. Hence, the data set became spotty and obscure.

There were also technical issues with the smart meter system, which not always would deliver its meter data on a regular 24 hour schedule. That is not a problem for the monthly or bi-monthly billing, but when used in a game designed for 24 hour cycles, it disturbed the game experience. This behaviour was not foreseen and there was no play element designed for that contingency.

9.2 Some Successes

The participants played AAPW on their own mobile phones, as the research team did not have the resources to buy phones for loan. This could potentially have caused problems, but the download and installation process ran smoothly, and there were few calls to the helpdesk. The number of active players on the server was as high as needed and expected.

The first time the participants logged in to AAPW, they were asked to enter the smart meter ID for their household; a number that can be found on the household's electricity bill. The team had expected that this would be a stumbling block. However it worked well; there were only few calls to the helpdesk about the issue. However, we cannot exclude the possibility that some players gave up at this point and did not bother to call. The drop out rate at this point could not be traced. Furthermore, a study performed by the designer team shows that there also was a decrease in the electricity consumption in participating households, and that this effect lasted for a while after the game (Larsson, 2011). This indicates a raise in the earlier mentioned category awareness of electricity, in the participating families.

10. SOLUTIONS AND RECOMMENDATIONS

AAPW was a joint project for the team from the Stockholm of Economics, the Energy Board of Mälardalen, and the Energy Design Studio of the Interactive Institute. Although no project is currently running, the partners consider further development of the game and a potential commercialisation route. An obstacle is that the implementation is by now quite antique. AAPW would need to be updated to modern mobile device technology if it is to be used again. Further issues pertain to the run-time management of AAPW: the original game required a managing staff and if it is going to be run at a large scale, then a more work-time efficient and perhaps wholly automated back-end system would need to be developed.

Though the AAPW project is finished, the core issue of visualising electricity and raising the awareness threshold of electricity consumption very much remains in society and hence remains open to new research inspired by AAPW. Furthermore, AAPW presents a direction for other societal issues where raising awareness of invisible connections is at focus. Issues where the actions of one person very much impacts the lives of others, and if progress in any direction is to be made then what needs to be impacted is the inter-personal relations. When taking action on such issues, it can be kept in mind that social expansion is a viable design option and that it can be used to raise awareness with both players and non-players.

REFERENCES

Abrahamse, W., Steg, L., Vlek, C., & Rothengatter, T. (2005). A review of interventions studies aimed at household energy conservation. *Journal of Environmental Psychology*, *25*(3), 273–291. doi:10.1016/j.jenvp.2005.08.002

Albarracin, D., Johnson, B. T., & Zanna, M. P. (2005). *Handbook of attitudes*. Hoboken, NJ: Lawrence Erlbaum Associates.

Bang, M., Gustafsson, A., & Katzeff, C. (2007). *Promoting new patterns in household energy consumption with pervasive learning games* (pp. 55–63). Lecture Notes in Computer Science Berlin: Springer. doi:10.1007/978-3-540-77006-0_7

Bang, M., Torstensson, C., & Katzeff, C. (2006). The Power House: A persuasive computer game designed to raise awareness of domestic energy consumption. In *First International Conference on Persuasive Computing for Well-being* (LNCS) (pp. 123 -132). Berlin: Springer.

Benford, S., Rowland, D., Flintham, M., Drozd, A., Hull, R., & Reid, J. et al. (2005). Life on the edge: Supporting collaboration in location-based experiences. In *Proceedings of the SIGCHI Conference on Human Factors in Computing Systems*. Portland, OR: ACM. doi:10.1145/1054972.1055072

Chaiken, S. (1987). The heuristic model of persuasion. In M. P. Zanna, J. M. Olson, & C. P. Herman (Eds.), *Social influence the Ontario symposium* (5th ed.). Hillsdale, NJ: Lawrence Erlbaum Associates.

Chaiken, S., Lieberman, A., & Eagly, A. (1989). Heuristic and systemic information processing within and beyond the persuasion context. In J. Uleman, & J. Bargh (Eds.), *Unintended thought* (p. 212). New York, NY: Guilford.

Chaiken, S., Pomerantz, E. M., & Giner-Sorolla, R. (1995). Structural consistency and attitude strength. In R. E. Petty, & J. A. Krosnick (Eds.), *Atitude strength, antecedents and consequences* (pp. 387–412). Mahwah, NJ: Lawrence Erlbaum.

Chen, S., & Chaiken, S. (1999). The heuristic- Systematic model in its broader context. In S. Chaiken, & Y. Trope (Eds.), *Dual-process theories in social psychology*. New York, NY: Guilford Press.

Colvert, A. (2009). Alternate reality gaming in primary school settings. In *Breaking New Ground: Innovation in Games, Play, Practice and Theory: Proceedings of DiGRA 2009*. London. DIGRA

Davies, H. (2007). Place as media in pervasive games. In *Proceedings of the 4th Australasian Conference on Interactive Entertainment*. Melbourne, Australia: Creativity ad Cognition Studios Press. Retrieved from http://dl.acm.org/ft_gateway.cfm?id=1367963&type=pdf&CFID=82449371&CFTOKEN=37911479

de Jong, A., Balksjö, T., & Katzeff, C. (2013). *Challenges in energy awareness: A Swedish case of heating consumption in households*. Paper presented at the ERSCP-EMSU Conference Bridges for a Sustainable Future. Istanbul, Turkey.

Dolak, D. (2012). *How to brand and market a commodity*. European Union Amazon Media EU.

Düüs-Henriksen, T. (2008). Extending experiences of learning games: Or why learning games should be neither fun nor educational or realistic. In O. Leino, H. Wirman, & A. Fernandez (Eds.), *Extending experiences: Structure, analysis and design of computer game player experience*. Rovaniemi, Finland: Lapland University Press.

Egenfeldt-Nielsen, S. (2007). *Beyond edutainment exploring the educational potential of videogames*. London, UK: Continuum International Publishing Group Ltd.

Evans, E. (2008). Character, audience agency and transmedia drama. *Media, Culture & Society March, 30*(2), 197-213. doi:10.1177/0163443707086861

Gardner, P. (2001). *Games with a day job: Putting the power of games to work*. Retrieved from http://www.gamasutra.com/view/feature/3071/games_with_a_day_job_putting_the_.php

Gee, J. P. (2003). What video games have to teach us about learning and literacy. *Computers in Entertainment, 1*(1), 20-20. http://doi.acm.org/10.1145/950566.950595

Gustafsson, A. (2010). *Positive persuasion - Designing enjoyable energy feedback experiences in the home*. (Doctoral dissertation). Retrieved from GUPEA 21-maj-2010

Gustafsson, A., & Bang, M. (2008). Evaluation of a pervasive game for domestic energy engagement among teenagers. In *Proceedings of the ACM SIGCHI International Conference on Advances in Computer Entertainment Technology* (ACE 2008). ACM. doi:10.1145/1501750.1501804

Gustafsson, A., Bang, M., & Svahn, M. (2009). Power explorer – A casual game style for encouraging long term behaviour change among teenagers. In *Proceedings of the ACM SIGCHI International Conference on Advances in Computer Entertainment Technology* (ACE 2009). Athens, Greece. ACM. doi:doi:10.1145/1690388.1690419

Katzeff, C. (2010, April 26). Engaging design for energy conservation in households. *Metering International Magazine,* 62-63.

Kruglanski, A., & Stroebe, W. (2005). The influence of beliefs and goals on attitudes: Issues of structure, function, and dynamics. In *The handbook of attitudes*. Mahwah, NJ: Erlbaum.

Lange, F. (2003). *Brand choice in goal-derived categories: What are the determinants?* (Doctoral Dissertation, EFI). Retrieved from http://hhs.diva-portal.org/smash/get/diva2:221410/FULLTEXT01

Larsson, C. (2011). *Slutrapport projekt – Unga utforskar energi. (Official No. Prj. nr: 32255-1)*. Energikontoret i Mälardalen.

Linderoth, J. (2009). Its not hard, just requires that you have no life: Computer games and the illusion of learning. *Digital Kompetanse*, *4*(1), 4–19.

Marcell, K., Agyeman, J., & Rappaport, A. (2004). Cooling the campus: Experiences from a pilot study to reduce electricity use at tufts university, USA, using social marketing methods. *International Journal of Sustainability in Higher Education*, *5*(2), 169–189. doi:10.1108/14676370410526251

Montola, M. (2011). A ludological view on the pervasive mixed-reality game research paradigm. *Personal and Ubiquitous Computing*, *15*(1), 3–12. doi:10.1007/s00779-010-0307-7

Montola, M., Stenros, J., & Waern, A. (Eds.). (2009). *Pervasive games: Theory and design*. Morgan Kaufman.

Paavilainen, J., Korhonen, H., Saarenpää, H., & Holopainen, J. (2009). Player perception of context information utilization in pervasive mobile games. In *Breaking New Ground: Innovation in Games, Play, Practice and Theory: Proceedings of DiGRA 2009*. London, UK: DIGRA.

Palazzi, C. E., Roccetti, M., & Marfia, G. (2010). Realizing the unexploited potential of games on serious challenges. *Computers in Entertainment*, *8*(4), 23. doi:10.1145/1921141.1921143

Perloff, M. R. (2010). *The dynamics of persuasion: Communication and attitudes in the 21st century* (4th ed.). Taylor & Francis.

Raessens, J. (2007). Playing history reflections on mobile and location based learning. In T. Hug (Ed.), *Didactics of microlearning: Concepts, discourses and examples*. Münster, Germany: Waxmann Verlag.

Reis, S., & Correia, N. (2012). Playing with the weather. In *Entertainment Computing - ICEC 2012 (LNCS)* (Vol. 7522, pp. 172–184). Berlin: Springer. doi:10.1007/978-3-642-33542-6_15

Robert, H. S. (1978). The twin rivers program on energy conservation in housing: Highlights and conclusions. *Energy and Building*, *1*(3), 207–242. doi:10.1016/0378-7788(78)90003-8

Solomon, S. D., & Qin, M. M. Z. Chen, M. M., & Avery, K. B. (2007). Contribution of working group I to the fourth assessment report of the intergovernmental panel on climate change. Cambridge, UK: Cambridge University Press.

Svahn, M. (2014). *Social expansion in marketing, focusing on pervasive-persuasive games for impacting energy consumption. (Unpublished PhD)*. Stockholm, Sweden: Stockholm School of Economics.

Svahn, M., & Lange, F. (2009). Marketing the category of pervasive games. In M. Montola, J. Stenros, & A. Waern (Eds.), *Pervasive games, theory and design*. Morgan Kaufman. doi:10.1016/B978-0-12-374853-9.00011-8

Swedish Energy Agency. (2012). *Energiläget 2012*. Energimyndigheten.

Szulborski, D. (2005). *This is not a game: A guide to alternate reality gaming*. Retrieved from http://www.lulu.com/product/hardcover/this-is-not-a-game-a-guide-to-alternate-reality-gaming/244023?productTrackingContext=search_results/search_shelf/center/1

Torstensson, C. (2005). *En förstudie för att utreda förutsättningarna för forskningsprojektet young energy*. Eskilstuna, Sweden: Interactive Institute.

van Houwelingen, J. H., & van Raaij, F. W. (1989). The effect of goal-setting and daily electronic feedback on in-home energy use. *The Journal of Consumer Research*, *16*(1), 98–105. doi:10.1086/209197

Waern, A., & Denward, M. (2009). On the edge of reality: Reality fiction in sanningen om marika. In *Breaking New Ground: Innovation in Games, Play, Practice and Theory: Proceedings of DiGRA 2009*. London, UK. DIGRA.

Walz, S. P., & Ballagas, R. (2007). Pervasive persuasive: A rhetorical design approach to a location-based spell-casting game for tourists. In *DiGRA 2007 - Situated Play: Proceedings of the 3rd International Conference of the Digital Games Research Association*. Tokyo: DIGRA.

World Energy Outlook 2011. (2011). Paris: International Energy Agency.

Chapter 11
The ASPIRE Program: Using Game-Based Learning to Reach Massive Audiences

Peter Christiansen
University of Utah, USA

EXECUTIVE SUMMARY

The ASPIRE Program is a science outreach program that was designed with the goal of teaching basic physics and math to middle school students and encouraging them to take an early interest in science. Our main tool in achieving this goal is a series of online games and activities that are designed to supplement classroom learning. The use of videogames as a teaching tool has enabled ASPIRE to reach thousands of students per day, while maintaining an average staff of only two or three employees. Although the games themselves are online, much of the success of ASPIRE can be attributed to connections with educators made through more traditional outreach activities. These connections serve as both a source of feedback for improving pro-learning behavioral effects in players and as a means of raising awareness for the games themselves.

BACKGROUND

Although public outreach is a significant part of many grant-funded research projects, limited resources often make it difficult to achieve high impact results on a large scale. Limited funds and staff can limit outreach efforts to small groups or create a reliance on large numbers of volunteers. The latter can also be unsustainable as

DOI: 10.4018/978-1-4666-6206-3.ch011

Copyright ©2014, IGI Global. Copying or distributing in print or electronic forms without written permission of IGI Global is prohibited.

volunteers may experience high turnover and often require additional management in order to be effective. Thus, for many research groups, particularly smaller ones, designing sustainable, high impact outreach activities is an elusive goal.

The Astrophysics Science Project Integrating Research & Education (ASPIRE) is the outreach branch of the Telescope Array Project, an National Science Foundation (NSF) funded experiment studying ultra-high energy cosmic rays. For over 15 years, ASPIRE has utilized a program centered around web-based videogames and interactive lessons capable of reaching audiences of over 7,000 users in a single day. By utilizing game-based learning techniques and by working in collaboration with state and local education organizations, ASPIRE is able to reach massive audiences while maintaining an average staff of two or three people.

The ASPIRE Program was conceived in 1997 by Professor Gene Loh and other researchers from the High Resolution Fly's Eye Experiment (HiRes), the precursor to the Telescope Array Project. At the time, HiRes was the largest cosmic ray detector in the world and was on the cutting edge of astrophysics research. With considerable support from the NSF, ASPIRE was originally envisioned on a large scale, enlisting the help of 26 teachers from the Utah school system, as well as a number of artists and engineers. This was also a time when the idea of the Internet as an educational tool was growing in popularity across the United States. In Utah, public middle and high schools had finally achieved 100 percent Internet connectivity, making online lessons a natural choice for ASPIRE.

The first ASPIRE lessons were created in Java, which was a new technology at the time. This provided a number of benefits, including a high level of interactivity as compared to most web-based technologies of the time, as well as relatively reliable cross-platform functionality. The latter was especially important because while computers were beginning to be taken much more seriously in public schools, the technology being used in school computer labs ranged from cutting edge computers to barely passable machines.

There were also some drawbacks to using Java as a platform. Although the project was, at this time, well funded and staffed with both programmers and artists, there was often difficulty in coordinating between the two groups. Artists had no access to code and programmers had no access to art tools. This decreased the efficiency of the development process, making the release of new lessons somewhat slower than it could have been.

In 2000, Julie Callahan, the current project coordinator of ASPIRE, was brought into the project, just as the project's three year funding cycle was coming to an end. During these three years, the economic climate in the US had changed substantially. The "Dot Com Bubble" had just burst, and budgets were being slashed across the country. When the new budget for the HiRes experiment was approved, the funding for ASPIRE was dramatically reduced, decreasing its staff from dozens to merely Julie and two programmers.

Despite being largely defunded, the ASPIRE program continued throughout the 2000s, though perhaps in a different direction from its original vision. Other than Julie, the program had no permanent employees – the other developers were generally students working part time and few stayed around longer than a year. Most were programmers, though several artists were hired during this time period, as well.

Moving from a large, structured staff to a small intermittent one required a much more flexible development cycle. Creating new games and activities was still a priority, though these projects would often be placed on hold when one developer left, only to be taken up by a new developer at a later date. Maintenance and updating of old lessons was also important, as web standards and aesthetics changed over the years. Even at times when the program lacked programmers to continue development, however, other outreach activities, such as public demonstrations, conferences, and school visits continued. These activities were important not only because they looked good on NSF reports, but because they helped to create connections and goodwill between ASPIRE and various other groups.

There were also more technical changes that took place during this time. It became clear that Java was not well suited to the needs of a long term project of this kind, as the language underwent a number of significant changes during its early years. Many of the early lessons, which were created in Java 1.0, began experiencing issues with deprecated code within a few years of their creation. Some of these programs were relatively complex for web applications in order to accurately model the physics being demonstrated in the lesson, as well as to create dynamic animations to accompany the simulation. This complexity made them prone to breaking as new versions of Java were released and also made them incredibly difficult, if not impossible, for new programmers to maintain after the original coders had moved on.

Due to these problems, ASPIRE switched from Java development to Macromedia (later Adobe) Flash. This switch was advantageous for a number of reasons. Most immediately, it solved the problem of code deprecation, as the Flash plugin to this day has maintained backwards compatibility with older versions all the way back to ActionScript 1.0. It also provided a development environment that combined art and animation tools with a coding tools, eliminating some of the barriers that had existed during the development of the first ASPIRE games. Since there was not always an artist on staff, Julie taught each of the incoming programmers how to use the basic drawing tools that Flash provided. Although some projects still required assistance from an artist, this allowed the development of new games to continue, if not a bit slowly, regardless of the current makeup of the ASPIRE staff.

SETTING THE STAGE

I joined the ASPIRE project in 2010. I had worked with Julie before and had referred a number of my former students to her as programmers. At that time, the ASPIRE website (aspire.cosmic-ray.org) already received a large amount of traffic, though there hadn't been any new lessons for several years. Additionally, the Simple Machines lesson, one of the few remaining (and most popular) Java activities, had lost a good deal of its functionality due to certain features being incompatible with the most recent versions of Java.

Our first priority was recreating the Simple Machines lesson, which in addition to having semi-functional Java-based activities, was still using late 90s HTML, complete with animated GIFs. While Julie began revising the design and layout of the web pages, I began creating Flash versions of the activities, along with Pace Sims, a programmer from the Entertainment Arts and Engineering program. Although we initially began programming different activities individually, we soon began specializing in certain tasks, with Pace taking the role of lead programmer and myself filling the roles of artist and designer.

In addition to fixing our broken lesson, we had a number of other goals for ASPIRE. The servers that hosted the online content needed to be updated, as did the site itself. Many of the site's original functionality, such as monitoring traffic, had long since ceased to function properly. We also hoped not only to update many of the current lessons, but to create new lessons for the site as well. Due to the nature of student employment, we had no idea how long we would have such a well-balanced team working together. Though our time constraints were very different from those encountered in the mainstream videogame industry, there was a definite sense of urgency about our goals for the project.

CASE DESCRIPTION: THE ASPIRE LABS

Learning Theories and Frameworks

The ASPIRE Model of curriculum development is loosely based on Gentry's (1994) model of Instructional Project Development and Management (IPDM). It also makes use of other theoretical frameworks, such as Bloom's Taxonomy of Educational Objectives (Bloom, Engelhart, Furst, Hill, & Krathwohl, 1956; Krathwohl, 2002). This model is used primarily when creating new content, however, it can also be used as a guide when revising existing material in order to improve it.

The IPDM model, as explained by Gentry, consists of fourteen interrelated, non-linear components (Gentry, 1994, pp. 3-4). These components are generally separated into supporting components, which are concerned with maintaining ongoing responsibilities such as management, facilities, and budgets, and development components, which are those aspects directly connected with the development of new educational material. It is these eight components that Gentry places into the latter category that are of most relevance to us.

Although the development of ASPIRE curriculum occasionally requires us to go through a process that touches on aspects of all eight components, this is usually unnecessary. Indeed, Gentry notes that depending on what point within the development process a designer begins, many components may have already been addressed (Gentry, 1994, p. 4). Since the IPDM model is generalized to be useful in any instructional development context, it is based on the assumption that you are starting completely from scratch. Thus, many of the components are focused on tasks such as assessing your educational goals and emphasizing the need for new instructional material. Rather than repeating this process for each new lesson, we generally base learning objectives and needs off of existing criteria that have been established by local or national educational organizations, but still match up with the goals of the ASPIRE program. This allows their work on curriculum and standards development to fulfill roll of several of the IPDM components. For example, since local Utah school districts are among our primary audiences, the goals of many of our lessons are based off of the Utah Core Standards, developed by the Utah State Board of Education. By using these learning objectives rather than developing our own, we are able to streamline the process of curriculum development considerably.

Beginning from established standards, our development process generally consists of five steps. The first step, which roughly corresponds to the IPDM design component, is to review our objectives. This involves reexamining our learning goals and defining objectives that will allow us to meet these goals. This is also the point in the development when the cognitive skills of Bloom's Taxonomy (knowledge, comprehension, application, analysis, synthesis, and evaluation) are discussed in relationship to these instructional objectives. The next step is the brainstorming-prototyping phase of development, when we begin to translate our abstract objectives into well-defined plans for activities and lessons. Following this is the development or production phase, where the actual programs are developed.

The final step in the development process is the revision or evaluation stage. This phase is perhaps the most critical for ensuring that the games promote desired learning behaviors in students who play them. Evaluating the program goes beyond simple bug testing to ensure proper functionality. It typically involves getting feedback from educators, administrators, and students in order to assure that the games are addressing the content areas that are relevant to the subject and that the game mechanics are effective at both conveying this information and encouraging the students to experiment and play with the mechanics on their own. As the ASPIRE program is technically an ongoing instructional design program, rather than a finite project (Greer, 1992, p. 2), revision and evaluation is an ongoing process of improving those lessons that are most in need of maintenance in order to meet our users needs.

The ASPIRE model of curriculum development can be understood as a very specific implementation of the IPDM model. While it is certainly possible that this model could be used effectively by other organizations, this is, perhaps, not the most useful way of understanding this example. Rather, I would suggest that the ASPIRE model is an example of a radical appropriation of a well-established instructional design model for a specific circumstance. While perhaps not directly applicable to the needs of other organizations, it provides a framework for understanding how generalized models can be used to fulfill very specific needs.

The Simple Machines Lab: Optimizing Workflow

While not the oldest lesson in the site, the Simple Machines lab had a distinct 1990s aesthetic. The character art used in the activities and elsewhere throughout the lesson was pre-rendered 3D models that didn't read particularly well in 2D. Many of these characters were animated, however, due to the low bandwidth requirements at the time they were created, most of these animations were somewhat jerky, especially those found in the many animated GIF images scattered throughout the lesson.

The more significant problem with the lesson, however, is that certain parts of the code no longer worked with the current version of Java. Although most of the applets retained their basic functionality, many dynamic animations and other visual components of the programs were displaying incorrectly or not at all. In addition to making the lesson look incomplete, it made the activities harder to use and much less engaging.

Although the original programmers of the Simple Machines lab had kept their original source code, this proved to be of little use in remaking the activities. While the programs were quite impressive for their time, they were essentially brilliant hacks created by talented programmers working by themselves. Within the code, variables were cryptic and comments were scarce. Going through old code gave little insight into how the original programs were created. We would have to begin again from scratch.

As previously mentioned, the division of responsibilities between the three ASPIRE employees for the Simple Machines lab changed substantially during the first few weeks of work. Nominally, both Pace and I had been hired on as programmers, leaving us with a rather engineering heavy team. We began work with a very flexible development approach. Both Pace and I began working on different activities in the lesson independently. Although neither programmer relied on the other, we shared a common office and would show each other what we were working on frequently.

By the time we had each completed our first small programs, we had begun to discover our relative strengths and weaknesses. Pace was by far the faster programmer. Although we were both able to complete our first programs with little difficulty, Pace could easily code twice as quickly as I could while still generating cleaner and more robust code. Art, on the other hand, came more naturally to me. Although I still considered myself much more of a programmer than an artist, I had a fair amount of experience with 2D animation, as well as experience in integrating artwork into ActionScript projects. By the time we had moved on to later activities in the Simple Machines lab, Pace had taken over nearly all of the core programming duties, while I was responsible for art and animation, as well as coding for user interface and several other secondary systems. Julie, meanwhile, had taken on the task of creating a new design for the web pages in which our Flash-based programs would be embedded, as well as updating them to comply with modern web standards.

Creating a more efficient division of responsibilities also required improvements in other areas. During the development of our first programs, Pace and I had been content to pass code and artwork back and forth by simply placing them on a USB drive or emailing them back and forth. As we began working together more and more, it became apparent that we would need a more reliable way of sharing files. We chose to set up a Subversion (SVN) system in order to both share our files more easily and make sure that we were both working on the same version of the program. Fortunately, TortoiseSVN, the Subversion client we decided to use, integrated well into Adobe Flash Builder, which we used for coding. Although there were one or two times when we lost work time due to a server failure or other SVN problem, the use of SVN greatly contributed to our ability to work together effectively as a team.

Although the Simple Machines lab was a fairly ambitious project that had been put off for years, we were able to complete the project in a matter of a few months. Much of this success was due to the fortuitous circumstances that brought such a well-balanced team together all at the same time, however, there were many aspects of ASPIRE as an organization that helped to facilitate the process. As the lesson had been around for over a decade, it had already received a considerable amount of scrutiny. Through connections with teachers who used the lessons, we were able to know before beginning the project exactly what aspects were most desirable from an educational perspective, as well as to have a general idea of the way that trends in education had shifted since the time the original lessons were created. Such a wealth of information on user needs and behaviors at the outset of a project would certainly be enviable to most game development studios. It seems reasonable, then, to suggest that the success of the Simple Machines lab was not just a product of a competent and well-balanced team, but also of an organizational structure that allowed us to hit the ground running without having to spend much time worrying about pre-production and planning.

Figure 1. The Simple Machines lessons present students with hypothetical situations in which simple machines can be used. It also generates experimental data that is difficult to measure in hands-on activities.

Shaping User Behavior

With the new Simple Machines lab up and running, we were finally able to turn our attention toward other tasks. Many of these related to the ASPIRE website itself. As part of an organization located on a University campus, we faced a number of unusual challenges. Our local network was located within the high performance computing center, behind a formidable number of firewalls. In order for our website to be accessible to the public, we needed two servers – one inside the firewall on our local network and a second public server outside the firewall. Updating and synchronizing between these two machines was handed by a number of somewhat complicated scripts, however, these scripts were also designed to allow easy access for ASPIRE employees so that they could update the site on a regular basis. The site was essentially run by a custom built PHP based content management system.

Unfortunately, by the time I joined ASPIRE, the servers were no longer running PHP properly. This both prevented many of the dynamic features that had originally been present in the site from working and made the process of updating the site much more difficult than it had originally been. We ultimately decided that the best solution would be to get two new servers and create a simple HTML site, as our needs did not actually require a dynamic, PHP-based system. The actual process of migrating to these new servers would take much longer than expected, but would eventually be completed. As with the Simple Machines lab, we took the opportunity to redesign the entire site using modern web standards and technology.

One of the dynamic features that had been present in the original site was a system for tracking the number of visitors. Knowing the amount of traffic that passed through was important for a number of reasons, not the least of which was to have something to show the NSF during their next evaluation. Although a comprehensive evaluation by a third party organization would have been the most accurate method of assessing the success of the ASPIRE program, such an evaluation would have been outside the limited budget of the project. Fortunately, the availability of free tools like Google Analytics not only replaced the functionality of the old PHP scripts, but allowed us to collect data on user traffic patterns that would have been impossible when ASPIRE was first created.

From our analytics, we learned a number of interesting facts about our site traffic. Perhaps most surprisingly, we discovered that despite the fact that that many of our lessons ranked quite high in Google searches, the largest percentage of our visitors arrived by direct link, suggesting that many teachers either had their students type in the full URL of a lesson or the lessons bookmarked. While it was very encouraging to know that our users liked our lessons enough to return year after year, it also meant that most of the users of the Simple Machines lab were skipping past the homepage and the links to the new lab, going straight to the old one. Although we debated creating a redirect page to send visitors directly to the new lessons, we decided instead to merely make a large notice on the front page of the Java lesson, directing people to move to the new lessons. After this, we began to see a small but steady shift in page views until the new lessons finally began to outrank the old ones. We also experienced a gradual but significant increase in overall traffic during this time, with our number of visitors surpassing 6000 per day toward the end of 2010.

Despite the fact that tools such as Google Analytics are geared toward advertising and business needs, they can be powerful tools for non-profit and educational sites as well. In fact, it is precisely because the users of such sites have distinct needs and motivations that analytics become important. Rather than blindly following conventional Internet wisdom, which may not apply in an educational context, analytics allow developers to make more reasonable assumptions about the habits of their visitors, allowing them to meet their needs more effectively.

Elementary Math: Balancing User Needs

Following the development cycle of the Simple Machines lab, we began work on two other projects. The first involved the rewriting of the Star Life Cycle lab, which at the time was the most popular lesson on the site. The second involved expanding our audience to include elementary school students by creating a new series of lessons.

In our discussions with educators and policy makers about our outreach program, a common theme began to emerge. Although our lessons were good for helping students understand various science topics, many middle-school aged students lacked the prerequisite mathematical knowledge to perform well in science classes. Thus, many schools and school districts had begun major pushes to get elementary aged students interested in math so that they would be more receptive to both math and science classes later in their schooling. In accordance with these needs, we decided to create a series of lessons to teach elementary school aged students about math.

As with most new lessons, we chose to base our learning objectives off of existing, widely accepted standards. To this end, we decided to match our math lessons to the Common Core (www.corestandards.org), a set of math and English language educational standards that are achieving widespread acceptance within the United States. We also held a number of meetings with representatives from local school districts and the state Office of Education in order to tailor our lessons to their needs.

One topic that was brought up repeatedly throughout these meetings was a desire for an evaluative component in our math games that teachers could use to identify areas where students struggled. While the benefits of such functionality seemed apparent, it also presented a number of major problems. Due to our small size and limited resources, maintaining any kind of a user database on our servers was out of the question. Perhaps even more problematic was the prospect of dealing with privacy concerns of a user base that was entirely made up of elementary school age children. Attempting to retain any kind of identifying information or even to manage the most basic forms of online interaction with users seemed impossible, yet in order for teachers to have access to any kind of useful information, there had to be some way to tie performance to an individual student.

Our solution to this problem was to create a "back end" to our math games that stored data about in-game performance locally on the host machine. This information would be tied to a user name selected by the player, though there would be no kind of account verification associated with this process. This meant that although the teacher could go to any computer and see the scores associated with any user name, there was nothing in the program to prevent two students from using the same user name or to prevent a single student from changing her name between plays. Such issues would have to be completely managed by the teacher. Also, while a single computer could track multiple users, the data would not follow a student from one computer to another. Thus, in order to track performance over any length of time, students would have to use the same machine every day. Although such constraints made the programs much less convenient for teachers to use, we agreed that the importance of protecting student privacy and preventing abuse of the system clearly outweighed the inconvenience.

Although much simpler than retaining data on centralized servers, creating a back end to track student progress locally proved to be a significant challenge, monopolizing most of our programming efforts. The games themselves, however, were fairly straightforward to create. The challenging part was not getting them to function, but rather to design them in such a way that they matched very closely to specific Common Core objectives while still being fun for the students.

As we looked at existing elementary math games, we found that a common design style was to simply present the player with traditional math problems and after a certain number of correct answers, reward her with a game. The game itself was usually unrelated to any learning objectives, serving simply as motivation to do math problems. Our goal was to avoid this style of design, making sure that the game itself was the learning tool, not just a carrot on a stick to get the players to push through boring material.

During this time, we were able to visit a number of classrooms and have students from our target age groups test our games. This proved to be incredibly helpful in the overall design process. In addition to getting feedback as to which games were fun and which were not, we were able to observe the way in which students used the games. This allowed us to see what parts of the user interface proved confusing or difficult to use, which mechanics didn't work quite as well as expected, and which aspects of the game seemed to most draw the students' attention. Of particular interest were the aspects of the games that promoted pro-learning behavioral change in students that were struggling to learn math through traditional means. Even small additions, such as more visual feedback when clicking on the right answer, often had a significant role in persuading children who were easily discouraged to continue trying.

At present, we have created about a half dozen math games targeted at first grade and kindergarten students. Some games have been instant hits with the students, while other continue to be revised and remade. In general, however, the teachers we have worked with continue to be enthusiastic about the games, so development continues.

Figure 2. The Elementary Math lessons integrated mathematical reasoning into the game mechanics, rather than keeping learning objectives and gameplay as separate components of the program

Star Life: Expanding Educational Objectives

At the same time that we were creating our elementary math lessons, our other major project was rewriting the Star Life Cycle lesson. This lesson was the most frequently used on the site, surpassing even the Simple Machines lab in terms of daily visits. Interestingly, it was also the lesson with the least interactive content. Nearly the entire lesson was static text and images. Additionally, the text itself, which had been contributed by a teacher working closely with the ASPIRE program, seemed to have been written with a very narrow focus in mind. While the lesson was supposedly focused on stellar evolution, nearly all the content dealt with the concept of hydrostatic equilibrium, glossing over or ignoring many other important concepts.

After reviewing the existing lesson, we decided that the best course of action would be to completely redesign it from scratch. This was no small task, as the lesson touched on a number of complex concepts in astrophysics, a field in which none of us at ASPIRE had formal training. Fortunately, since ASPIRE is the outreach program for one of the largest particle astrophysics experiments in the world, most of our colleagues were, in fact, astrophysicists. Even with this incredible resource, however, redesigning the curriculum took several months.

Besides simply clarifying and correcting the information presented in the lesson, our goal was to make it more interactive and engaging. To this end, we decided to add two new Flash activities to the lesson – an interactive Hertzsprung-Russel diagram and a simple game that would allow the students to apply their knowledge of stellar evolution by forming their own star. In addition to hopefully making the lesson more interesting for students, this would allow us to target more of the cognitive skills in Bloom's Taxonomy than merely presenting information for rote memorization.

Creating these two programs was a somewhat slow process. As Pace was busy working on the back end for the math games, the programming for the Star Life Cycle activities fell to me. Although many aspects of these programs were fairly straightforward, they required a great deal of work with color transformations in order to achieve the full spectrum of stars from huge blue supergiants to tiny red dwarfs. Additionally, since the base animations differed considerably in their original color palettes, much of the process of coding the color transformations was simply a matter of trial and error.

So far, responses to the new lesson have been positive. We have received feedback both from members of the Telescope Array project and the local planetarium. Hopefully, by adding more activities to the lesson and making it deal with a broader range of topics related to stellar evolution, it will be useful to a much wider audience.

Figure 3. The Star Formation Simulator takes the principles learned throughout the Star Life lessons and allows students to apply them by creating their own star

CURRENT CHALLENGES

Just as it has been since the early days of ASPIRE, our greatest challenge is still dealing with employee turnover. Pace, our lead programmer, eventually was hired by Electronic Arts and left the project. Although we have had the opportunity to work with a number of other students since his departure, we have yet to find another programmer that could fill his role on the team. As such, the new math lessons have been put on hold and several other ambitious projects have been returned to the back burner.

With no dedicated programmer, we are left with several options in regards to the math games. First, we could wait in hopes of hiring another programmer who can finish the back end. Alternatively, I could begin working on the back end myself, though this would most likely take a considerable amount of time and take me away from other, more immediate projects. A third option, which seems more practical at the moment, is to remove the user tracking components and release the games without the ability to view students' progress. Due to the high demand for this feature among the educators we met with, it is unfortunate to have to abandon all our work on it. Our hope, however, is to be able to eventually reintegrate that functionality at a point in the future when we have more programmer talent to spare. For now, at least, this option would allow us to begin releasing the nearly completed games while still giving us the flexibility to continue with other projects.

Although losing talented employees in such a small organization will always be a significant setback, ASPIRE has been set up in such a way that we are able to weather such difficulties without losing the ability to continue our mission. Even with our new content delayed, traffic to the site continues and even grows, occasionally passing 7000 visitors in a single day. We are still involved in more traditional outreach activities, such as school visits and public demonstrations, and are expanding into new outreach methods, such as web video, in order to reach new audiences. Hopefully some new programmers will come around, but for now, we are still moving forward, slowly but surely.

SOLUTIONS AND RECOMMENDATIONS

The ASPIRE program demonstrates the ability of web-based educational videogames to reach large audiences. It also demonstrates the degree of organizational flexibility that such a model will allow, particularly when games are based on reliable platforms that will be supported for years to come. Particularly for organizations where funding is variable and susceptible to changes in external factors, the ability to maintain a strong public presence even in difficult times is invaluable.

It should also be noted that although the ASPIRE website has maintained a steady stream of visitors ever since its creation in the late 1990s, its success is not simply the inevitable result of maintaining a consistent web presence. Rather, the success of ASPIRE's online endeavors are directly linked to its more traditional outreach practices. Although not as far reaching as our web games, classroom visits, educational conferences and other in-person outreach efforts have not only yielded positive results on their own, but contributed to the general awareness of the ASPIRE website. Indeed, with the ability to use Google Analytics to determine not just region, but individual cities from which our traffic comes, it is possible to see correlations between high traffic locations and areas where we have done in-person activities. Other anecdotal evidence, such as email from educators, seems to support the premise of local word-of-mouth communication driving much of our traffic to the site.

As an ongoing instructional design program, rather than a single instructional project, the lessons learned from the ASPIRE program are most applicable to organizations with long-term goals of sustainable development and impactful outreach. Even with short-term projects, however, the results of the ASPIRE Program highlight the importance that organizational flexibility and community connections can have in getting small, web-based educational games into the hands of their target audiences.

REFERENCES

Bloom, B. S., Engelhart, M. D., Furst, E. J., Hill, W. H., & Krathwohl, D. R. (1956). *Taxonomy of educational objectives: Handbook I: Cognitive domain.* New York: David McKay.

Gentry, C. G. (1993). *Introduction to instructional development: Process and technique*. Belmont, CA: Wadsworth Publishing Company.

Greer, M. (1992). *ID project management: Tools and techniques for instructional designers and developers*. Englewood Cliffs, NJ: Educational Technology.

Krathwohl, D. R. (2002). A revision of Bloom's taxonomy: An overview. *Theory into Practice, 41*(4), 212–218. doi:10.1207/s15430421tip4104_2

KEY TERMS AND DEFINITIONS

Adobe Flash: Formerly known as Macromedia Flash, it is a platform designed to create animations, games, and other small programs. These could then be delivered to users via the Adobe Flash Player, a browser plugin that allows Flash programs to be run across various browsers and operating systems.

Cosmic Ray: A high energy particle that enters earth's atmosphere from space, sometimes creating a shower of secondary particles upon impacting other molecules in the atmosphere.

Hertzsprung-Russel Diagram: A graph showing the relationship between stars' absolute magnitudes, or brightness, and their spectral types (red, yellow, white, blue), which are indications of their temperatures.

Java: A programming language designed to run across multiple platforms. Java could be used to write applets, which could be embedded in web pages.

PHP: A server-side scripting language designed for dynamically creating web content.

Stellar Evolution: The process by which a star changes throughout its life cycle, creating increasingly heavier elements through various forms of fusion.

Ultra High Energy Cosmic Ray: A rare cosmic ray with energy greater than 10^{18} eV, far beyond energies typical of common cosmic rays.

Chapter 12
A Hostile World: A Pervasive Urban Game to Sensitise and Foster a Cross-Cultural Reflection

Maresa Bertolo
Politecnico di Milano, Italy

Ilaria Mariani
Politecnico di Milano, Italy

EXECUTIVE SUMMARY

A Hostile World is a persuasive game designed for an urban context with a high level of multiethnic presence, a recurrent feature of the contemporary megalopolis. Our players are ordinary native citizens who are plunged into an alternative reality where they can realize how complex and demanding it is to deal with gestures and tasks of everyday life in a foreign context, trusting them to live a destabilizing experience that aims to increase the sensitivity, understanding, and empathy towards foreigners, soothing the existing multicultural tensions. The game is a quest-based system; quests recreate situations of everyday-life needs, from shopping to bureaucratic adventures; it's designed to be modular and its sessions may change in the number and quality of quests adapting to different cities, contexts, and targets. The authors identify its effectiveness through the analysis of data collected during and after actual gameplay.

DOI: 10.4018/978-1-4666-6206-3.ch012

Copyright ©2014, IGI Global. Copying or distributing in print or electronic forms without written permission of IGI Global is prohibited.

BACKGROUND

Politecnico di Milano is a scientific-technological university which trains engineers, architects and designers. Its focus is on quality and innovation of teaching, research and technological knowledge transfer. Research activity constitutes a path parallel to that formed by cooperation and alliances with the industrial system. The drive to internationalization sees Politecnico di Milano take part in the European and world network of leading technical universities and offers several exchange and double-degree programmes beside Bachelor of Science, Master of Science, Specialization Master's and Doctoral degree.

We are developing our research at the Department of Design, which elaborates and promotes research and education in design investigating various fields that, with a focus on the centrality of the individuals and of the community, study: the spaces in which people live and the objects they interact with; their ways and means to communicate, to move, to consume, to produce and dispose of tangible and intangible assets, to enjoy and increase the value of cultural and environmental heritage, to innovate social interactions. It also promotes and coordinates research and training initiatives, publications, events, repositories of knowledge and documentation centers. In collaboration with other departments, centers and institutions it carries out research activities and provides consultancy on contracts and agreements.

The Department is divided into research groups. We are part of the *Products, Strategies and Services* one which is theoretically and methodologically based on an idea of the project as a continuum of products, strategies, services and communications that affect the evolution of society, culture, technology and economy, with a special eye on innovation and sustainability. Our research on Game Design fits well in this context, since our work includes theoretical and applied research as well as teaching and thesis tutoring. In particular, we design projects for ludic activities and we are interested in the dynamics that concern both the urban area and the people as a result of the activities we offer. The project presented here comes from one of our ideas and has been developed as the output of a Master of Science thesis in Communication Design at the School of Design, Politecnico di Milano, submitted in 2013 by Lavinia Ierardi.

SETTING THE STAGE

We live in a historical moment in which our communities are urged to face and deal with complex and pressing socio-cultural issues. The contemporary western cities are hybrid and multifaceted realities, characterized by a fluid and dynamic nature that makes them able to change and transform as needed. Being the stage of social, technological and architectural change, cities influence to a large extent the life of their citizens.

In the last decades we have witnessed an unprecedented technological evolution that has resulted in a more than just significant impact on living spaces and how they are lived (Shepard, 2011; Flusser, 2004). The cities are spaces in which tangible and intangible entities co-exist; they are areas of prevailing knowledge crossed by a network of information with multiple access points. They are interwoven with a mobile and widespread pervasive technology that has profoundly changed the way in which people relate to each other, to the information itself, and especially to the space and time of everyday life. By nature, the city has always been stratified, but today it seems that this connotative trait has been brought to whole new levels.

Looking at all this with the eyes of a designer, we notice that there is a space of possibilities, full of urgencies and opportunities, in which many entities need to be designed by someone who is aware of the different dimensions involved. This complex system results to be extremely interesting for the research on the contiguous areas of Communication and Game Design, since more and more frequently the designer is asked to design both the interaction and the experience between the land itself and its inhabitants. Using instruments that are suitable to support social innovation, the designer has shown to be able to become an interpreter and *facilitator* of the dialoguing processes and social sharing (Meroni & Sangiorgi, 2011).

We focus on Game Studies, working on theoretical issues with attention to the possible design evolutions. In particular, we are interested in how the game can become a vehicle and facilitator for communication and socio-cultural innovation. In this sense, our goal is to stimulate city dwellers to look at the urban space as a place of diversity, meetings and exchanges. A place that has the shape of a complex and evolving organism in which values and cultures of diverse origins live side by side, with all the consequences that this entails. It is a city to experience, claiming back and reconquering lost spaces. Thanks to international research, the game has emerged as an artifact with a strong communicative impact, able to deal with challenging and serious issues, often adopting conceptual simplifications that can create *meaningful experiences*; experiences that can sensitize and persuade the player to reconsider positions, preconceptions and ideas.

Our proposed project is *A Hostile World*. It arises with the intent to lead the players to consider the difficulties faced by foreigners, and ultimately aims at increasing the level of empathy and at relaxing the intercultural tensions that so often are exacerbated by the reluctance to put ourselves *in the shoes of others*. As we will describe in greater detail later, it is a persuasive game designed to take place in urban environments with a highly multiethnic presence, such as contemporary metropolis are. An important aspect of the resulting experience is related to the almost total absence of digital elements. We intentionally detached from the trend to develop persuasive game with heavy digital components and connected to the network. Our desire is to reduce the mediation between players and the gaming environment, giving the feeling of *really* being *in the other reality*. That's why we wanted to work on a *concrete and physical* experience that is strongly linked to real time personal interaction. Although the game is currently characterized by a low digital content, it immerses its players into an Alternate Reality where they live an experience that has been described as intense, rewarding and fun but also unsettling and disturbing (Figure 4). The play sessions have shown that players actually indicate a greater sensitivity, understanding and empathy towards foreigners at the end of the gameplay.

CASE DESCRIPTION

This case study is an urban pervasive game that arises from the consideration that the play activity may lead players to go through meaningful play experiences. In so doing players can firstly interact with the context, and getting into its habits and uses allow a deeper understanding. The concept of *experience* is crucial to the Game Design, and persuasive games in particular. Playing, the *Homo Ludens* described by Huizinga dives in a *fictional world* that has rules, conventions and characteristics that separate it from the real world and the everyday life. The context in which the ludic activity takes place is typically called *'magic circle'*, a space whose boundaries have been studied in detail (Zimmerman, 2012; Walther, 2005; Juul, 2008) and that, especially when discussing about urban games, have a tendency to blur and become membrane. Its very existence defines a place that is *other*, in which the player distances himself from the ordinary. It's a *protected* environment where the player can go through the conflict with serenity and safety, living an experience whose consequences are particularly interesting to us. Indeed, we can look at the play activity as a safe way to experience reality (Crawford, 1984). In regards to this, Csikszentmihalyi, a well known psychology professor, has associated the act of experiencing to the freedom to make mistakes ([1992] 2009, p. 59): the play experience allows users to immerse themselves in an *other context* leading to a state of openness to possibilities and consequently resulting in a consideration of possible change and facilitation of learning. The immersion into the *ludic realty,* a reality that is set apart from the normal one, demands the player to trust and believe in the game world – just like a reader trusts his narrator and, when reading a story, is drawn into the narrative context. This immersion can cause the player to detach from his usual *frames of references* and momentarily adopt other frames (Sclavi, 2003). The concepts of safety experience, of immersion into a fictional world, and the resulting openness towards changes of positions are on the very ground of our case study.

A Hostile World is at the same time an urban and a persuasive game: it is able to persuade players to reconsider their own bias and preconceptions by diving them into a concrete and tangible experience. The focus is: *How did you feel going through that situation?*, and consequently to invite players to reflect on their inner preconceptions. This statement arises from the assumption that, quoting Huizinga, we may relate to the game both "as a contest for something or a representation of something. These two functions can unite in such a way that the game 'represents' a contest, or else becomes a contest for the best representation of something" (Huizinga, [1938] 1939, p. 19). The game is indeed a system of representation able to reflect and return in some traits the context and the culture it was born in. The act of playing within a particular playground context induces the player to overcome the magic circle boundaries and symbolically interact with the socio-cultural elements which are constitutive of the society itself (Salen & Zimmerman, 2004). Through the game and the resulting meaningful play experience, players are invited to interact with the context as a systems of values, habit and frameworks of reference within it (Flanagan, 2009; Juul, 2010; Iacovoni, 2006).

Game Goal: The Experience of Diversity

An important aspect of our research is the desire to design and create games that are able to improve people's lives. We are convinced that *play* is a major element of people's lives and society (Huizinga 1938; Fink 1957), and is to be performed without the need for a secondary *justification*. We also believe that the ability to create, through his own games, meaningful gameplay experiences that are intense and pleasant at the same time, is a high achievement for a game designer. At the same time, however, we believe that the game can offer experiences that prompt players to think and reconsider their actions or their points of view. Observing the society, the people and the city, we notice numerous types of problems, many of which could be solved, or eased by prompting those that are involved to reflect, think differently and *put themselves in the shoes of others*. This is why we strive to design games that encourage people to temporarily change their frame of reference, and bring back in the real world those ideas experienced during the gameplay. This is the approach *A Hostile World* was born from. Milan is rich in culture, art, a city able to captivate visitors and offer ongoing interesting experiences to those who live in it. At the same time, just like all other large cities, Milan faces a very diverse set of problems: traffic, alienation, stress; like many other Italian and international cities, it is a place of great multicultural and multiethnic presence, and it is stage of scenes of misunderstanding, tensions and all those complications that arise from the coexistence of people with different patterns of interpretation.

A Hostile World

A Hostile World addresses native citizens, accustomed to living in a well-known environment, to using their mother tongue, to understanding what other people mean and what to expect from the environment and the events. It targets people who are not used to interact with the '*different*'. Our player doesn't belong to a specific age group, or to a professional category and doesn't have particular characteristics. He is a *normal person,* and as such we are interested in him as a target, because during the gameplay we want to *make him different*. We want to plunger him in an alternate reality where he may realize how complex and demanding it is to deal with gestures and tasks of the everyday life in a foreign context. The project immediately immerses its players into a fictional world in which they are the different ones, the strangers. The language in which everything is described or communicated is an unknown one, as we will describe shortly. Players move around a familiar and well-known environment, but their sense of safety is in a way destabilized: they do not understand what they're told, nor what they need to do to solve the assigned missions, therefore living the experience of foreigners in their own city. The game sessions have demonstrated the validity of our design assumptions: players testified the experience as intense, deep, often characterized by elements of great unpleasantness, but almost paradoxically it was perceived in a positive way. Our players lived a destabilizing experience, they felt different, often powerless and frustrated (Figure 4), but they were able to sense what it means *to be different,* to be *foreigners in a hostile world,* and they then brought the memory of this feeling in their daily life outside the game.

A Hostile World manages to get the player involved in immersive and proactive experiences, and – by stimulating meditation on this subject - it suggests him a critical review of his interpretative habits, mainly thanks to making him *walk into somebody else's shoes*, even if it was *only* in a game. This project has thus demonstrated its ability to stimulate a critical thinking that could lead to reconsideration of views and habits, to *persuasion*. This allows us as designers to observe how the gaming experience once again is able to help develop an openness to change. In addition, it allows us to analyze a fascinating aspect of the relationship between humans and the game: our propensity to declare ourselves happy and satisfied to have experienced something negative and stressful (Figure 3). At the end of this case study we will analyze this typical paradox of games in the light of the data collected during the sessions of *A Hostile World.*

Through this project we do not want to manage the coexistence of the various actors that populate our multi-ethnic metropolis, nor do we want to propose a solution to the many problems related to their living side by side. Instead we want to stimulate the understanding of the *others*, thanks to the interdisciplinary analysis and design approach that characterize the areas of Game Design and Communication Design. Our aim is to make the player aware of a topic of considerable social interest, by creating experiences in the city that help generating a sense of *identification* with some of the issues felt by ethnic and racial minorities.

The Game

A Hostile World is an urban persuasive game in which players are asked to solve simple missions. To succeed in those assigned tasks they have to move in the game environment, recognize some important places, and interact with those who play the role of non-playing characters – henceforth called Actors – in order to obtain either information or key objects – Tokens – that prove the accomplishment of the mission.

The purpose of the game is to make people live the experience of being alienated foreigners in a foreign land, and force them to interact with an environment that looks normal, but in fact is not. This is why we chose to develop all of the game elements in a foreign language: Actors speak a language unknown to the players, and the signs and the components of the game use a language little widespread too, a *foreign* language. The language we chose is Esperanto: an artificial language, created in the nineteenth century as an auxiliary language that lives alongside the individual national ones. The idea was to respect the diversity of cultures and languages of all peoples, and to facilitate dialogue, understanding and peace through a simple but expressive language, which belonged to humanity and not to a specific population. Its simple structure, without complicated grammar rules and exceptions, makes it easy to learn. It also has its own international heritage, expressed both in the presence of multi-lingual influences in its vocabulary, and a rich artistic production. Two million people circa, a limited but significant number, speak it: it is known all over the world, even though it is hard to run into someone who actually uses it. Esperanto has many similarities with other languages; when hearing or reading it, even without knowing its vocabulary, people have the bizarre sensation of understanding and not understanding at the same time. A feeling which links well together with our desire to create an experience where people are in their city, in an area well known to them, but they interact with people and objects that are difficult to understand: *everything seems normal, but in fact it isn't.*

A Hostile World can have different settings, all of them characterized by simple, everyday elements. If the setting is the university – as in our test sessions – the missions may require you to buy a sandwich (Token *Sandviĉo*) at the bar (*Restoracio*), or to get a patch in the infirmary, or again, to obtain permission to access the library or use a lab. We've also considered other settings such as the health care environment, where the tasks are related to obtaining a health care card, prescriptions for medications, appointments with specialists or booking medical exams. Another setting is the bureaucratic environment, with missions related to gaining a VISA, identity documents, or signing of a lease.

Regardless of the setting, the missions have a structure composed of:

- **Starting Point:** The place where the mission is assigned. Most of the times it coincides with the Reception, but it could also be any of the other place of the game. For example, a traffic policeman Karactero can assign the mission *Pay a fine;*
- **Goal:** The Token that has to be collected. Each mission states where to find the specific object. In the sessions conducted we used cards with icons or short texts as Tokens;
- **Closing Point:** The place where players demonstrate that the mission was accomplished. This place depends on the mission: it may be the Reception, but it could also be another point of the game. For example, the payment receipt of the fine may have to be reported to the traffic policeman, or delivered to a different Karatctero.

Missions can be made up of two or more sub-missions. For example the task *Organise a dinner* may comprise the sub-missions *Invite friends, Buy the food*, and *Tell the neighbors,* each of which falls within the structure described above.

A Hostile World is designed in such a way that it can be played in spaces of different sizes and types. The game area may be relatively small, such as a square, a campus (as in the game sessions analyzed later in the case study), a museum complex, or it can instead overlap entire city neighborhoods. The game space is a normal, ordinary environment reinterpreted, in which players live experiences that are different from those of someone who is using the same space but is alien to the game. Regardless of where it is staged, the space of *A Hostile World* consists of a set of game places, called Game-points. There are no preferential pathways to connect the Game-points, since – as we'll see later – their research and identification are important elements of the gaming experience. The players spend part of their time exploring the environment looking for the locations they must reach in that step of the game. We can therefore imagine the gaming space as a constellation of zero dimension places (the Game-points), reachable through paths that depend on both the place where the game is actually played, and the ability of the players to recognize them.

The main Game-point where the game is managed from, is called Reception. This is where people access the world of the game: they are greeted by human trained personnel who equips them with a badge (to be applied in a visible position on the chest) with the inscription *Ludanto* that identifies them as players, and with directions for the accomplishment of a mission.

The other Game-points are the places where players have to travel to, or move through, to solve their missions. Game-points are either mobile or fixed, but in any case they imply the presence of a person (a game Actor) with whom the player has to interact. The mobile Game-points are simply Actors who roam in the game space, playing the roles of policemen, janitors or passersby; they have a *Karactero* badge, which identifies them as potential characters to interact with. The fixed Game-points are specific places, with the constant presence of a Karactero and, as we'll see shortly, they have different functions depending on the missions they are connected to. The main ludic activities happens in these places – whether fixed or mobile. These are the locations where players come into contact with Actors in order to solve their missions. The area between Game-points is intentionally not designed: *A Hostile World* uses the surrounding environment, so that players walk on real pathways and sidewalks. As they move around they have to understand how to proceed and search for the next Game-point.

In-game items can be grouped into two main categories: *spatial elements* and *mission items*.

- *Spatial elements* are all the signs that identify the game locations. For example, they can be those signs with icons and texts (in Esperanto, obviously) indicating the nature of a point of the game, or the direction to reach it, as in Figure 1. Their quantity and size depend on the nature and size of the space where the game is staged. We point out the importance of strong visual consistency, which we achieved by using bright yellow paper and a simple, very straightforward and recognizable graphic layout.
- *Mission items* are the cards describing the nature of the mission, and those that the player gets as evidence of the accomplished task. These are graphically recognizable as belonging to the game; they are small, easy to handle even while the player is walking, gesticulating, drawing or miming.

A Hostile World requires the presence of trained staff: Actors are people who know Esperanto, or have studied a set of phrases and terms selected as essential terminology and vocabulary needed in order to allow dialoguing whit players. For example the person appointed as a Karactero to perform the role of the guardian of an administrative building must know the terms for directions, stairs, elevator, offices, and at the same time can completely ignore the medical, culinary or artistic jargon.

During play testing, we had the chance to see that a full knowledge of Esperanto is not necessary for Actors, since they were able to do their task without any issues after studying only the phrase book especially prepared for them and consisting on average of about twenty sentences, terms, and idiomatic expressions.

We believe it is crucial to give players the chance to experience the feeling people get when finding themselves in a context that is *adverse* because it is *different*. We investigated in particular those behaviors that people adopt, even if they are just playing, when experiencing a situation of linguistic incompatibility. At the center of the project there are questions such as *How do we behave and what feeling do we experience when, facing a different and unknown language, we need to make ourselves understood? To which extent can we detach from our frames in an attempt to understand the other? How much are we willing to put ourselves to the test?* In a sense, we can say that *A Hostile World* is aimed at pushing players to investigate their limitations and their cultural flexibility by putting themselves to the test and inventing the most diverse communication expedients they can come up with. According to the sociologist Goffman (1959, p. 83), a representation succeeds when the observers believe that the actors are sincere – whether really sincere or insincere, but anyway genuinely convinced of their own part – and therefore persuasive. The role played by the Actors inside of *A Hostile World* is very significant in order to maintain the persuasiveness of the ludic representation. The sincerity of the environment recreated, the honesty of the actions and of the emotions rely on the ability of the actors to interpret their roles. The actors of the game sessions conducted by us did not have a complete and total mastery of the language; they were indeed people to whom Esperanto was totally unknown except for the phrase books supplied by us; even so the play experiences were intense and significant.

Management and Organizational Concerns

From an organizational point of view, each game session needs the definition of missions that are specifically selected for the target audience they are addressed to: the tone of voice of the missions and their difficulty varies greatly whether we are targeting junior high schools or university students or members of a cultural association.

Another significant decision that has crucial consequences on the gameplay concerns the setting which on one hand serves as a connecting element between the different missions and the space in which they occur, on the other hand helps motivating the players who are interested in performing the tasks. Different settings also require different amount and types of missions. Using the examples proposed previously, bureaucratic environments will require missions in which players are sent to search for documents such as a residence permit or a labor card, moving between an articulated and complex system of offices that bounce them from one to another. But if we consider the health care environment we can expect missions such as the booking of special blood tests, getting a prescription for a specific drug that treats certain symptoms or claiming a health care card. It is clear that the level of complexity and difficulty of each mission can be increased, giving players the idea of being in a more or less hostile and adverse environment. In this regard, the definition of the setting should take into account the everyday tasks in which the target player is usually involved, and the designer can decide to either move towards that direction, or the opposite one and thus to immerse the players in worlds they are not familiar with. However, several elements must be taken into account when deciding the amount of missions and sub-missions, such as the budget, the amount of people in the team and the time available to structure the game, the amount of players involved and the expected duration of the game. The location where the game takes place and the time available for completing the individual missions are central elements of the planning: a spacious and structurally complex setting requires physical energy and time to be explored.

In regards to the practical organization, each session requires the production of printed material containing the game instructions and other elements in support of each mission: a set of directions for the Actors, to help them during the ludic representation they will take part in; the production and allocation of special signage that will direct the players. The entire gaming activity is based on a careful and conscious management of both people and the roles required: this means that the game can't in any way be automated and played permanently, as it requires a strong and vital human component. *A Hostile World* is not a game that may be played at will: it has to be arranged.

Learning Theories and Frameworks

Working on this project we focused on the *frames of references*, the role they play in our daily lives and how, through the act of playing, we are able to stand back from them and change our point of view, even if just temporarily. Growing up in a community, learning a language, we create our set of complex hierarchies of implicit assumptions that are given for granted in that mileu, and are a safe ground that allows us to understand the environment, the people and the events (Sclavi 2003, p. 30). As human beings we have absorbed and interiorised a social legacy that is the result of a system of thoughts and choices passed down over the years. This heritage belongs to us and it manifests itself not only in our actions and reactions, but also in our own way of thinking, feeling and believing. It is interesting to note that if someone moves far from the identity schemes to which he grew up, he is perceived by the others as different.

Think of how we react when we see a space that is used in what *we consider an improper way.* We might be astonished and amazed, or stunned and confused, but also agitated and upset, or even distressed and frightened, because often we do not understand and are frightened by what is different. *A Hostile World* aims at generating a gaming experience that takes place in an urban space and connects people with the concept of diversity. For this reason, in order to understand this project it is important to focus on the cultural interpretation of the city environment. In particular, it is interesting to recall the analysis proposed by Christopher Alexander in his fascinating book *The Timeless way of Building* (1979, p. 72): any space, whether urban or not, public or private, can be interpreted differently depending on where the observer comes from and on his cultural context. For example in New York just as in Milan, the sidewalk is primarily a place of transit, on which people move quickly, often bumping into each other. In India or China instead, the sidewalk is often considered a place to sit, talk, play, and even sleep. Our players are driven to feel temporarily as *foreigners in a foreign land*, finding themselves in a familiar urban environment, which – because of the gameplay – is not the same. The habit and knowledge that certain spaces are linked to certain types of tasks is part of each of us: our daily life consists of places where certain actions are performed, actions that we could call ordinary, common, everyday actions. Throughout this project the ordinary acquires new and unexpected features, urging the players to face unfamiliar and unsettling situations. The *patterns of events* (Alexander, 1979) are reconsidered. Think about how in a foreign city local customs may appear *bizarre* and incomprehensible: we lack shared frames of reference. In such conditions, even the gestures of everyday life may become obstacles to tackle and cope with. Acts of racism and discrimination against the different emerge as a symptom of a plurality of cultures in our cities. Our project intends to stimulate an attitude of openness and willingness to understand, by trying to make a difference (in more than one way) on those territories where the presence of the *other* is often perceived in a negative way. *A Hostile World* wants to influence the player not only during the game, but also in the period that follows. Thanks to the significant experience lived playing, a shift of meaning happens from the fictional world to the real world and the different is not so different any more (Bertolo & Mariani, 2014).

CURRENT CHALLENGES

A Hostile World is a project with a well defined structure and a clear objective: we organized game sessions, during which we studied both the players and their behaviors through observation and ethnographic interviews, according to a hermeneutic approach supported by the collection of data and the filling of evaluation questionnaires on the game and the experience. At the same time, we've had the opportunity to verify its effectiveness and identify the potentials and problems that – like in any game – emerge only when it is really played.

The use of Esperanto was very appreciated by our players. It helped to create an experience that was really democratic for all: none of the players knew it and therefore no one was favored. This language requires training of the actors involved, not only in the form of guidance on how to perform the different roles, such as the attitude to embrace (hostile-friendly-neutral). The Actors who do not speak Esperanto must be provided with a phrase book of expressions. These expressions have been in several cases well assimilated by Actors and players, who have used them as sayings or inserts into phrases. At the end of the game, almost all of them greeted us with *Saluton!* and *Danko tre multa!*

When choosing the area to play in, be sure that the Game-points are easy to reach with the means available to players: usually our participants move on foot and within well known and small spaces (3 to 5 km^2). However, it is not unconceivable that means such as bicycles, skateboards, etc. become part of the game. In these cases the size of the playgrounds must take this into account and indeed, enhance the experience. Our gaming sessions were held with a multiethnic target mainly from the university – students, researchers, teachers of our campus; playing in a well-known and clearly circumscribed area has allowed players to experience the feeling of being lost generated by the game without worrying about the real surroundings. This is a choice that makes the player feel in a state of full safety, in a safe space. It allows the player to focus on the ludic experience, living it to the fullest, rather than feel limited in the exploration and the experience itself. Think at the fear of getting lost that could grasp anyone who is in an unknown space, maybe a big city, where they do not have their bearings: it is a situation we have to take into account, and that has to be designed in a conscious way, for example by limiting visually the playground area, thus avoiding the player exiting the game space without realizing it, and providing maps and tools that will enable him to tranquilly get lost.

In this scenario, the designer must allocate the Game-points of the mission in places that are easily accessible but not clearly in sight. The engaging in their search and the gratification that comes when discovering them are in fact part of the ludic experience. In this sense, the coexistence of urban spaces and points with ludic meanings constitutes a problem: it is essential to use signage elements that are clearly visible and distinct from ordinary urban ones, able to emerge from the visual urban noise.

As previously mentioned, the project consciously relies on non-digital game material. This is partly due to the fact that otherwise it could create a filtering of the players on the basis of the technological equipment that they possess. Another reason for this choice is our desire to enhance the feeling of being in the shoes of people who often are not equipped with smart mobile devices and leave the players wondering "What if I didn't have it?". It was very interesting to see how our players often did not think about getting help during the game by using their devices to access search engines and automatic translators (Figure 2, right). Some said they decided not to use them because they thought it would be an unfair shortcut, despite the intentional absence of any instructions about it. At the same time we saw players use an online translator openly, and others hide behind parked cars to search for the translation of a word.

We deliberately didn't provide rules concerning the possibility to play in a group or individually, but we noticed a strong tendency to aggregate in groups (Figure 2, left). Some of the players who started the experience by themselves, then aggregated with others claiming later that they felt united by a feeling of alienation and disorientation and legitimated in the creation of mutual support groups. We read this behavior as an appeal to the collective intelligence in order to collaborate and solve more quickly the game missions.

The analysis of the data collected during and after the game sessions gave us satisfactory results. Most of the players said they liked the game (Figure 3), describing it as a stimulating and very positive experience. This is a particularly interesting aspect, as it comes as a further confirmation of how a game can lead people to live substantially difficult and negative experiences, but leave them a positive sense of satisfaction. Observing Figure 4 we can see that in the questionnaires the players used many and significant *negative* keywords to describe their experiences: difficulty, confusion and frustration. Yet, there are also numerous positive ones, and the percentage of *'Yes'* to the question 'Did you like it?' was almost full (Figure 3). The gaming experience was generally perceived as a very strong one, and generated the desired feeling of empathy towards those who really live in a hostile world like a foreign and incomprehensible land. The players have left statements like 'It is difficult to shake this experience off', 'Fortunately, the game then ends', indicating a state of total *frustrating fun* and great dedication to the completing of the missions. We saw players get hold of an Actor and have him point the way step by step, while others protested and called him back to order; people got depressed and literally beat their head against the wall in frustration or feigned an illness and fell to the ground trying to make themselves understood by the Actor-pharmacist (Figure 1). Players worked hard, using their creative abilities, sometimes even with behaviors that looked hilarious to the observers: the mission *Scholarship* led many people to try to explain it by pointing to their bag and miming the act of flipping through a book, hoping that – due to the fact that in Italian a scholarship is called 'bag of study' – the Actor would understand the concept (and of course, being a very good italian Actor, he did not).

During the game sessions, *A Hostile World* proved to be capable of providing the experience we were looking for in the design phase: the facing the difficulty to perform simple tasks in an environment where the available information is incomprehensible. An experience that aroused empathy with the stranger, and was intense, full of difficulties but still remembered with pleasure. Numerous players canceled previous commitments to take part in more missions, and asked us to keep them updated on upcoming sessions. What our players experienced is the almost unconscious crossing from taking part in a leisure activity to experiencing a powerful rhetorical tool: while playing they not only learned some words of Esperanto, but they also got to better understand how *different* people live their daily lives, especially when there are no more frames of reference and shared vocabularies.

SOLUTIONS AND RECOMMENDATIONS

Although it has been played during several sessions, *A Hostile World* project presents several opportunities for further development and the positive response of the players during and after the sessions encourages us not only to plan more of them, but also to invest our energy in evolutions of the game.

The very nature of *A Hostile World* is such that every game session requires planning that takes into account – once the budget is known – the type and size of the spaces, the quantity and typology of players, and the key issues of the host city. We are drafting a *Manual of a Hostile World,* a set of guidelines that can help people to put in place the game. It is a multimedia document that illustrates the nature of the project and provides both methodological and practical tools for the creation of the different elements of the game. It contains a set of missions already organized, and instructions and tips for creating new ones, an archive of templates for the production of game materials, as well as a photographic-video archive of the sessions carried out previously to serve as historical memory and as inspiration and guidance.

The creation of an itinerant team doesn't compete with the creation of the manual. Once a group of people has organized a session of *A Hostile World* and has developed its own experience in the various roles involved, that team may arrange other session both in their city and in other places. We do not exclude the emerging of expert groups who are able to organize the game as one of their activities.

Figure 1. On the left: the game signage, as indications (above) and in a Game-point; in the center, printed materials; on the right: players interacting with an Actor (above) and a player miming trying to get understood

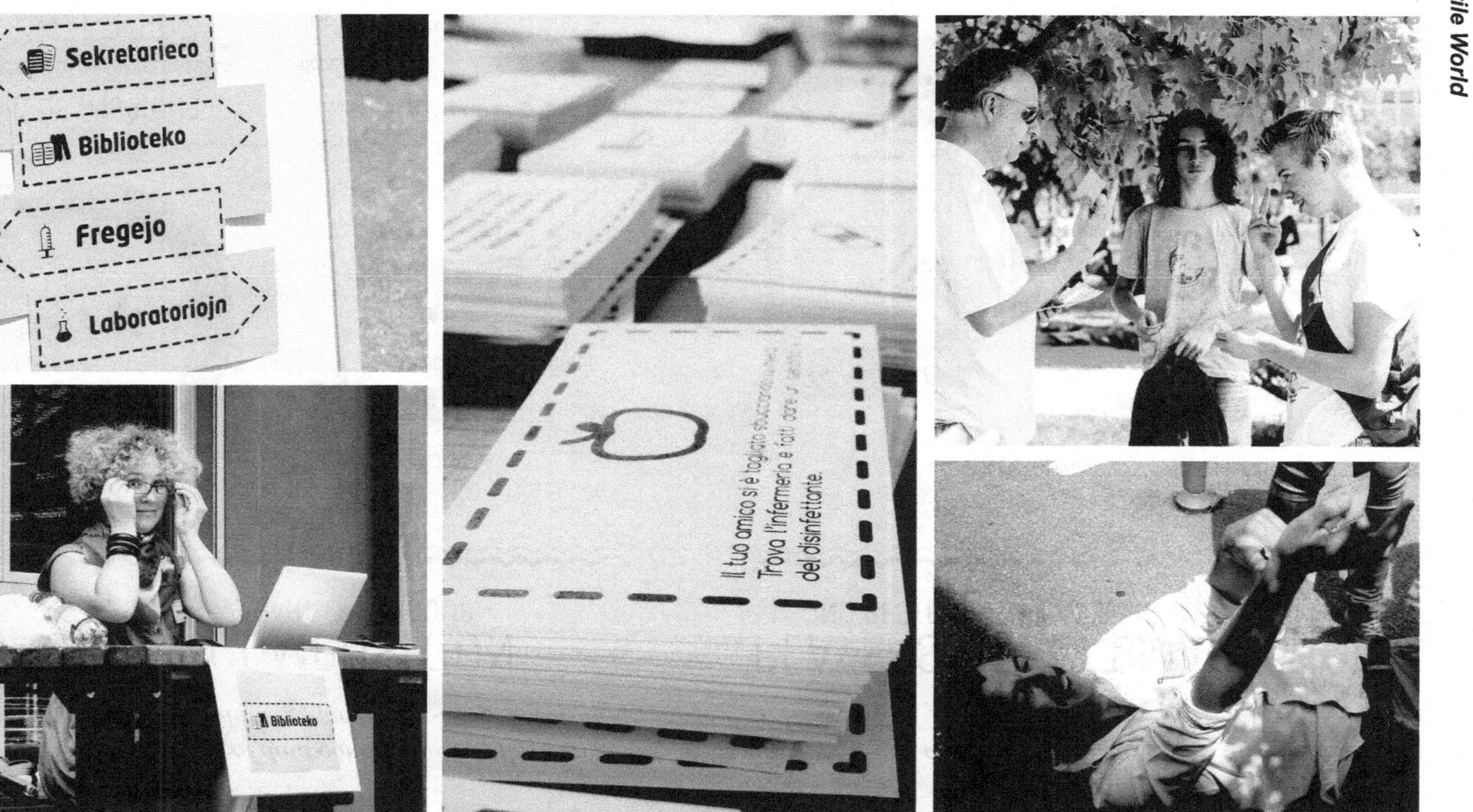

Figure 2. Visual analysis of data concerning (on the left) gameplay as single or group activity and (on the right) the use of mobile communication devices during gameplay

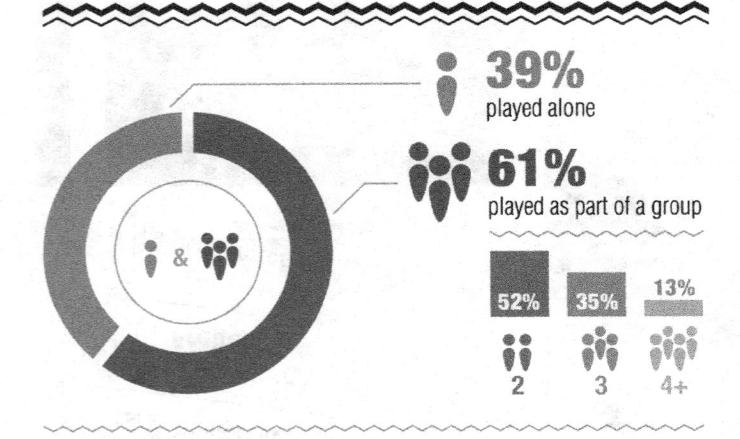

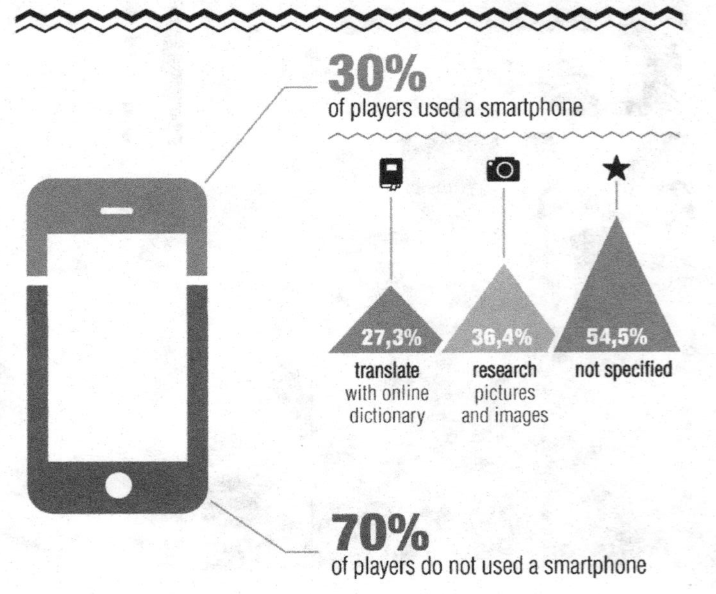

Figure 3. Visual representation of players responses concerning gameplay appreciation

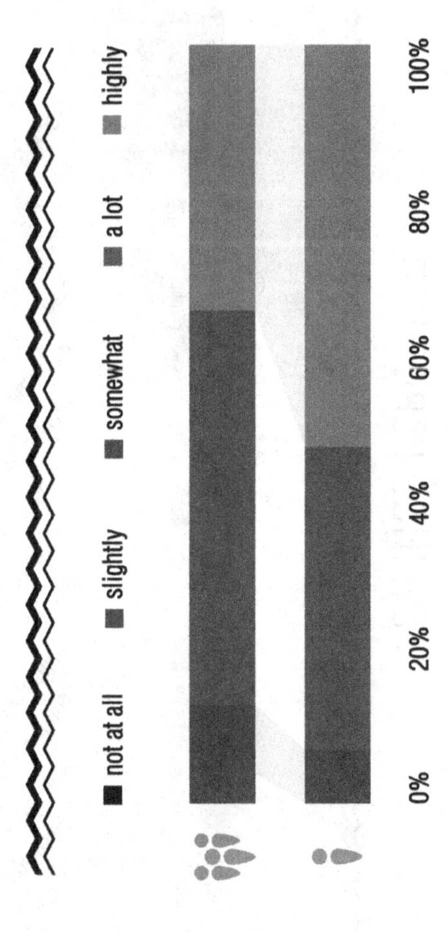

Figure 4. Visual analysis of data concerning emotions and feelings stated by players right after game sessions

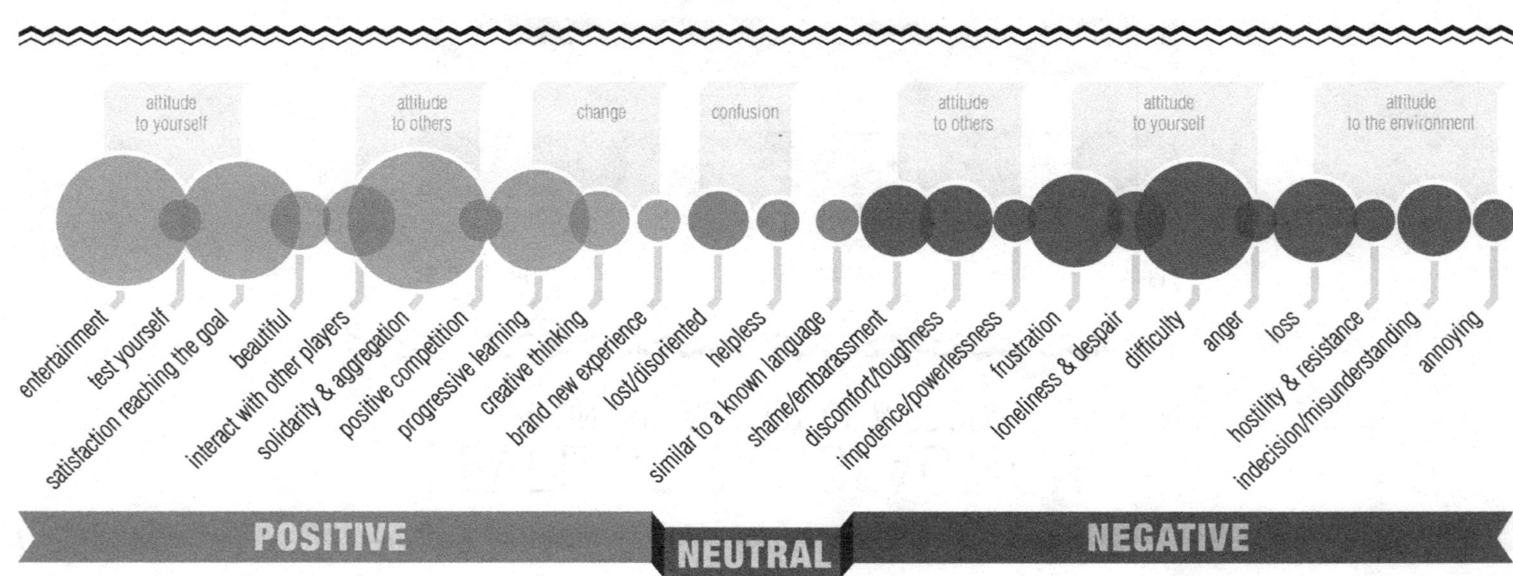

REFERENCES

Alexander, C. (1979). *The timeless way of building*. New York: Oxford University Press.

Bateson, G. (1956). *The Message "This is Play"*. New York: Macy Foundation.

Bertolo, M., & Mariani, I. (2014). *Game Deisgn. Gioco e giocare tra teoria e progetto*. Milano: Pearson.

Bogost, I. (2007). *Persuasive games: The expressive power of videogames*. Cambridge, MA: MIT Press.

Castells, M. (2004). *The network society: A cross-cultural perspective*. London: Edward Elgar. doi:10.4337/9781845421663

Crawford, C. (1984). *The art of computer game design*. New York: Osborne/McGraw-Hill.

Csikszentmihalyi, M. (2008). *Flow: The psychology of optimal experience*. New York: HarperCollins.

De Luca, V., & Bertolo, M. (2012). Urban games to design the augmented city. *Eludamos: Journal for Computer Game Culture*, *6*(1), 71–83.

Fink, E., Saine, U., & Saine, T. (1968). The oasis of happiness: Toward an ontology of play. *Yale French Studies*, *41*(41), 19–30. doi:10.2307/2929663

Flanagan, M. (2009). *Critical play: Radical game design*. Cambridge, MA: MIT Press.

Flusser, V. (2004). *La cultura dei media*. Milano: Bruno Mondadori.

Galbiati, M., & Piredda, F. (2012). *Visioni urbane: Narrazioni per il design della città sostenibile*. Milano: Franco Angeli.

Goffman, E. (1959). *The presentation of self in everyday life*. New York: Garden City.

Gregory, P. (2003). *New scapes: Territories of complexity*. London: Springer.

Huizinga, J. (1949). *Homo ludens*. London: Routledge & Kegan Paul.

Iacovoni, A. (2004). *Game zone: Playgrounds between virtual scenarios and reality*. Basel: Birkhäuser.

Juul, J. (2008). The magic circle and the puzzle piece. In *Proceedings of the Philosophy of Computer Games*. Academic Press.

Juul, J. (2010). *A casual revolution: Reinventing video games and their players*. Cambridge, MA: MIT Press.

Mariani, I., & Ciancia, M. (2013). The urban space as a narrative and ludic playground. In *Proceedings of CIDAG International Conference in Design and Graphic Arts*. ISEC – Instituto Superior de Educação e Ciências, IPT – Instituto Politécnico de Tomar.

McGonigal, J. (2011). *Reality is broken: Why games make us better and how they can change the world*. New York: Penguin Press.

Meroni, A., & Sangiorgi, D. (2011). *Design for services*. Farnham, MA: Gower Publishing, Ltd.

Montola, M., Stenros, J., & Waern, A. (2009). *Pervasive games: Theory and design*. Burlington, MA: Morgan Kaufmann Publishers.

Salen, K., & Zimmerman, E. (2004). *Rules of play: Game design fundamentals*. Cambridge, MA: MIT Press.

Sclavi, M. (2003). *Arte di ascoltare e mondi possibili: Come si esce dalle cornici di cui siamo parte*. Milano: Mondadori Bruno.

Shepard, M. (2011). *Sentient city: Ubiquitous computing, architecture, and the future of urban space*. New York: Architectural League of New York.

Szulborski, D. (2005). *This is not a game: A guide to alternate reality gaming. Macungie*. New Fiction Publishing.

Walther, B. K. (2005). Reflections on the methodology of pervasive gaming. In *Proceedings of the 2005 ACM SIGCHI International Conference on Advances in Computer Entertainment Technology*, (pp. 176-179). ACM. doi:10.1145/1178477.1178501

Zimmerman, E. (2012). Jerked around by the magic circle: Clearing the air ten years later. *Gamasutra Blog*. Retrieved from http://www.gamasutra.com/view/feature/6696/

KEY TERMS AND DEFINITIONS

Diversity: Presence of features that differentiate a thing or a person from another that is part of the same type.

Empathy: Psychological process that allows a person to identifies with someone else by understanding and sharing his/her emotional situation.

Esperanto: The most common among auxiliary languages, Esperanto is an artificial international language designed by the Polish physician L.L. Zamenhof (who published the first grammar in 1887) . Aiming at the maximum internationality, its grammar is characterized by a remarkable simplicity.

Frames of References: As human beings we have absorbed and interiorised a social legacy that is the result of a system of thoughts and choices passed down over the years. This heritage manifests itself not only in our actions and reactions, but also in our own way of thinking, feeling and believing.

Fun: It is an experience of pleasure; with respect to games it is particularly difficult to define fun: there are too many different ways to enjoy playing games. Physically, fun is a neurochemical reward that encourages people to keep doing what they are doing.

Hostile: To be adverse to something or someone, or to show dislike and distrust.

Meaningful Experience: Games can convey a "meaning" in the sense of providing information and have a communicative function; as well as they can provoke a "significant" experience, that is intense, deeply emotional. The play activity is "significant" when the player performs actions that mean something during the gameplay. However, players can live meaningful experiences that sometime cause a shift of meaning from the fictional world to the real one.

Multiculturalism: Is the cultural diversity of communities and the policies that promote this diversity. As a descriptive term, it is the simple fact of cultural diversity and refers to the demographic presence of several cultural or ethnic groups within a society. As a prescriptive term, multiculturalism encourages ideologies and policies that promote the diversity and the desire amongst people to express their own identity.

Pattern: A recurrent set of ideas and habits, as well as a recurrent way of usage for places or objects. According to Alexander (1979), each pattern expresses a relation between a certain context with a certain system of forces occurring, a problem and a solution. The pattern can refer to events, spaces and languages.

Sensitize: Make someone aware of something by awakening the attention and interest aiming at enhancing a certain participation. It is generally used to refer to the problems of a social nature.

Chapter 13
Strategies and Tactics in Digital Role-Playing Games:
Persuasion and Social Negotiation of the Natural Order Doctrine in *Second Life's Gor*

Christophe Duret
Université de Sherbrooke, Canada

EXECUTIVE SUMMARY

In this chapter, the authors examine the strategies and tactics of persuasion used by the players in the Gorean role-playing games organized in Second Life, which is a video gaming adaptation of the series of novels The Chronicles of Gor by John Norman, games in which a doctrine is both defended and contested. These strategies and tactics fall into six different categories of closure conveyed in Gorean role-playing games: institutionalized enculturation, informal enculturation, hermeneutical closure, sociotechnical closure, narrative closure, and legal closure. The chapter shows that the PRPG-VE is an inadequate medium when it comes to conveying a persuasive message to a target, but it can be useful in the context of a media critical education program.

DOI: 10.4018/978-1-4666-6206-3.ch013

Copyright ©2014, IGI Global. Copying or distributing in print or electronic forms without written permission of IGI Global is prohibited.

BACKGROUND[1]

As a communicologist, I am particularly interested in sociocriticism, reception theories in general and hermeneutics in particular. It is with this dual focus that I feel particularly suited to the uncovering of social meanings embodied in video games and which brings me to focus on the study of role playing in *Second Life*. I have defined these games elsewhere (Duret, 2014) as "Participatory Role-playing Games in Virtual Environment" (henceforth "PRPG-VE"). Of the PRPG-VE, it is the Gorean role-playing games which have caught my attention, for two reasons: firstly, they are the videoludic adaptation of a work of fiction where the ludic activities they embody concretely reflect the appropriation of the work by the players; secondly, they convey an explicit *doctrine* (which already exists in the original work) which the players defend or challenge, and which they either apply literally or subvert in their role play. Thus, it is a role-playing game with a *subverted doctrine* that I present here. It is not as a player or as a member of an organization or a producer, that I will approach this atypical object, but as a researcher attracted by its complexity, for its uniqueness and for the enlightenment that it offers regarding role playing and video games in general.

SETTING THE STAGE

I will discuss the Gorean role-playing games in the light of the transmission and challenge of a doctrine in a participatory environment in which persuasion is no longer the privilege of a single author, but of the players. To do this, I shall first describe the Gorean role-playing games and their characteristics as subverted doctrine. Then I shall describe them as texts, after which I shall elaborate a typology that will allow me to classify the strategies and tactics deployed by players to support their perspective on the doctrine conveyed by the game. Finally, in conclusion, I shall show how my analysis of the strategies and tactics of the players is enlightening for anyone interested in games as a medium of persuasion and as a learning tool. The case study I present here is based on an exploratory study conducted in the summer of 2012 on Gorean role-playing games. I later analyzed the documents produced by the players in an extrinsic play [2] situation (posts published on the forums *Gor-SL.com* and *The Gorean Forum*, as well as blogs, fanzines, online journals, wikis, game scenarios, game narratives, etc.). In addition, during the same summer I dedicated my time to a non-participatory observation role in virtual Gorean environments.

CASE DESCRIPTION

The Gorean role-playing games are participatory role-playing games in virtual world organized on *Second Life* and inspired by the fictional universe of the science fiction novels *The Chronicles of Gor* by John Norman. The setting for these games is the planet Gor which was summed up by Sixma (2009) as a "barbaric planet [...] where men are bold masters and women are either frigid mistresses or sexual slaves" (p. 5).

The participatory role-playing games in virtual environment are multiplayer role playing games in which players take on a role through an avatar in multi-user virtual environments (MUVE) which may or may not have been specifically designed for such activities. The PRPG-VE are representative of the participatory culture, as understood by Jenkins (2006), insofar as the players contribute to the development of the avatar scripts[3] and accessories as well as the game mechanics, the virtual environment, the rules of the game and the game world (the diegetic frame) of the role-play[4].

According to the site statistics for the Gorean Meter Support Portal (2012a; 2012b), dated July 15, 2013, there were 9,516 active avatars in 250 Gorean sims[5] between July 2 and 15, 2013. Although there is no official census of the number of players who engage in Gorean role-playing games in *Second Life*, it could have been have more than 70,000 in 2009, as reported by Au (2009).

The Gorean community includes two types of players characterized by their position with respect to the reading of *The Chronicles of Gor*, which is either literal or distanced (Duret, 2014). On the one hand, we find players who want to remain faithful to the novels of Norman in their role play (literal reading position). They call themselves "By the Books" or "BtB". On the other hand, there are players who distance themselves from the content of the novels to introduce outside elements into Gorean role-playing games. These are the "Gor Evolved" or "GE". As mentioned in the Gorean game manual *A Brief Guide to Gorean Roleplay in Second Life* (Ghiardie, 2010), the conflict between the two groups mainly concerns the role of women in Gorean role-playing games. The "BtB" say that in the context of the planet Gor, women are not equal to men and cannot be permitted to carry weapons. The "GE" instead state that it is realistic to imagine that the Goreans may have evolved socially to the point where women are considered equal to men and are thus considered capable of fighting alongside them. This view is contrary to that of the author, who writes:

There are no 'female warriors' on Gor. Gor is on the whole an honest, male-dominated realistic world [...] There are panther girls and talunas on Gor. They are unhappy, frustrated, disturbed women, half alienated from their sex. They tend to run in dangerous feline packs. Once captured and subdued it is said they make excellent slaves. 'Bring me into the collar if you can!' Amazon women/Mrs. Conan the Barbarian does not belong in the Gorean world. (Norman, 2001, non-paginated document)

Description of the Game as Subverted Doctrine

The Chronicles of Gor assume the specific characteristics of *romans à thèse* (thesis novels), which Suleiman (1983) defined as a type of novel which explicitly puts forth "recognized body of doctrine or system of ideas" (p. 1) and which has an "unambiguous dualistic system of values [...] rules of action addressed to the reader [and a] doctrinal intertext" (p. 54). The doctrine defended by Norman is the law of natural order, in which, according to the words of the author, "[t]he male sex is naturally dominant, and the female dominance-responsive" (quoted in Smith, 1996, non-paginated document), but which carries the evolutionary paradigm in a broader sense. It is the justification for why slavery is prevalent in Gorean society. In the framework of Gorean role-playing games, the law of natural order both structures the ludic experience for the players and also divides the community. In effect, the "BtB" players defend the law of natural order, while the "GE" question the accuracy and relevance of this doctrine. They challenge it by playing the roles of female warriors grouped into clans: the Panthers and the Talunas.

The Gorean Role-Playing Games Taken as Texts

In role-playing games, the text is understood by Padol (1996) as any material (I will add: any *semiotic* material) that attains the ludic interface between at least two players, or their descriptions (game settings and the action), diagrams, documents and settings provided by the game master[6], the reactions of the players and the characters faced with the game master's descriptions and the interactions between the characters played by the participants. The text is therefore not the result of a game, but the game itself. In the context of PRPG-VE, the ludic interface corresponds to its virtual environment and the navigator which allows circulation.

The definition given by Padol is applicable to PRPG-VE. Considering the latter as texts, it is possible to raise a sociocritical perspective in order to understand how the doctrine of the law of natural order is challenged or defended by the players. However, this definition is incomplete. Both role playing games and video games consist of a formal structure, which like the cybertext of Aarseth (1997), produces, consumes and organizes verbal signs and forms of expression, similarly to how it produces sequences of events when the player interacts with it. This structure is the instance that simulates the game world of PRPG-VE and the events that occur through it.

Therefore, it is appropriate to designate two types of texts: virtual text and actualized text. The first is the sum of possible texts or texts that have not yet occurred (and which may never occur). This textual level governs the simulation. It constitutes a field of possibilities structured, on the one hand by all the relevant social, formal and diegetic game system factors that program the actualized texts and, on the other hand by the meanings and motivations that the players, both as players and as individuals, bring with them (e.g. attitudes, values, needs). The actualized texts are concrete materializations of the virtual text and are constitutive of the game world.

Beyond this conceptualization of PRPG-VE, the modes through which sociality[7] leaves a footprint in the text PRPG-VE should not be overlooked, as without these the transactions of the players at the doctrine level of the game would appear anecdotal. These modalities are similar to morphogenesis. According to Cros (2003; 2006), a text has a dialectical structure due to social discourse (conveyors of mental structures and ideological formations of a social formation) and contradictory intertexts[8], which represent the frame and in which the encounter is constitutive of conflicting spaces which will result in "semiotic islets" that Duchet (1979) calls "sociograms." This structure is reproduced in the same way at all levels of the text, but appears heterogeneously. The text, in accordance with the principle of morphogenesis thus develops patterns and laws of repetition. Also according to Cros, the text is divisible into two phenomena: genotext and phenotext. The genotext consists of elements that reflect a moral value, a social value, or an abstract concept, and which are structured in polarized terms. It is a "virtual space where the initial structures program the process of semiotic productivity" (Cros, 2003, p. 55). The phenotexts are actualizations of the genotextual programming through the process of writing (or, in the framework of PRPG-VE: by the production of actualized texts using simulation). They are "dedicated to achieving the syntax of preprogrammed messages on all textual levels, depending on the specificity of each one" through the genotext (p. 55). From the sociocritical perspective, the virtual text corresponds to the genotext and the actualized texts, to the phenotexts.

Sociality enters the texts in the form of sociograms. Sociograms are "socio-discursive concretion[s]" (Cros, 2003, p. 39). According to Angenot (1989), this is a "Gordian knot of entangled ideologemes" that thematize and interpret a "doxic object" (p. 103). Duchet more precisely defines it as an "unclear, unstable, conflictual set of partial representations centered around a core of mutual interaction" (quoted in Duchet, Herschberg Pierrot, & Neefs, 1994, p. 118). This set is unclear, says Robin (1993), because of its indeterminate random nature. It is unstable because it is moving, just like the image of the socio-historical reality that underlies it and because the meaning that it holds is transformed by the work of the fiction. It is conflictual because it reflects opposed discourses. It consists of partial representations because the social discourse on which it rests cannot be grasped in its entirety. The conflictual core of this set consists of a stereotype, a maxim, a lexicalized sociolect, a cultural cliché, an abstract concept, etc., such that it is ideologically marked without ever constituting a monolithic doctrine. An example of a sociogram, given by Robin (1993), is that of "literature" in Balzac's *Lost Illusions*. The novel, in effect, contrasts the contradictory figures of poet and journalist at a time when literature was overwhelmed by the rapid development of the daily press.

In the context of the Gorean role-playing games, the central sociogram is natural order. It is inscribed in the text at the levels of the social organization of the game, of its formal structure and of its diegetic frame.

At the level of the social organization of the game, the relevance of the socio-discursive content that feed the sociogram of natural order is measured in terms of the real world: gender relations, relations of submission and domination, slavery, etc[9]... The values and attitudes of the players linked to the socio-discursive content of the sociogram emerge in the social organization of the game. For example, on the discussion thread *Gorean women: haven't we seen them before?,* the player Sayrax Wiefel raised some ethical issues related to the law of natural order likely to lead other players to reflect on their own relationship with a doctrine that undermines the achievements of feminism:

It is something I struggle with everyday, I enjoy this setting yet understand fully what am doing is wrong on some moral level. I go through bouts of self doubt with long breaks from this place [the gorean sims] at times. IT is why I turned to building and expanding my realm to outside gor, feeling it sorta eating me on the inside and wondering what effect it is having on my moral bearing as a person (Gor-SL, 2011a).

At the level of the formal structure, the conflict which reflects the sociogram of natural order is present in the rules (rules implemented by computer or applied by the players) and the parameters governing the simulation, as well as in the sociotechnical systems of the game. Millerand, Proulx & Rueff (2010) describe as "sociotechnical" "the social relations embodied in the design of the technical system, a design that constrains and makes some uses more possible than others" (p. 4). For example, in the Gorean role-playing games, the antagonism between "BtB" and "GE" is translated into combat scripts. The latter allow women to bear arms and the damage they inflict is equal to that inflicted by men. These scripts, in a "GE" perspective, challenge the law of natural order by making women equal to men rather than complementary. In contrast, the "BtB" players prohibit female avatars from bearing arms in their "sims" and express, in the form of regulations, the idea that women are physically unfit for combat.

Indeed, the sociogram of natural order as embodied within the diegetic frame is born from the meeting of the game's actualized texts, from the *Chronicles of Gor* and from their respective intertext. Thus, the Darwinian considerations that dot the novels of Norman echo a more general intertext: that of the outcomes of the evolutionary paradigm and its manifestations such as social Darwinism and evolutionary psychology. For example, in *Tarnsman of Gor*, the main character and narrator writes:

For whatever reason, the larl will always prefer ruining a hunt, even one involving a quarry of several animals, to allowing a given animal to move past it to freedom. Though I suppose this is purely instinctive on the larl's part, it does have the effect, over a series of generations, of weeding out animals which, if they survived, might transmit their intelligence, or perhaps their erratic running patterns, to their offspring (Norman, 2007a [1966], p. 174).

Interviews with Norman on the complementarity of the sexes also raise that of intertextuality. For example, Norman (quoted in Smith 1996) speaks of slavery which he depicts in his novels as "a celebration of the glory of nature and the reality of dimorphic sexuality [...] Ultimately, of course, the male is the master, and the female is the slave" (non-paginated document). However, feminist texts of the sixties to which the *Chronicles of Gor* constitute a reactionary antithesis are also constitutive of intertextuality.

Strategies and Tactics

The conflicts that relate to the doctrine of natural order, as we have seen, take several forms and leave detectable traces in the multilevel analysis of the text. According to Fiske (1987), "(a) text is the site of struggles for meaning that reproduce the conflicts of interest between the producers and consumers of the cultural commodity. A program is produced by the industry, a text by its readers" (p. 14). The *Chronicles of Gor* act as both programs and cultural commodity. Therefore, the ludic activity of members of the Gorean community illustrates, or even embodies, socially differentiated readings in the form of texts, since, as noted by Fiske, "one program can stimulate the production of many texts according to the social conditions of its reception" (p. 14). Yet conflicts between the players in the Gorean sims also replicate the struggles between producers and consumers. However, the antagonism not only places an actor in the cultural industry in opposition to the consumers, but also the consumers / readers in opposition to other consumers / readers, "BtB" opposing "GE". This conflict between readers to competing interpretative postures should not be seen as oppositions of equal strength as PRPG-VE are characterized by a hierarchical power structure: firstly, there are the simple players, and secondly, the administrators and moderators of the sims. This imbalance demonstrates, as in the producer / consumer relationship, the ownership of the sims, in virtue of which the administrators (or delegated moderators) have the last word in arbitration, since (unless power-sharing with other players) they have the power to exclude players and invalidate game sequences that violate the rules they have set, denying them any possibility to officially participate in the diegetic part of the game.

To account for interpretive conflicts that divide the players, the balance of power, and the attempts of each to impose their interpretative posture, I will use here the concepts of "strategy" and "tactics" of Michel de Certeau (1990), after which I will establish a typology of strategies and tactics deployed by the players in the Gorean role-playing games.

According to de Certeau (1990), strategy is employed by the dominant order, which includes the producers of the culture industry. It involves a calculation of the balance of power, a vantage point from which to manage uncertainty and the movements of opponents, as well as a fixed operations base to build on the achievements. Tactics, in turn, are used by the dominated, the consumers and users of media content that have neither independence nor their own space. They are forced to act in the other's territory according to externally imposed laws; they have to "make do." As Macherey (2005) summarized, tactics are "this 'art' which 'plays' on the flaws of the system and which, without leaving the system, creates room for maneuver" (non-paginated document).

It thus becomes possible to distinguish the strategies of the administrators and moderators of the "sims" (the producers) and the tactics of the simple players (the users). However, these strategies and tactics are manifested in different ways, depending on how the videoludic experience is organized by the player, according to the frame used by the player. We have revealed five frames constitutive of the video gaming experience, all closely interdependent, from where the strategies and tactics of players are applied: the ludic, diegetic, social, metaludic and hermeneutical frames.

Role play, says Fine (1983), consists of both the realms of discourse and of social worlds. Social worlds are the "bounded set [...] of social conventions" (p. 183) which, according to Goffman (1974), form the constitutive frames of the experience through which the individual gives meaning to their perceptions. Goffman distinguishes two types of framework, namely primary and secondary frameworks. The primary frameworks pertain to everyday experiences. They allow individuals to make sense of events, to place them, classify them, etc… On the one hand, there are the social frameworks, which take place in ordinary interpersonal relationships, and on the other, there are the natural frameworks, which refer to natural forces. The secondary frameworks include, meanwhile, keyed primary frameworks. For example, murder falls within the primary framework. When simulated on the stage of a theater, it is keyed into a secondary framework. The keys are fivefold: make-believe (e.g. fantasies, theater), contests, ceremonials, technical redoings (e.g. fire drills) and regroundings (e.g. the carnival). Role play appears, therefore, as make-believe.

When we defined the PRPG-VE (Duret, 2014), we carefully discriminated many of their components: the social organization, the formal structure and the game world. These three components require players to shift between several frames during the game to understand them. These frames reflect both the gaming experience in the PRPG-VE and that which is experienced during the events of extrinsic play related to the PRPG-VE. Drawing inspiration from Fine (1983) and Harviainen (2009) and enhancing their analyses of role play games in terms of frame analysis, we distinguish three frames that echo these components, in the primary framework of the game as a social organization and the ludic and diegetic frames as well as two additional primary frameworks, or meta-ludic and hermeneutic frames. All five are constitutive of the videoludic experience in situations of intrinsic play and of extrinsic play.

As a social organization, the primary framework of the game is the frame in which the players are considered as individuals. This frame allows the players to consider the experience of past situations in which they have placed any frame constitutive of intrinsic play in parentheses, situations where social interaction of the players takes place as opposed to interaction between the characters. For example, two individuals who flirt in "OOC" ("out of character") mode are part of such a frame.

The ludic frame corresponds to the experience of the player through their interactions with the formal structure of the game. This reflects the mental schemes used by the participant in order to interpret in formal terms the situations and events which happened in the PRPG-VE. In this context, a sword injury received to the shoulder by a character is not interpreted as a mutilation or war injury, but rather as the loss of X hit points.

The metaludic frame is that which is used by the players in their interactions with other players in order to regulate, comment, apply or modify the formal structure of the PRPG-VE. The discussions in the forums dedicated to Gorean role-playing games demonstrate the importance of metaludic activity. This is where disputes between players are settled, where they ardently defend their style of play (mimetic, optimizer-agonistic[10]), look for solutions to settle technical problems, etc... This framework is also applied in game situations when two players negotiate the application of a rule they interpret differently.

The diegetic frame corresponds to the experience of the player as a character in their interactions with the game world and the constitutive events. In this frame, the player interprets situations and events that occur in the game in diegetic terms. For example, in a fantasy role-playing game hosted on *Second Life*, a participant may in diegetic terms interpret the intrusion of a non-player user entering the virtual environment whose avatar is dressed as a twenty-first century person as a divine apparition or as someone attending a carnival.

The hermeneutical (or metatextual) frame is twofold. On the one hand, it allows the player to enter the videoludic experience from a holistic perspective, in other words, make it coherent as a complete multifaceted experience that falls into different frames, an experience which can be talked about later through narratives. This sequencing of the videoludic experience goes through the process of emplotment (*mise en intrigue*)[11]. Furthermore, this frame makes it possible to account for the dialectical relationship between the player seen as an individual and the PRPG-VE seen simultaneously as a social organization, formal structure and game world (1), in terms of migration of meaning, or, what an individual brings to the game (a view of the world, attitudes and values, etc..) (2), and what the game brings to the individual as a text within Ricœurian hermeneutics (Ricoeur, 1986) (3).

These five frames are the theater of strategies and tactics of the players, which can be classified into one of six types of strategies and tactics: institutionalized enculturation, informal enculturation, hermeneutical closure, sociotechnical closure, narrative closure and legal closure. These are inspired by the concepts of "discursive closure", Deetz (1991) and "preferred meaning", Hall (1980; 1989).

The concept of "discursive closure" reflects the methods used in a discursive field given in order to silence or eliminate the conflicts and contradictions. As a concept of "preferred meaning" it refers to the fact that the significance is "polysemic in its intrinsic nature" and that "[i]t is caught in and constituted by the struggle to 'prefer' one among many meanings as the dominant" (Hall 1989, p. 47). If the producer of media content encodes a program to give it a preferred interpretation, the inherent polysemy of the signs does not guarantee that its reception, or its reading, will be concordant with the dominant meaning. What is retained, therefore, of the connection between these two concepts and those of "strategy" and "tactic" in the context of Gorean role-playing games, is that the doctrine of natural order conveyed by *The Chronicles of Gor* and adopted by some players is the subject of an oppositional reading, that there is conflict within the Gorean community to impose a particular reading of this doctrine and that this imposition, far from being frontal, goes through subtle processes that aim to define the meanings related to it, by reducing the polysemy of *The Chronicles of Gor* to extract a monosemic reading.

Now we examine the six types of strategies and tactics mentioned above:

Institutionalized enculturation and informal enculturation: enculturation is the process by which an individual gradually acquires the characteristics and norms of a group or culture (Oxford Dictionary, 2013). In the Gorean community, that means attempts on the part of some members of the various constitutive groups ("BtB", "GE", "lifestylers", "BDSM", etc…) to educate players to make them adopt their interpretative stance on *The Chronicles of Gor* and / or the Gor game world. The ways of acquiring these characteristics and standards pass through either a formal structure (institutionalized), or an informal structure (*ad hoc*).

Institutionalized enculturation refers to the various schools created in the Gorean "sims" themselves for use by the players so they can learn the habits, customs and norms of a particular group. For example, the *Gorean University of Scrolls*, closed in 2011, specialized in "BtB" education of the players (characteristics specific to different castes of Gorean society, particularities of free companionship relationships[12], etc...). Most institutions aim to educate slaves to their role. Thus, the *Gorean Pleasure Silk University* and the *Mur Slave Academy* specialize in teaching the role of slave. This directly concerns the doctrine of natural order, since here players learn submission to a master, the rules to be obeyed, the positions and behaviors to adopt, the techniques to serve their master, quotes from the Norman novels that enlighten the role of a slave, the philosophy to be adopted by a slave in order to fulfill their role, etc…

Informal enculturation, meanwhile, refers to attempts on the part of some players to educate other players in game situations or in the Gorean forums. For example, Kaitlin, a "BtB" player, in order to preserve a literal reading of *The Chronicles of Gor*, aims to train players who have either read little or nothing of the books by illustrating "in character" situations. Thus, she writes: "How do we bring back Gor? Educate people on what is really inside those pages through observable RP since we know most can't be bothered to pick one up (The Gorean Forums, 2012)."Another player, lil.niosaki, demonstrates the efficiency of this process by writing: "When I first started Gor… I had some amazing teachers... who showed me the ropes by their amazing RPs... they were out there every day RPing and including others in their RP that it become infectious" (The Gorean Forums, 2012). Finally, Babybear Serenity, rather than trying to change the habits of players in game situations, opted for the publication of an educational post on a forum. Here, the aim is to ensure that women not subjected to slavery, (the free women), wear an opaque veil to hide their hair and face, according to the Gorean custom:

I recently found myself in a gorean city where almost every free woman, high caste and low caste were wearing sheer veils. I did my best to 'deal with it in RP', but the insults fell on deaf ears. […] I fear that in this instance I wasn't confident there was much I could do to give them education without creating drama […] I then decided to form an educational post on my House of Seginus forum. I could perhaps direct people (Gor-SL, 2011b).

Institutionalized enculturation is about strategy, since trainers define and enforce the rules that trainees must follow and also manage the space in which these courses are held. Informal enculturation is an instrument of preferred closure for tacticians, since it can occur in areas that they do not own. However, it can also form part of the strategies for administrators and moderators of Gorean "sims".

Hermeneutical closure describes the exegetical work of players in the framework of extrinsic play, activities from which a personal interpretation of the novels of Norman is put forward. These exegeses are posted on forums, blogs, notecards and fanzines. Some are presented in the form of online encyclopedias or take the form of collections of quotes of excerpts from *The Chronicles of Gor* and classified by themes. Their influence on Gorean role-playing games is extensive, since many players, for the most part "GE", have read little or none of *The Chronicles of Gor* novels and they rely on these exegeses to build their character and their role play. They unwittingly adopt the perspective of the authors of these exegesis, without being able to develop their own interpretation.

Here is an example of such exegeses: when players who want to play a female warrior legitimize the existence of such roles in *The Chronicles of Gor*, they usually refer to the character of Tarna, a woman placed at the head of an army in the *Tribesmen of Gor* novel. A "GE" reader questions the relevance of the law of natural order since Tarna illustrates that it is possible for a woman to fight alongside men. In contrast, according to the work of exegesis presented on the blog *The Tahari Desert of Gor* (Sheraka, 2009) and the 15th essay in *Luther's Gorean Essays* (n.d.), Tarna rather reinforces the law of natural order, as the reason why she is at the head of an army is that she is manipulated by her husband, a warlord who uses her for political reasons: to humiliate his opponents, dishonoring them and forcing them to fight an army led by a woman, etc…).

Hermeneutical closure is independent of the power hierarchy that divides the players, since it exerts itself outside the Gorean role-playing games. It can therefore be strategic and tactical.

Sociotechnical closure describes, meanwhile, the interpretations of the work of Norman embodied in the technical devices of Gorean sims. In the context of Gorean role-playing games, these devices are software and pass through a procedural rhetoric (Bogost, 2007). The procedural rhetoric allows a video game designer to implement dynamic models with the aid of rules such that the designer can express their ideas and use this as a way to convince the players. It is particularly useful in illustrating conceptual order systems because of its symbolic nature. Frasca (2003), meanwhile, talks about simulation rhetoric, since it is through simulation that the point of view of the author is conveyed. The author cites the example of the game *The Sims*, in which the rules allow kissing between members of the same sex. Homosexuality, in these conditions, is an avenue that the character can take and which is opened up by the designer of the game.

Sociotechnical closure thus appears to be a way of enclosing the meaning of a work or of putting forward a doctrine by allowing players to take certain actions that support a specific interpretation or a particular doctrine, or preventing contradictory actions. For example, as we saw earlier, combat scripts developed for the Gorean role-playing games allow women to fight and the damage they inflict is equal to that inflicted by male combatants. These scripts question the law of natural order by making a female combatant equal to that of a male combatant. This is a strategic action in that, firstly, it is the producers who parameterize the devices and not their users and, secondly, it is the administrators/moderators who decide whether or not to accept the use of these devices in their "sim".

Legal closure concerns the rules that are not implemented in the game software that govern the Gorean sims and which support a point of view regarding the doctrine of natural order. For example, the rules governing the Gorean sim *The Great Tahiri* stipulate that women cannot carry weapons except for a short dagger which must remain hidden, that raids by the Panthers are prohibited, as are female outlaws, mercenaries and pirates (regulations 2.e and 5.a). The dress code for free women is also regulated. They must thus be "fully covered, veiled. No skin and hair showing. Misbehavior will result in a collar[13]" (regulation 5.d).

Finally, *narrative closure* refers to the emplotment of the events that occurred during a game session. These events are subject to selection on the part of the players / authors and are presented according to their own perspective. Therefore, the narratives that result from this selection and this perspective carry their attitudes and values which underpin the interpretive posture of the players. Moreover, the game narratives can reintegrate into the diegesis of the game, if only unofficially, events that occurred in a game situation that administrators and moderators may have invalidated because they resulted in behavior contrary to the regulations of their "sim". However, these regulations, as I mentioned, also carry a perspective on the doctrine of natural order. In doing so, the reintegration of invalidated events represents a tactic for players without power over the definition and application of the rules, a tactic that allows the diffusion of their interpretive posture. Note, however, that narrative closure is an instrument that can also be employed by the strategists, since, as in the case of hermeneutical closure, it takes place outside the Gorean sims, where the hierarchy of players no longer exists. These game narratives may be scriptural, but they can also be represented in the form of videos (machinima) or sequences of still images (snapshots) from the events in question.

CURRENT CHALLENGES

As I have demonstrated in this chapter, preferred meanings can be put forward in the context of PRPG-VE and the polysemy of the text can be reduced by using strategies and tactics. However, because of the social and participatory nature of the PRPG-VE, it is difficult to envisage that a doctrine, unambiguous as it is, can be conveyed without being questioned, challenged or subverted, following the specific beliefs, attitudes and values of each player. Of course, all the strategies and tactics highlighted here are tools available to game designers who want to convey messages in a non-frontal manner. Moreover, these tools can be effective in a non-participatory context when the designer is solely responsible for determining the content, the rules and the simulation, since it is the ability of the players to put forward their views in a gaming framework that makes the exposure of a persuasive message problematic.

SOLUTIONS AND RECOMMENDATIONS

While it appears that PRPG-VE do not provide an adequate medium for persuasion, they can however serve as a case study for critical media education (Masterman, 1993) (Piette, 1996; 2006) in an educational framework, following the approach of the "non-transparency" of the media (Masterman, 1993). The latter considers the constructed nature of reality conveyed by the media and its influence on the representation of reality and social imagination. Following Piette (1996), it is thus possible to approach the study of the media from the angle of production, persuasion (professional rhetoric, ideologies, values and ideals conveyed in media content) and from the point of view of the reception of media messages. Using PRPG-VE for illustrative purposes would therefore allow the implementation of a program of critical education aimed at the videoludic medium.

ACKNOWLEDGMENT

The research on which this article is based has received financial support from the Social Sciences and Humanities Research Council (SSHRC) and the *Fonds québécois de recherche Société et culture* – FQRSC- (Quebec Fund for Research, Society and Culture).

REFERENCES

Aarseth, E. J. (1997). *Cybertext: Perspectives on Ergodic Literature*. Baltimore, MD: Johns Hopkins University Press.

Acculturation. (2013). In *Oxford dictionary*. Retrieved July 20, 2013, from http://oxforddictionaries.com/definition/english/acculturate?q=acculturation#acculturate__8

Angenot, M. (1986). *Le cru et le faisandé: Sexe, discours social et littérature à la Belle époque*. Bruxelles: Éditions Labor.

Angenot, M. (1989). *1889, Un état du discours social*. Longueuil: Le Préambule.

Bogost, I. (2007). *Persuasive games: The expressive power of videogames*. Cambridge, MA: The MIT Press.

Certeau, M. (Ed.). (1990). L'invention du quotidien: Vol. 1. *Arts de faire*. Paris: Éditions Gallimard.

CRIST. (n.d.). *Manifeste*. Retrieved October 20, 2011, from http://www.site.sociocritique-crist.org/p/manifeste.html

Cros, E. (2003). *La sociocritique*. Paris: L'Harmattan.

Cros, E. (2006). Spécificités de la sociocritique. *La sociocritique d'Edmond Cros*. Retrieved April 25, 2012, from http://sociocritique.fr/spip.php?article6

Deetz, S. A. (1991). *Democracy in an age of corporate colonization: Developments in communication and the politics of everyday life*. New York: State University of New York Press.

Duchet, C. (1971). Pour une socio-critique ou variations sur un incipit. *Littérature*, (1), 5-14.

Duchet, C. (1979). Positions et perspective. In C. Duchet (Ed.), *Sociocritique* (pp. 3–8). Paris: Nathan.

Duchet, C. (1986). La manœuvre du bélier: Texte, intertexte et idéologies dans L'Espoir. *Revue des Sciences Humaines*, (204): 107–131.

Duchet, C., Herschberg Pierrot, A., & Neefs, J. (1994). Sociocritique et génétique, entretien avec Anne Herschberg Pierrot et Jacques Neefs. *Genesis, Manuscrits, Recherche, Invention*, (6), 117-127.

Duchet, C., & Maurus, P. (2011). *Un cheminement vagabond: Nouveaux entretiens sur la sociocritique*. Paris: Honoré Champion.

Duret, C. (2013). Jeux de rôle participatifs en environnement virtuel et herméneutique conflictuelle: L'expérience ludique des joueurs dans les jeux de rôle goréens. *Communication. Lettres et Sciences du Langage*, *7*(1), 4–20.

Duret, C. (2014). *Les jeux de rôle participatifs en environnement virtuel: Définition et enjeux théoriques*. (Master's thesis). Université de Sherbrooke, Sherbrooke, Canada.

Encyclopaedia Universalis. (2013). *Acculturation*. Retrieved July 25, 2013, from http://www.universalis.fr/encyclopedie/acculturation/

Fine, G. A. (1983). *Shared fantasy: Role-playing games as social worlds*. Chicago: University of Chicago Press.

Fiske, J. (1987). *Television culture*. London: Routledge.

Frasca, G. (2003). Simulation versus narrative: Introduction to ludology. *Ludology.org*. Retrieved September 20, 2011, from http://www.ludology.org/articles/VGT_final.pdf

Genette, G. (1972). *Figures III*. Paris: Éditions du Seuil.

Goffman, E. (1974). *Frame analysis: An essay on the organization of experience*. New York: Harper & Row.

Gor-SL. (2011a). *Gorean women: Haven't we seen them before?* Retrieved August 31, 2012, from http://www.gor-sl.com/index.php/topic,9435.msg79290.html#msg79290

Gor-SL. (2011b). *Dear Dove*. Retrieved August 31, 2012, from http://www.gor-sl.com/index.php/topic,9896.msg83954.html#msg83954

Gorean Forums. (2012). *Bringing Gor back to Gor*. Retrieved July 25, 2012, from http://www.goreanforums.net/viewtopic.php?f=230&t=6840&start=20#p162775

Hall, S. (1980). Encoding/decoding. In S. Hall, D. Hobson, A. Love, & P. Willis (Eds.), *Culture, media, language* (pp. 128–138). London: Hutchinson.

Hall, S. (1989). Ideology and communication theory. In B. Dervin, L. Grossberg, B. O'Keefe, & E. Wartella (Eds.), *Rethinking communication I: Paradigm dialogues*. Newbury Park, CA: Sage.

Harviainen, J. T. (2009). A hermeneutical approach to role-playing analysis. *International Journal of Role-Playing*, *1*(1).

Jenkins, H. (2006a). *Convergence culture: Where old and new media collide*. New York: New York University Press.

Luther's Gorean Essays. (n.d.). *Female Warriors #15, 5.0 Version*. Retrieved July 25, 2013, from http://www.gor-now.net/delphius2002/id31.htm

Macherey, P. (2005). *Michel de Certeau et la mystique du quotidien*. Retrieved July 25, 2013, from http://stl.recherche.univ-lille3.fr/seminaires/philosophie/macherey/macherey20042005/macherey06042005.html

Maingueneau, D. (1976). *Initiation aux méthodes de l'analyse du discours*. Paris: Hachette.

Masterman, L. (1993). The media education revolution. *Canadian Journal of Educational Communication*, *22*(1), 5–14.

Millerand, F., Proulx, S., & Rueff, J. (2010). Introduction. In F. Millerand, S. Proulx, & J. Rueff (Eds.), *Web social: Mutation de la communication* (pp. 15–32). Québec, Canada: Presses de l'Université du Québec.

Montola, M. (2003). Role-playing as interactive construction of subjective diegeses. In M. Gaden, L. Thorup, & M. Sander (Eds.), As LARP grows up: The book from knudepunkt (pp. 82-89). Frederiksberg: Projektgruppen KP03.

Murray, J. (1998). *Hamlet on the holodeck*. Cambridge, MA: The MIT Press.

Padol, L. (1996). *Playing stories, telling games: Collaborative storytelling in role-playing games*. Retrieved April 25, 2013, from http://www.recappub.com/games.html

Piette, J. (1996). *Éducation aux médias et fonction critique*. Montreal, Canada: L'Harmattan.

Piette, J. (2006). La démarche d'enseignement en éducation aux médias. *Revue Vie pédagogique*, (140), 1-5.

Ricoeur, P. (1983). Temps et récit: Vol. 1. *L'intrigue et le récit historique*. Paris: Éditions du Seuil.

Ricoeur, P. (1986). *Du texte à l'action: Essais d'herméneutique II*. Paris: Éditions du Seuil.

Robin, R. (1993). Pour une socio-poétique de l'imaginaire social. *Discours social/Social Discourse, 5* (1-2), 7-32.

Second Life Wiki. (2009). *Category: LSL Script/fr*. Retrieved July 30, 2013, from http://wiki.secondlife.com/wiki/

Sheraka. (2009). Who is Tarna? *The Tahari Desert of Gor*. Retrieved July 25, 2013, from http://tahari.wordpress.com/2009/09/30/who-is-tarna

Siang Ang, C., Zaphiris, P., & Wilson, S. (2010). Computer games and sociocultural play: An activity theoretical perspective. *Games and Culture*, *5*(4), 354–380. doi:10.1177/1555412009360411

Sixma, T. (2009). The Gorean community in second life: Rules of sexual inspired role-play. *Journal of Virtual Worlds Research*, *1*(3).

Suleiman, S. R. (1983). *Authoritarian fictions: The ideological novel as a literary genre*. New York: Columbia University Press.

KEY TERMS AND DEFINITIONS

Actualized Text (*cf.* Virtual Text): Concrete materializations of the virtual text. Constitutive of the game world.

MUVE: Online, multi-user virtual environments (e.g. *Second Life, IMVU*).

Participatory Role-Playing Games in Virtual Environment (PRPG-VE): Multiplayer role playing games in which players take on a role through an avatar in multi-user virtual environments (MUVE) which may or may not have been specifically designed for such activities. The PRPG-VE are representative of the participatory culture, as the players contribute to the development of the avatar scripts and accessories as well as the game mechanics, the virtual environment, the rules of the game and the game world of the role-play.

Sociogram: "Unclear, unstable, conflictual set of partial representations centered around a core of mutual interaction" (Duchet, quoted in Duchet, Herschberg Pierrot, & Neefs, 1994, p. 118).

Strategy (*cf.* Tactics): Employed by the dominant order, which includes the producers of the culture industry. It involves a calculation of the balance of power, a vantage point from which to manage uncertainty and the movements of opponents, as well as a fixed operations base to build on the achievements.

Tactics (*cf.* Strategy): Used by the dominated, the consumers and users of media content that have neither independence nor their own space. They are forced to act in the other's territory according to externally imposed laws.

Virtual Text (*cf.* Actualized Text): The sum of possible texts or texts that have not yet occurred. This textual level governs the simulation. It constitutes a field of possibilities structured, on the one hand by all the relevant social, formal and diegetic game system factors that program the actualized texts and, on the other hand by the meanings and motivations that the players, both as players and as individuals, bring with them (e.g. attitudes, values, needs).

ENDNOTES

1. Translation by Brian Cleveland.
2. Siang Ang, Zaphiris and Wilson (2010) make the distinction between intrinsic play and extrinsic play. Intrinsic play is contained within the predetermined limits of the game structure, while extrinsic play extends beyond these limits; it is manifested, for example, in forums where players share tips to improve their game performance.
3. In *Second Life*, the scripts are lines of programming that achieve certain effects: changing the appearance of an object, how it moves, its interaction and exchanges with avatars, etc... (Second Life Wiki, 2009).
4. "The diegesis" writes Genette (1972), "is the spatiotemporal universe designated by the narrative" (p. 280) or, in the context of role play, the space-time universe in which the characters move. Montola (2003) defines the diegesis as the amount of information in the game world, the laws that govern its fictional reality, the verbalizations of the players and game masters during the game, as well as their thoughts, their emotions and their actions. Diegesis thus represents the elements of the game world.
5. A "sim" or "simulator" in *Second Life* is a three-dimensional virtual space hosted on a server. The "sims" are leased to users who may develop and administer them.
6. The administrators/moderators in the context of the PRPG-VE.
7. The term "sociality" falls within the definition of social and historical order in the framework of sociocritic (e.g. CRIST, n.d.; Cros, 2003; Duchet, 1971; 1979; 1986; Duchet & Maurus, 2011; Robin, 1993).
8. Maingueneau (1976) describes intertextuality as a "set of relationships with other texts manifesting within a text."
9. In the hermeneutic scope of Gorean role-playing games: Duret (2013).
10. See Duret (2014)
11. Emplotment is a way to order and understand the experience which, without its assistance, is fundamentally discordant. This experience is ordered in stories (fiction, historical chronicles, articles, etc...) through the process of triple *mimesis* (Ricoeur, 1983).
12. Gorean equivalent of marriage.
13. To be "collared", in the Gorean context, means the fitting of a collar that marks enslavement.

Related References

To continue our tradition of advancing information science and technology research, we have compiled a list of recommended IGI Global readings. These references will provide additional information and guidance to further enrich your knowledge and assist you with your own research and future publications.

Aboolian, R., Sun, Y., & Leu, J. (2012). Deploying a Zone-Based Massively Multiplayer Online Game on a Congested Network. [IJISSCM]. *International Journal of Information Systems and Supply Chain Management*, *5*(1), 38–57. doi:10.4018/jisscm.2012010103

Abrams, S. S. (2013). The Dynamics of Video Gaming: Influences Affecting Game Play and Learning. In I. Management Association (Ed.), Digital Literacy: Concepts, Methodologies, Tools, and Applications (pp. 684-697). Hershey, PA: Information Science Reference. doi:10.4018/978-1-4666-1852-7.ch035

Adamo-Villani, N., & Dib, H. (2014). Online Virtual Learning Environments: A Review of Two Projects. [IJSSOE]. *International Journal of Systems and Service-Oriented Engineering*, *4*(1), 1–20. doi:10.4018/ijssoe.2014010101

Adams, S. S., & Holden, J. (2011). Games, Ethics and Engagement: Potential Consequences of Civic-Minded Game Design and Gameplay. In K. Schrier, & D. Gibson (Eds.), *Designing Games for Ethics: Models, Techniques and Frameworks* (pp. 291–311). Hershey, PA: Information Science Reference.

Ahamer, G. (2012). A Four-Dimensional Maxwell Equation for Social Processes in Web-Based Learning and Teaching: Windrose Dynamics as GIS (Games' Intrinsic Spaces). [IJWLTT]. *International Journal of Web-Based Learning and Teaching Technologies*, *7*(3), 1–19. doi:10.4018/jwltt.2012070101

Ahamer, G. (2012). The Web-Supported Negotiation Game "Surfing Global Change": Rules, History and Experiences. [IJOPCD]. *International Journal of Online Pedagogy and Course Design*, *2*(2), 60–85. doi:10.4018/ijopcd.2012040105

Ahamer, G. (2013). Joyfully Map Social Dynamics when Designing Web-Based Courses. [IJWLTT]. *International Journal of Web-Based Learning and Teaching Technologies*, *8*(2), 19–57. doi:10.4018/jwltt.2013040102

Akilli, G. K. (2014). Games and Simulations: A New Approach in Education? In J. Bishop (Ed.), *Gamification for Human Factors Integration: Social, Education, and Psychological Issues* (pp. 272–289). Hershey, PA: Information Science Reference.

Aldrich, C., & DiPietro, J. C. (2011). An Overview of Gaming Terminology: Chapters I – LXXVI. In I. Management Association (Ed.), Gaming and Simulations: Concepts, Methodologies, Tools and Applications (pp. 24-44). Hershey, PA: Information Science Reference. doi:10.4018/978-1-60960-195-9.ch102

All, A., Van Looy, J., & Castellar, E. P. (2013). An Evaluation of the Added Value of Co-Design in the Development of an Educational Game for Road Safety. [IJGBL]. *International Journal of Game-Based Learning*, *3*(1), 1–17. doi:10.4018/ijgbl.2013010101

Anagnostou, K. (2011). How has the Internet Evolved the Videogame Medium? In M. Cruz-Cunha, V. Varvalho, & P. Tavares (Eds.), *Business, Technological, and Social Dimensions of Computer Games: Multidisciplinary Developments* (pp. 448–462). Hershey, PA: Information Science Reference.

Anagnostou, K. (2011). Research Note: Narration vs. Simulation. [IJGCMS]. *International Journal of Gaming and Computer-Mediated Simulations*, *3*(2), 67–77. doi:10.4018/jgcms.2011040105

Anagnostou, K., & Pappa, A. (2011). Video Game Genre Affordances for Physics Education. [IJGBL]. *International Journal of Game-Based Learning*, *1*(1), 59–74. doi:10.4018/ijgbl.2011010105

Anderson, J. L. (2014). Games and the Development of Students' Civic Engagement and Ecological Stewardship. In J. Bishop (Ed.), *Gamification for Human Factors Integration: Social, Education, and Psychological Issues* (pp. 199–215). Hershey, PA: Information Science Reference.

Andrews, S. S., Stokrocki, M., Jannasch-Pennell, A., & DiGangi, S. A. (2012). The Development of a Personal Learning Environment in Second Life. In M. Thomas (Ed.), *Design, Implementation, and Evaluation of Virtual Learning Environments* (pp. 219–236). Hershey, PA: Information Science Reference.

Annetta, L. A., Holmes, S., & Cheng, M. (2012). Measuring Student Perceptions: Designing an Evidenced Centered Activity Model for a Serious Educational Game Development Software. In R. Ferdig, & S. de Freitas (Eds.), *Interdisciplinary Advancements in Gaming, Simulations and Virtual Environments: Emerging Trends* (pp. 165–182). Hershey, PA: Information Science Reference.

April, K. A., Goebel, K. M., Blass, E., & Foster-Pedley, J. (2012). Developing Decision-Making Skill: Experiential Learning in Computer Games. [IJISSC]. *International Journal of Information Systems and Social Change*, *3*(4), 1–17. doi:10.4018/jissc.2012100101

Aranda, D., & Sánchez-Navarro, J. (2011). How Digital Gaming Enhances Non-Formal and Informal Learning. In P. Felicia (Ed.), *Handbook of Research on Improving Learning and Motivation through Educational Games: Multidisciplinary Approaches* (pp. 395–412). Hershey, PA: Information Science Reference.

Ardito, C., & Lanzilotti, R. (2011). An EUD Approach to the Design of Educational Games. [IJDET]. *International Journal of Distance Education Technologies*, *9*(4), 25–40. doi:10.4018/ijdet.2011100103

Arora, P., & Itu, S. (2012). Arm Chair Activism: Serious Games usage by INGOs for Educational Change. [IJGBL]. *International Journal of Game-Based Learning*, *2*(4), 1–17. doi:10.4018/ijgbl.2012100101

Arriaga, P., Esteves, F., & Fernandes, S. (2013). Playing for Better or for Worse?: Health and Social Outcomes with Electronic Gaming. In M. Cruz-Cunha, I. Miranda, & P. Gonçalves (Eds.), *Handbook of Research on ICTs for Human-Centered Healthcare and Social Care Services* (pp. 48–69). Hershey, PA: Medical Information Science Reference.

Arroyo-Palacios, J., & Romano, D. M. (2012). Bio-Affective Computer Interface for Game Interaction. In R. Ferdig, & S. de Freitas (Eds.), *Interdisciplinary Advancements in Gaming, Simulations and Virtual Environments: Emerging Trends* (pp. 249–265). Hershey, PA: Information Science Reference.

Arya, A., Chastine, J., Preston, J., & Fowler, A. (2013). An International Study on Learning and Process Choices in the Global Game Jam. [IJGBL]. *International Journal of Game-Based Learning*, *3*(4), 27–46. doi:10.4018/ijgbl.2013100103

Asbell-Clarke, J., Edwards, T., Rowe, E., Larsen, J., Sylvan, E., & Hewitt, J. (2012). Martian Boneyards: Scientific Inquiry in an MMO Game. [IJGBL]. *International Journal of Game-Based Learning*, *2*(1), 52–76. doi:10.4018/ijgbl.2012010104

Aubrecht, M. (2012). Games in E-learning: How Games Teach and How Teachers Can Use Them. In H. Wang (Ed.), *Interactivity in E-Learning: Case Studies and Frameworks* (pp. 179–209). Hershey, PA: Information Science Reference.

Ausburn, L. J. (2012). Learner Characteristics and Performance in a First-Person Online Desktop Virtual Environment. [IJOPCD]. *International Journal of Online Pedagogy and Course Design*, *2*(2), 11–24. doi:10.4018/ijopcd.2012040102

Ausburn, L. J., Ausburn, F. B., & Kroutter, P. J. (2013). Influences of Gender and Computer Gaming Experience in Occupational Desktop Virtual Environments: A Cross-Case Analysis Study. [IJAVET]. *International Journal of Adult Vocational Education and Technology*, *4*(4), 1–14. doi:10.4018/ijavet.2013100101

Bachvarova, Y., & Bocconi, S. (2014). Games and Social Networks. In T. Connolly, T. Hainey, E. Boyle, G. Baxter, & P. Moreno-Ger (Eds.), *Psychology, Pedagogy, and Assessment in Serious Games* (pp. 204–219). Hershey, PA: Information Science Reference.

Badman, C., & DeNote, M. (2013). Are Wii Having Fun Yet? In Y. Baek, & N. Whitton (Eds.), *Cases on Digital Game-Based Learning: Methods, Models, and Strategies* (pp. 25–49). Hershey, PA: Information Science Reference.

Baek, Y. (2011). Principles of Educational Digital Game Structure for Classroom Settings. In I. Management Association (Ed.), Gaming and Simulations: Concepts, Methodologies, Tools and Applications (pp. 229-239). Hershey, PA: Information Science Reference. doi:10.4018/978-1-60960-195-9.ch113

Baggio, B., & Beldarrain, Y. (2011). Groups, Games & Community. In Anonymity and Learning in Digitally Mediated Communications: Authenticity and Trust in Cyber Education (pp. 262-286). Hershey, PA: Information Science Reference. doi:10.4018/978-1-60960-543-8.ch012

Related References

Balzert, S., Pannese, L., Walter, M., & Loos, P. (2012). Serious Games in Business. In M. Cruz-Cunha (Ed.), *Handbook of Research on Serious Games as Educational, Business and Research Tools* (pp. 539–558). Hershey, PA: Information Science Reference.

Barab, S. A., Gresalfi, M., Dodge, T., & Ingram-Goble, A. (2012). Narratizing Disciplines and Disciplinizing Narratives: Games as 21st Century Curriculum. In R. Ferdig, & S. de Freitas (Eds.), *Interdisciplinary Advancements in Gaming, Simulations and Virtual Environments: Emerging Trends* (pp. 18–31). Hershey, PA: Information Science Reference.

Barlow, M. (2011). Game Led HCI Improvements. In M. Cruz-Cunha, V. Varvalho, & P. Tavares (Eds.), *Business, Technological, and Social Dimensions of Computer Games: Multidisciplinary Developments* (pp. 126–145). Hershey, PA: Information Science Reference.

Bates, M. I., Brown, D., Cranton, W., & Lewis, J. (2011). Formulating a Serious-Games Design Project for Adult Offenders with the Probation Service. [IJGBL]. *International Journal of Game-Based Learning*, *1*(4), 1–10. doi:10.4018/ijgbl.2011100101

Becker, K. (2011). The Magic Bullet: A Tool for Assessing and Evaluating Learning Potential in Games. [IJGBL]. *International Journal of Game-Based Learning*, *1*(1), 19–31. doi:10.4018/ijgbl.2011010102

Bekebrede, G., Harteveld, C., Warmelink, H., & Meijer, S. (2013). Beauty or the Beast: Importance of the Attraction of Educational Games. In C. Gonzalez (Ed.), *Student Usability in Educational Software and Games: Improving Experiences* (pp. 138–160). Hershey, PA: Information Science Reference.

Benson, P., & Chik, A. (2011). Towards a More Naturalistic CALL: Video Gaming and Language Learning. [IJCALLT]. *International Journal of Computer-Assisted Language Learning and Teaching*, *1*(3), 1–13. doi:10.4018/ijcallt.2011070101

Benson, P., & Chik, A. (2012). Towards a More Naturalistic CALL: Video Gaming and Language Learning. In B. Zou (Ed.), *Explorations of Language Teaching and Learning with Computational Assistance* (pp. 75–88). Hershey, PA: Information Science Reference.

Berland, M., & Lee, V. R. (2011). Collaborative Strategic Board Games as a Site for Distributed Computational Thinking. [IJGBL]. *International Journal of Game-Based Learning*, *1*(2), 65–81. doi:10.4018/ijgbl.2011040105

Bertozzi, E., Krilov, L. R., & Walker, D. (2013). Successful Game Development Partnerships between Academics and Physicians: Two Case Studies. [IJGCMS]. *International Journal of Gaming and Computer-Mediated Simulations*, *5*(3), 97–107. doi:10.4018/jgcms.2013070107

Bezio, K. M. (2014). Friends and Rivals: Loyalty, Ethics, and Leadership in BioWare's "Dragon Age II". In D. Hickey, & J. Essid (Eds.), *Identity and Leadership in Virtual Communities: Establishing Credibility and Influence* (pp. 145–169). Hershey, PA: Information Science Reference.

Bidarra, J., Rothschild, M., Squire, K., & Figueiredo, M. (2013). The AIDLET Model: A Framework for Selecting Games, Simulations and Augmented Reality Environments in Mobile Learning. [IJWLTT]. *International Journal of Web-Based Learning and Teaching Technologies*, *8*(4), 50–71. doi:10.4018/ijwltt.2013100104

Black, R. W., & Reich, S. M. (2011). Affordances and Constraints of Scaffolded Learning in a Virtual World for Young Children. [IJGBL]. *International Journal of Game-Based Learning*, *1*(2), 52–64. doi:10.4018/ijgbl.2011040104

Blasko, D. G., Lum, H. C., White, M. M., & Drabik, H. B. (2014). Individual Differences in the Enjoyment and Effectiveness of Serious Games. In T. Connolly, T. Hainey, E. Boyle, G. Baxter, & P. Moreno-Ger (Eds.), *Psychology, Pedagogy, and Assessment in Serious Games* (pp. 153–174). Hershey, PA: Information Science Reference.

Bösche, W., & Kattner, F. (2011). Fear of (Serious) Digital Games and Game-Based Learning?: Causes, Consequences and a Possible Countermeasure. [IJGBL]. *International Journal of Game-Based Learning*, *1*(3), 1–15. doi:10.4018/ijgbl.2011070101

Bottino, R. M., Ott, M., & Tavella, M. (2014). Serious Gaming at School: Reflections on Students' Performance, Engagement and Motivation. [IJGBL]. *International Journal of Game-Based Learning*, *4*(1), 21–36. doi:10.4018/IJGBL.2014010102

Boyle, E. (2014). Psychological Aspects of Serious Games. In T. Connolly, T. Hainey, E. Boyle, G. Baxter, & P. Moreno-Ger (Eds.), *Psychology, Pedagogy, and Assessment in Serious Games* (pp. 1–18). Hershey, PA: Information Science Reference.

Boyle, E., Terras, M. M., Ramsay, J., & Boyle, J. M. (2014). Executive Functions in Digital Games. In T. Connolly, T. Hainey, E. Boyle, G. Baxter, & P. Moreno-Ger (Eds.), *Psychology, Pedagogy, and Assessment in Serious Games* (pp. 19–46). Hershey, PA: Information Science Reference.

Related References

Breiter, A., & Kolo, C. (2011). Electronic Gaming in Germany as Innovation in Education. In I. Management Association (Ed.), Gaming and Simulations: Concepts, Methodologies, Tools and Applications (pp. 426-439). Hershey, PA: Information Science Reference. doi:10.4018/978-1-60960-195-9.ch207

Brown, D. (2013). Overcoming the Barriers to Uptake: A Study of 6 Danish Health-Based Serious Games Projects. [IJGBL]. *International Journal of Game-Based Learning*, *3*(3), 1–9. doi:10.4018/ijgbl.2013070101

Bruzzone, A. G., Frascio, M., Longo, F., Massei, M., & Nicoletti, L. (2013). An Innovative Serious Game for Education and Training in Health Care. [IJPHIM]. *International Journal of Privacy and Health Information Management*, *1*(2), 132–146. doi:10.4018/ijphim.2013070109

Bryant, J. A., Akerman, A., & Drell, J. (2012). Wee Wii: Preschoolers and Motion-Based Game Play. In R. Ferdig, & S. de Freitas (Eds.), *Interdisciplinary Advancements in Gaming, Simulations and Virtual Environments: Emerging Trends* (pp. 61–77). Hershey, PA: Information Science Reference.

Budiharto, W., Ricky, M. Y., & Rachmawati, R. N. (2013). The Novel Method of Adaptive Multiplayer Games for Mobile Application using Neural Networks. [IJMCMC]. *International Journal of Mobile Computing and Multimedia Communications*, *5*(1), 10–24. doi:10.4018/jmcmc.2013010102

Burton, A. M., Liu, H., Battersby, S., Brown, D., Sherkat, N., Standen, P., & Walker, M. (2011). The Use of Motion Tracking Technologies in Serious Games to Enhance Rehabilitation in Stroke Patients. [IJGBL]. *International Journal of Game-Based Learning*, *1*(4), 60–73. doi:10.4018/ijgbl.2011100106

Caperton, I. H. (2012). Toward a Theory of Game-Media Literacy: Playing and Building as Reading and Writing. In R. Ferdig, & S. de Freitas (Eds.), *Interdisciplinary Advancements in Gaming, Simulations and Virtual Environments: Emerging Trends* (pp. 1–17). Hershey, PA: Information Science Reference.

Caponetto, I., Earp, J., & Ott, M. (2013). Aspects of the Integration of Games into Educational Processes. [IJKSR]. *International Journal of Knowledge Society Research*, *4*(3), 11–21. doi:10.4018/ijksr.2013070102

Cavanaugh, C. (2011). Augmented Reality Gaming in Education for Engaged Learning. In I. Management Association (Ed.), Gaming and Simulations: Concepts, Methodologies, Tools and Applications (pp. 45-56). Hershey, PA: Information Science Reference. doi:10.4018/978-1-60960-195-9.ch103

Chaka, C. (2012). Second Life and World of Warcraft: Harnessing Presence Learning. In H. Yang, & S. Yuen (Eds.), *Handbook of Research on Practices and Outcomes in Virtual Worlds and Environments* (pp. 236–254). Hershey, PA: Information Science Reference.

Chamberlin, B., Maloney, A., Gallagher, R. R., & Garza, M. L. (2013). Active Video Games: Potential for Increased Activity, Suggestions for Use, and Guidelines for Implementation. In S. Arnab, I. Dunwell, & K. Debattista (Eds.), *Serious Games for Healthcare: Applications and Implications* (pp. 191–212). Hershey, PA: Medical Information Science Reference.

Chamberlin, B., Trespalacios, J., & Gallagher, R. (2012). The Learning Games Design Model: Immersion, Collaboration, and Outcomes-Driven Development. [IJGBL]. *International Journal of Game-Based Learning*, *2*(3), 87–110. doi:10.4018/ijgbl.2012070106

Chang, C., & Chin, Y. (2011). Predicting the Usage Intention of Social Network Games: An Intrinsic-Extrinsic Motivation Theory Perspective. [IJOM]. *International Journal of Online Marketing*, *1*(3), 29–37. doi:10.4018/ijom.2011070103

Charij, S., & Oikonomou, A. (2013). Using Biometric Measurement in Real-Time as a Sympathetic System in Computer Games. [IJGBL]. *International Journal of Game-Based Learning*, *3*(3), 21–42. doi:10.4018/ijgbl.2013070103

Charlier, N., & De Fraine, B. (2012). Game-Based Learning in Teacher Education: A Strategy to Integrate Digital Games into Secondary Schools. [IJGBL]. *International Journal of Game-Based Learning*, *2*(2), 1–12. doi:10.4018/ijgbl.2012040101

Charnock, D., & Standen, P. J. (2013). Second-Hand Masculinity: Do Boys with Intellectual Disabilities Use Computer Games as Part of Gender Practice? [IJGBL]. *International Journal of Game-Based Learning*, *3*(3), 43–53. doi:10.4018/ijgbl.2013070104

Chen, K., Chen, J. V., & Ross, W. H. (2012). Antecedents of Online Game Dependency: The Implications of Multimedia Realism and Uses and Gratifications Theory. In K. Siau (Ed.), *Cross-Disciplinary Models and Applications of Database Management: Advancing Approaches* (pp. 176–208). Hershey, PA: Information Science Reference.

Cheney, A. W., & Bronack, S. C. (2011). Presence Pedagogy as Framework for Research in Virtual Environments. [IJGCMS]. *International Journal of Gaming and Computer-Mediated Simulations*, *3*(1), 79–85. doi:10.4018/jgcms.2011010106

Chib, A. (2011). Promoting Sexual Health Education via Gaming: Evidence from the Barrios of Lima, Peru. In P. Felicia (Ed.), *Handbook of Research on Improving Learning and Motivation through Educational Games: Multidisciplinary Approaches* (pp. 895–912). Hershey, PA: Information Science Reference.

Coelho, H. (2011). Computer Games and Libraries. In M. Cruz-Cunha, V. Varvalho, & P. Tavares (Eds.), *Computer Games as Educational and Management Tools: Uses and Approaches* (pp. 52–66). Hershey, PA: Information Science Reference.

Colman, J., & Gnanayutham, P. (2014). Assistive Technologies for Brain-Injured Gamers. In G. Kouroupetroglou (Ed.), *Assistive Technologies and Computer Access for Motor Disabilities* (pp. 28–56). Hershey, PA: Medical Information Science Reference.

Connolly, T., & Makriyannis, E. (2012). OERopoly: Collaborative Learning about Open Educational Resources through Game-Playing. In A. Okada, T. Connolly, & P. Scott (Eds.), *Collaborative Learning 2.0: Open Educational Resources* (pp. 391–409). Hershey, PA: Information Science Reference.

Connolly, T., & Stansfield, M. (2011). From E-Learning to Games-Based E-Learning. In I. Management Association (Ed.), Gaming and Simulations: Concepts, Methodologies, Tools and Applications (pp. 1763-1773). Hershey, PA: Information Science Reference. doi:10.4018/978-1-60960-195-9.ch706

Correia, P., & Carrasco, P. (2012). Serious Games for Serious Business: Improving Management Processes. In M. Cruz-Cunha (Ed.), *Handbook of Research on Serious Games as Educational, Business and Research Tools* (pp. 598–614). Hershey, PA: Information Science Reference.

Cox, A. D., Eno, C. A., & Guadagno, R. E. (2012). Beauty in the Background: A Content Analysis of Females in Interactive Digital Games. [IJICST]. *International Journal of Interactive Communication Systems and Technologies*, *2*(2), 49–62. doi:10.4018/ijicst.2012070104

Crespo, S., Melfi, V., Fisch, S. M., Lesh, R. A., & Motoki, E. (2011). Television, Games, and Mathematics: Effects of Children's Interactions with Multiple Media. [IJGCMS]. *International Journal of Gaming and Computer-Mediated Simulations*, *3*(3), 1–18. doi:10.4018/jgcms.2011070101

Cuccurullo, S., Francese, R., Passero, I., & Tortora, G. (2013). A 3D Serious City Building Game on Waste Disposal. [IJDET]. *International Journal of Distance Education Technologies*, *11*(4), 112–135. doi:10.4018/ijdet.2013100108

Cummings, J. J., & Ross, T. L. (2011). Optimizing the Psychological Benefits of Choice: Information Transparency and Heuristic Use in Game Environments. [IJGCMS]. *International Journal of Gaming and Computer-Mediated Simulations*, *3*(3), 19–33. doi:10.4018/jgcms.2011070102

DaCosta, B., Nasah, A., Kinsell, C., & Seok, S. (2011). Digital Propensity: An Investigation of Video Game and Information and Communication Technology Practices. In P. Felicia (Ed.), *Handbook of Research on Improving Learning and Motivation through Educational Games: Multidisciplinary Approaches* (pp. 1148–1173). Hershey, PA: Information Science Reference.

Daloukas, V., Rigou, M., & Sirmakessis, S. (2012). Is there a Place for Casual Games in Teaching and Learning?: The Snakes and Ladders Case. [IJGBL]. *International Journal of Game-Based Learning*, *2*(1), 16–32. doi:10.4018/ijgbl.2012010102

Das, S. (2014). Levelling (Up) the Playing Field: How Feminist Gamers Self-Identify and Learn in Online Communities. In V. Venkatesh, J. Wallin, J. Castro, & J. Lewis (Eds.), *Educational, Psychological, and Behavioral Considerations in Niche Online Communities* (pp. 81–100). Hershey, PA: Information Science Reference.

de Freitas, S. (2013). Towards a New Learning: Play and Game-Based Approaches to Education. [IJGBL]. *International Journal of Game-Based Learning*, *3*(4), 1–6. doi:10.4018/ijgbl.2013100101

de Freitas, S., & Griffiths, M. (2011). Massively Multiplayer Online Role-Play Games for Learning. In I. Management Association (Ed.), Gaming and Simulations: Concepts, Methodologies, Tools and Applications (pp. 779-793). Hershey, PA: Information Science Reference. doi:10.4018/978-1-60960-195-9.ch312

De Grove, F., Van Looy, J., & Mechant, P. (2013). Learning to Play, Playing to Learn: Comparing the Experiences of Adult Foreign Language Learners with Off-the-Shelf and Specialized Games for Learning German. [IJGBL]. *International Journal of Game-Based Learning, 3*(2), 22–35. doi:10.4018/ijgbl.2013040102

Deale, D. F., Key, S. S., Regina, M., & Pastore, R. (2012). Women and Gaming. [IJGCMS]. *International Journal of Gaming and Computer-Mediated Simulations, 4*(1), 86–89. doi:10.4018/jgcms.2012010105

del Blanco, Á., Torrente, J., Moreno-Ger, P., & Fernández-Manjón, B. (2014). Enhancing Adaptive Learning and Assessment in Virtual Learning Environments with Educational Games. In I. Management Association (Ed.), K-12 Education: Concepts, Methodologies, Tools, and Applications (pp. 578-597). Hershey, PA: Information Science Reference. doi:10.4018/978-1-4666-4502-8.ch034

Demirbilek, M. (2011). The Use of Electronic Games in Distance Learning as a Tool for Teaching and Learning. In I. Management Association (Ed.), Gaming and Simulations: Concepts, Methodologies, Tools and Applications (pp. 1233-1250). Hershey, PA: Information Science Reference. doi:10.4018/978-1-60960-195-9.ch421

Denham, A. R. (2013). Strategy Instruction and Maintenance of Basic Multiplication Facts through Digital Game Play. [IJGBL]. *International Journal of Game-Based Learning, 3*(2), 36–54. doi:10.4018/ijgbl.2013040103

Denholm, J. A., Protopsaltis, A., & de Freitas, S. (2013). The Value of Team-Based Mixed-Reality (TBMR) Games in Higher Education. [IJGBL]. *International Journal of Game-Based Learning, 3*(1), 18–33. doi:10.4018/ijgbl.2013010102

Dickinson, A. R., & Hui, D. (2013). Enhancing Intelligence, English and Math Competencies in the Classroom via e@Leader Integrated Online Edutainment Gaming and Assessment. In M. Khosrow-Pour (Ed.), *Cases on Assessment and Evaluation in Education* (pp. 277–303). Hershey, PA: Information Science Reference.

Drachen, A., & Canossa, A. (2012). Analyzing User Behavior in Digital Games. In A. Lugmayr, H. Franssila, P. Näränen, O. Sotamaa, J. Vanhala, & Z. Yu (Eds.), *Media in the Ubiquitous Era: Ambient, Social and Gaming Media* (pp. 1–28). Hershey, PA: Information Science Reference.

Driver, P. (2012). Pervasive Games and Mobile Technologies for Embodied Language Learning. [IJCALLT]. *International Journal of Computer-Assisted Language Learning and Teaching, 2*(4), 50–63. doi:10.4018/ijcallt.2012100104

Duarte, I. D., & Valente de Andrade, A. M. (2011). Games and Advertising. In M. Cruz-Cunha, V. Varvalho, & P. Tavares (Eds.), *Business, Technological, and Social Dimensions of Computer Games: Multidisciplinary Developments* (pp. 366–382). Hershey, PA: Information Science Reference.

Dubbels, B. (2011). Cognitive Ethnography: A Methodology for Measure and Analysis of Learning for Game Studies. [IJGCMS]. *International Journal of Gaming and Computer-Mediated Simulations*, *3*(1), 68–78. doi:10.4018/jgcms.2011010105

Dubbels, B. (2013). Gamification, Serious Games, Ludic Simulation, and other Contentious Categories. [IJGCMS]. *International Journal of Gaming and Computer-Mediated Simulations*, *5*(2), 1–19. doi:10.4018/jgcms.2013040101

Duin, H., Cerinšek, G., Fradinho, M., & Taisch, M. (2012). Serious Gaming Supporting Competence Development in Sustainable Manufacturing. In M. Cruz-Cunha (Ed.), *Handbook of Research on Serious Games as Educational, Business and Research Tools* (pp. 47–71). Hershey, PA: Information Science Reference.

Duin, H., Cerinšek, G., Fradinho, M., & Taisch, M. (2013). Serious Gaming Supporting Competence Development in Sustainable Manufacturing. In I. Management Association (Ed.), Industrial Engineering: Concepts, Methodologies, Tools, and Applications (pp. 766-790). Hershey, PA: Engineering Science Reference. doi:10.4018/978-1-4666-1945-6.ch043

Dunbar, N. E., Wilson, S. N., Adame, B. J., Elizondo, J., Jensen, M. L., & Miller, C. H. et al. (2013). MACBETH: Development of a Training Game for the Mitigation of Cognitive Bias. [IJGBL]. *International Journal of Game-Based Learning*, *3*(4), 7–26. doi:10.4018/ijgbl.2013100102

Dunwell, I., & Jarvis, S. (2013). A Serious Game for On-the-Ward Infection Control Awareness Training: Ward Off Infection. In S. Arnab, I. Dunwell, & K. Debattista (Eds.), *Serious Games for Healthcare: Applications and Implications* (pp. 233–246). Hershey, PA: Medical Information Science Reference.

Durga, S., & Squire, K. (2011). Productive Gaming and the Case for Historiographic Game-Play. In I. Management Association (Ed.), Gaming and Simulations: Concepts, Methodologies, Tools and Applications (pp. 1124-1141). Hershey, PA: Information Science Reference. doi:10.4018/978-1-60960-195-9.ch415

Dyer, R. (2013). Games in Higher Education: Opportunities, Expectations, and Challenges of Curriculum Integration. In S. de Freitas, M. Ott, M. Popescu, & I. Stanescu (Eds.), *New Pedagogical Approaches in Game Enhanced Learning: Curriculum Integration* (pp. 38–59). Hershey, PA: Information Science Reference.

Edwards, R. (2013). Pogo Chat. In M. Garcia-Ruiz (Ed.), *Cases on Usability Engineering: Design and Development of Digital Products* (pp. 378–404). Hershey, PA: Information Science Reference.

El Ali, A., & Ketabdar, H. (2013). Magnet-Based Around Device Interaction for Playful Music Composition and Gaming. [IJMHCI]. *International Journal of Mobile Human Computer Interaction*, *5*(4), 56–80. doi:10.4018/ijmhci.2013100103

Ellcessor, E., & Duncan, S. C. (2013). Forming The Guild: Star Power and Rethinking Projective Identity in Affinity Spaces. In P. Felicia (Ed.), *Developments in Current Game-Based Learning Design and Deployment* (pp. 188–202). Hershey, PA: Information Science Reference.

Erb, U., Moura de Araújo, L., Klein, L., Königschulte, A., & Simonow, N. (2012). Serious Games for Exhibition Contexts: Limitations and Design Decisions. In M. Cruz-Cunha (Ed.), *Handbook of Research on Serious Games as Educational, Business and Research Tools* (pp. 708–729). Hershey, PA: Information Science Reference.

Evans, M. (2011). I'd Rather Be Playing Calculus: Adapting Entertainment Game Structures to Educational Games. In P. Felicia (Ed.), *Handbook of Research on Improving Learning and Motivation through Educational Games: Multidisciplinary Approaches* (pp. 153–175). Hershey, PA: Information Science Reference.

Evans, M., Jennings, E., & Andreen, M. (2011). Assessment through Achievement Systems: A Framework for Educational Game Design. [IJGBL]. *International Journal of Game-Based Learning*, *1*(3), 16–29. doi:10.4018/ijgbl.2011070102

Evans, M., Jennings, E., & Andreen, M. (2013). Assessment through Achievement Systems: A Framework for Educational Game Design. In P. Felicia (Ed.), *Developments in Current Game-Based Learning Design and Deployment* (pp. 302–315). Hershey, PA: Information Science Reference.

Evans, M. A. (2011). Procedural Ethos: Confirming the Persuasive in Serious Games. [IJGCMS]. *International Journal of Gaming and Computer-Mediated Simulations*, *3*(4), 70–80. doi:10.4018/jgcms.2011100105

Evett, L., Ridley, A., Keating, L., Merritt, P., Shopland, N., & Brown, D. (2011). Designing Serious Games for People with Disabilities: Game, Set and Match to the Wii™. [IJGBL]. *International Journal of Game-Based Learning*, *1*(4), 11–19. doi:10.4018/ijgbl.2011100102

Farmer, L. S. (2011). Gaming in Adult Education. In V. Wang (Ed.), *Encyclopedia of Information Communication Technologies and Adult Education Integration* (pp. 687–706). Hershey, PA: Information Science Reference.

Faust, K., Meyer, J., & Griffiths, M. D. (2013). Competitive and Professional Gaming: Discussing Potential Benefits of Scientific Study. [IJCBPL]. *International Journal of Cyber Behavior, Psychology and Learning*, *3*(1), 67–77. doi:10.4018/ijcbpl.2013010106

Feinberg, J. R., Schewe, A. H., Moore, C. D., & Wood, K. R. (2013). Puttering, Tinkering, Building, and Making: A Constructionist Approach to Online Instructional Simulation Games. In R. Hartshorne, T. Heafner, & T. Petty (Eds.), *Teacher Education Programs and Online Learning Tools: Innovations in Teacher Preparation* (pp. 417–436). Hershey, PA: Information Science Reference.

Ferri, G. (2013). Rhetorics, Simulations and Games: The Ludic and Satirical Discourse of Molleindustria. [IJGCMS]. *International Journal of Gaming and Computer-Mediated Simulations*, *5*(1), 32–49. doi:10.4018/jgcms.2013010103

Findley, M. R. (2011). The Relationship between Student Learning Styles and Motivation during Educational Video Game Play. [IJOPCD]. *International Journal of Online Pedagogy and Course Design*, *1*(3), 63–73. doi:10.4018/ijopcd.2011070105

Fratto, V. A. (2011). Enhance Student Learning with PowerPoint Games: Using Twenty Questions to Promote Active Learning in Managerial Accounting. [IJICTE]. *International Journal of Information and Communication Technology Education*, *7*(2), 13–20. doi:10.4018/jicte.2011040102

Frederico, C. (2013). Nutrition Games. In S. Arnab, I. Dunwell, & K. Debattista (Eds.), *Serious Games for Healthcare: Applications and Implications* (pp. 167–190). Hershey, PA: Medical Information Science Reference.

Freedman, K. (2014). The Art of Gaming: Knowledge Construction in Visual Culture Learning Communities. In V. Venkatesh, J. Wallin, J. Castro, & J. Lewis (Eds.), *Educational, Psychological, and Behavioral Considerations in Niche Online Communities* (pp. 1–13). Hershey, PA: Information Science Reference.

Fromme, J., Jörissen, B., & Unger, A. (2011). (Self-) Educational Effects of Computer Gaming Cultures. In I. Management Association (Ed.), Gaming and Simulations: Concepts, Methodologies, Tools and Applications (pp. 1251-1269). Hershey, PA: Information Science Reference. doi:10.4018/978-1-60960-195-9.ch501

Gajadhar, B. J., deKort, Y. A., & IJsselsteijn, W. A. (2011). Rule of Engagement: The Presence of a Co-Player Does Not Hinder Gamers' Focus. In R. Ferdig (Ed.), *Discoveries in Gaming and Computer-Mediated Simulations: New Interdisciplinary Applications* (pp. 147–162). Hershey, PA: Information Science Reference.

Garcia-Ruiz, M. A., Tashiro, J., Kapralos, B., & Martin, M. V. (2011). Crouching Tangents, Hidden Danger: Assessing Development of Dangerous Misconceptions within Serious Games for Healthcare Education. In I. Management Association (Ed.),Gaming and Simulations: Concepts, Methodologies, Tools and Applications (pp. 1712-1749). Hershey, PA: Information Science Reference. doi:10.4018/978-1-60960-195-9.ch704

Garrett, M., & McMahon, M. (2012). Computer-Generated Three-Dimensional Training Environments: The Simulation, User, and Problem-Based Learning (SUPL) Approach. In R. Ferdig, & S. de Freitas (Eds.), *Interdisciplinary Advancements in Gaming, Simulations and Virtual Environments: Emerging Trends* (pp. 183–201). Hershey, PA: Information Science Reference.

Garrett, M., & McMahon, M. (2012). The SUPL Approach: A Conceptual Framework for the Design of 3D E-Simulations Based on Gaming Technology within a Problem-Based Learning Pedagogy. In D. Holt, S. Segrave, & J. Cybulski (Eds.), *Professional Education Using E-Simulations: Benefits of Blended Learning Design* (pp. 233–254). Hershey, PA: Business Science Reference.

Gatzidis, C. (2013). First-Person Shooter Game Engines and Healthcare: An Examination of the Current State of the Art and Future Potential. In S. Arnab, I. Dunwell, & K. Debattista (Eds.), *Serious Games for Healthcare: Applications and Implications* (pp. 76–89). Hershey, PA: Medical Information Science Reference.

Gazzard, A. (2012). Re-coding the Algorithm: Purposeful and Appropriated Play. In A. Lugmayr, H. Franssila, P. Näränen, O. Sotamaa, J. Vanhala, & Z. Yu (Eds.), *Media in the Ubiquitous Era: Ambient, Social and Gaming Media* (pp. 200–214). Hershey, PA: Information Science Reference.

Ghuman, D., & Griffiths, M. (2012). A Cross-Genre Study of Online Gaming: Player Demographics, Motivation for Play, and Social Interactions Among Players. [IJCBPL]. *International Journal of Cyber Behavior, Psychology and Learning*, *2*(1), 13–29. doi:10.4018/ijcbpl.2012010102

Göbel, S., & Mehm, F. (2013). Personalized, Adaptive Digital Educational Games using Narrative Game-Based Learning Objects. In K. Bredl, & W. Bösche (Eds.), *Serious Games and Virtual Worlds in Education, Professional Development, and Healthcare* (pp. 74–84). Hershey, PA: Information Science Reference.

Göbel, S., & Mehm, F. (2014). Personalized, Adaptive Digital Educational Games using Narrative Game-Based Learning Objects. In I. Management Association (Ed.),K-12 Education: Concepts, Methodologies, Tools, and Applications (pp. 281-291). Hershey, PA: Information Science Reference. doi:10.4018/978-1-4666-4502-8.ch016

Goggins, S., Schmidt, M., Guajardo, J., & Moore, J. L. (2011). 3D Virtual Worlds: Assessing the Experience and Informing Design. [IJSODIT]. *International Journal of Social and Organizational Dynamics in IT*, *1*(1), 30–48. doi:10.4018/ijsodit.2011010103

Gogolin, G., Gogolin, E., & Kam, H. (2014). Virtual Worlds and Social Media: Security and Privacy Concerns, Implications, and Practices. [IJALR]. *International Journal of Artificial Life Research*, *4*(1), 30–42. doi:10.4018/ijalr.2014010103

Gomez, J. (2014). A Match Made in "Outer Heaven:": The Digital Age Vis-à-Vis the Bomb in Guns of the Patriots. In G. Verhulsdonck, & M. Limbu (Eds.), *Digital Rhetoric and Global Literacies: Communication Modes and Digital Practices in the Networked World* (pp. 248–282). Hershey, PA: Information Science Reference.

Gopin, E. (2014). Finding and Evaluating Great Educational Games. In Z. Yang, H. Yang, D. Wu, & S. Liu (Eds.), *Transforming K-12 Classrooms with Digital Technology* (pp. 83–97). Hershey, PA: Information Science Reference.

Griffiths, M., Hussain, Z., Grüsser, S. M., Thalemann, R., Cole, H., Davies, M. N., & Chappell, D. (2011). Social Interactions in Online Gaming. [IJGBL]. *International Journal of Game-Based Learning*, *1*(4), 20–36. doi:10.4018/ijgbl.2011100103

Griffiths, M., Hussain, Z., Grüsser, S. M., Thalemann, R., Cole, H., Davies, M. N., & Chappell, D. (2013). Social Interactions in Online Gaming. In P. Felicia (Ed.), *Developments in Current Game-Based Learning Design and Deployment* (pp. 74–90). Hershey, PA: Information Science Reference.

Griffiths, M., Kuss, D. J., & Ortiz de Gortari, A. B. (2013). Videogames as Therapy: A Review of the Medical and Psychological Literature. In M. Cruz-Cunha, I. Miranda, & P. Gonçalves (Eds.), *Handbook of Research on ICTs and Management Systems for Improving Efficiency in Healthcare and Social Care* (pp. 43–68). Hershey, PA: Medical Information Science Reference.

Groff, J., Howells, C., & Cranmer, S. (2012). Console Game-Based Pedagogy: A Study of Primary and Secondary Classroom Learning through Console Video Games. [IJGBL]. *International Journal of Game-Based Learning, 2*(2), 35–54. doi:10.4018/ijgbl.2012040103

Gupta, M., Jin, S., Sanders, G. L., Sherman, B. A., & Simha, A. (2012). Getting Real About Virtual Worlds: A Review. [IJVCSN]. *International Journal of Virtual Communities and Social Networking, 4*(3), 1–46. doi:10.4018/jvcsn.2012070101

Gwee, S., San Chee, Y., & Tan, E. M. (2011). The Role of Gender in Mobile Game-Based Learning. [IJMBL]. *International Journal of Mobile and Blended Learning, 3*(4), 19–37. doi:10.4018/jmbl.2011100102

Gwee, S., San Chee, Y., & Tan, E. M. (2013). The Role of Gender in Mobile Game-Based Learning. In D. Parsons (Ed.), *Innovations in Mobile Educational Technologies and Applications* (pp. 254–271). Hershey, PA: Information Science Reference.

Hainey, T., Connolly, T., Stansfield, M., & Boyle, L. (2011). The Use of Computer Games in Education: A Review of the Literature. In P. Felicia (Ed.), *Handbook of Research on Improving Learning and Motivation through Educational Games: Multidisciplinary Approaches* (pp. 29–50). Hershey, PA: Information Science Reference.

Hainey, T., Connolly, T. M., Chaudy, Y., Boyle, E., Beeby, R., & Soflano, M. (2014). Assessment Integration in Serious Games. In T. Connolly, T. Hainey, E. Boyle, G. Baxter, & P. Moreno-Ger (Eds.), *Psychology, Pedagogy, and Assessment in Serious Games* (pp. 317–341). Hershey, PA: Information Science Reference.

Hamlen, K. R., & Gage, H. E. (2011). Negotiating Students' Conceptions of 'Cheating' in Video Games and in School. [IJGCMS]. *International Journal of Gaming and Computer-Mediated Simulations*, *3*(2), 44–56. doi:10.4018/jgcms.2011040103

Hanghøj, T. (2013). Game-Based Teaching: Practices, Roles, and Pedagogies. In S. de Freitas, M. Ott, M. Popescu, & I. Stanescu (Eds.), *New Pedagogical Approaches in Game Enhanced Learning: Curriculum Integration* (pp. 81–101). Hershey, PA: Information Science Reference.

Haring, P., Chakinska, D., & Ritterfeld, U. (2011). Understanding Serious Gaming: A Psychological Perspective. In P. Felicia (Ed.), *Handbook of Research on Improving Learning and Motivation through Educational Games: Multidisciplinary Approaches* (pp. 413–430). Hershey, PA: Information Science Reference.

Harwood, T. (2011). Convergence of Online Gaming and E-Commerce. In B. Ciaramitaro (Ed.), *Virtual Worlds and E-Commerce: Technologies and Applications for Building Customer Relationships* (pp. 61–89). Hershey, PA: Business Science Reference.

Harwood, T. (2012). Emergence of Gamified Commerce: Turning Virtual to Real. [JECO]. *Journal of Electronic Commerce in Organizations*, *10*(2), 16–39. doi:10.4018/jeco.2012040102

Haskell, C. (2013). 3D GameLab: Quest-Based Pre-Service Teacher Education. In Y. Baek, & N. Whitton (Eds.), *Cases on Digital Game-Based Learning: Methods, Models, and Strategies* (pp. 302–340). Hershey, PA: Information Science Reference.

Hauge, J. B., Boyle, E., Mayer, I., Nadolski, R., Riedel, J. C., & Moreno-Ger, P. et al. (2014). Study Design and Data Gathering Guide for Serious Games' Evaluation. In T. Connolly, T. Hainey, E. Boyle, G. Baxter, & P. Moreno-Ger (Eds.), *Psychology, Pedagogy, and Assessment in Serious Games* (pp. 394–419). Hershey, PA: Information Science Reference.

Hauge, J. B., Hoeborn, G., & Bredtmann, J. (2012). Challenges of Serious Games for Improving Students' Management Skills on Decision Making. In M. Cruz-Cunha (Ed.), *Handbook of Research on Serious Games as Educational, Business and Research Tools* (pp. 947–964). Hershey, PA: Information Science Reference.

Hawreliak, J. (2013). "To Be Shot at Without Result": Gaming and the Rhetoric of Immortality. In R. Luppicini (Ed.), *Handbook of Research on Technoself: Identity in a Technological Society* (pp. 531–553). Hershey, PA: Information Science Reference.

Hayashi, T., & Ohsawa, Y. (2013). Processing Combinatorial Thinking: Innovators Marketplace as Role-Based Game Plus Action Planning. [IJKSS]. *International Journal of Knowledge and Systems Science*, *4*(3), 14–38. doi:10.4018/ijkss.2013070102

Heeter, C., Lee, Y., Magerko, B., & Medler, B. (2011). Impacts of Forced Serious Game Play on Vulnerable Subgroups. [IJGCMS]. *International Journal of Gaming and Computer-Mediated Simulations*, *3*(3), 34–53. doi:10.4018/jgcms.2011070103

Heeter, C., Lee, Y., Magerko, B., & Medler, B. (2013). Impacts of Forced Serious Game Play on Vulnerable Subgroups. In R. Ferdig (Ed.), *Design, Utilization, and Analysis of Simulations and Game-Based Educational Worlds* (pp. 158–176). Hershey, PA: Information Science Reference.

Heeter, C., Sarkar, C. D., Palmer-Scott, B., & Zhang, S. (2012). Engineering Sociability: Friendship Drive, Visibility, and Social Connection in Anonymous Co-Located Local Wi-Fi Multiplayer Online Gaming. [IJGCMS]. *International Journal of Gaming and Computer-Mediated Simulations*, *4*(2), 1–18. doi:10.4018/jgcms.2012040101

Heinrichs, W. L., Davies, D., & Davies, J. (2013). Virtual Worlds in Healthcare: Applications and Implications. In S. Arnab, I. Dunwell, & K. Debattista (Eds.), *Serious Games for Healthcare: Applications and Implications* (pp. 1–22). Hershey, PA: Medical Information Science Reference.

Heitmann, M., & Tidten, K. (2011). New Business Models for the Computer Gaming Industry: Selling an Adventure. In M. Cruz-Cunha, V. Varvalho, & P. Tavares (Eds.), *Business, Technological, and Social Dimensions of Computer Games: Multidisciplinary Developments* (pp. 401–415). Hershey, PA: Information Science Reference.

Heng, L. P., Wen, L. D., & Huey, T. H. (2012). Regulation of Violence in MMORPG. In R. Sharma, M. Tan, & F. Pereira (Eds.), *Understanding the Interactive Digital Media Marketplace: Frameworks, Platforms, Communities and Issues* (pp. 349–367). Hershey, PA: Information Science Reference.

Herro, D. (2013). Elements of Game Design: Developing a Meaningful Game Design Curriculum for the Classroom. In Y. Baek, & N. Whitton (Eds.), *Cases on Digital Game-Based Learning: Methods, Models, and Strategies* (pp. 240–255). Hershey, PA: Information Science Reference.

Herro, D. (2014). Elements of Game Design: Developing a Meaningful Game Design Curriculum for the Classroom. In I. Management Association (Ed.), K-12 Education: Concepts, Methodologies, Tools, and Applications (pp. 269-280). Hershey, PA: Information Science Reference. doi:10.4018/978-1-4666-4502-8.ch015

Hersh, M. A., & Leporini, B. (2013). An Overview of Accessibility and Usability of Educational Games. In C. Gonzalez (Ed.), *Student Usability in Educational Software and Games: Improving Experiences* (pp. 1–40). Hershey, PA: Information Science Reference.

Hersh, M. A., & Leporini, B. (2014). An Overview of Accessibility and Usability of Educational Games. In I. Management Association (Ed.), Assistive Technologies: Concepts, Methodologies, Tools, and Applications (pp. 63-101). Hershey, PA: Information Science Reference. doi:10.4018/978-1-4666-4422-9.ch005

Hoffman, B. (2011). Mobile Gaming: Exploring Spaces and Places. In M. Cruz-Cunha, V. Varvalho, & P. Tavares (Eds.), *Business, Technological, and Social Dimensions of Computer Games: Multidisciplinary Developments* (pp. 185–199). Hershey, PA: Information Science Reference.

Hoge, B. (2014). GBL as PBL: Guidelines for Game-Based Learning in the Classroom and Informal Science Centers. In Z. Yang, H. Yang, D. Wu, & S. Liu (Eds.), *Transforming K-12 Classrooms with Digital Technology* (pp. 58–82). Hershey, PA: Information Science Reference.

Holden, C. L., & Sykes, J. M. (2011). Leveraging Mobile Games for Place-Based Language Learning. [IJGBL]. *International Journal of Game-Based Learning*, *1*(2), 1–18. doi:10.4018/ijgbl.2011040101

Hollins, P., & Whitton, N. (2011). From the Games Industry: Ten Lessons for Game-Based Learning. [IJVPLE]. *International Journal of Virtual and Personal Learning Environments*, *2*(2), 73–82. doi:10.4018/jvple.2011040107

Hollins, P., & Whitton, N. (2013). From the Games Industry: Ten Lessons for Game-Based Learning. In M. Thomas (Ed.), *Technologies, Innovation, and Change in Personal and Virtual Learning Environments* (pp. 79–88). Hershey, PA: Information Science Reference.

Holloway, A. (2013). Better Birth through Games: The Design of the Prepared Partner and Digital Birth. [IJGCMS]. *International Journal of Gaming and Computer-Mediated Simulations*, *5*(3), 43–71. doi:10.4018/jgcms.2013070104

Holmes, S., Thurmond, B., Annetta, L. A., & Sears, M. (2012). Serious Educational Games (SEGs) and Student Learning and Engagement with Scientific Concepts. In L. Lennex, & K. Nettleton (Eds.), *Cases on Inquiry through Instructional Technology in Math and Science* (pp. 464–486). Hershey, PA: Information Science Reference.

Hoppenbrouwers, S., Schotten, B., & Lucas, P. (2012). Towards Games for Knowledge Acquisition and Modeling. In R. Ferdig, & S. de Freitas (Eds.), *Interdisciplinary Advancements in Gaming, Simulations and Virtual Environments: Emerging Trends* (pp. 281–299). Hershey, PA: Information Science Reference.

Hoppenbrouwers, S., Weigand, H., & Rouwette, E. (2011). Exploring Dialogue Games for Collaborative Modeling. In N. Kock (Ed.), *E-Collaboration Technologies and Organizational Performance: Current and Future Trends* (pp. 292–317). Hershey, PA: Information Science Reference.

Hromek, R. (2013). Facilitation of Trust in Gaming Situations. In C. Gonzalez (Ed.), *Student Usability in Educational Software and Games: Improving Experiences* (pp. 161–173). Hershey, PA: Information Science Reference.

Hsu, P., & Tsai, C. (2012). Evaluating the Virtual Products for Online Games via the Grey Relational Analysis. [IJEA]. *International Journal of E-Adoption*, *4*(3), 39–47. doi:10.4018/jea.2012070103

Hua Chen, V. H., & Lirn Duh, H. B. (2011). Socializing in the Online Gaming Community: Social Interaction in World of Warcraft. In I. Management Association (Ed.), Virtual Communities: Concepts, Methodologies, Tools and Applications (pp. 145-160). Hershey, PA: Information Science Reference. doi:10.4018/978-1-60960-100-3.ch111

Huang, W. D., & Tettegah, S. Y. (2014). Cognitive Load and Empathy in Serious Games: A Conceptual Framework. In J. Bishop (Ed.), *Gamification for Human Factors Integration: Social, Education, and Psychological Issues* (pp. 17–30). Hershey, PA: Information Science Reference.

Hubal, R., & Pina, J. (2012). Serious Assessments in Serious Games. [IJGCMS]. *International Journal of Gaming and Computer-Mediated Simulations*, *4*(3), 49–64. doi:10.4018/jgcms.2012070104

Hudlicka, E. (2011). Affective Gaming in Education, Training and Therapy: Motivation, Requirements, Techniques. In P. Felicia (Ed.), *Handbook of Research on Improving Learning and Motivation through Educational Games: Multidisciplinary Approaches* (pp. 482–511). Hershey, PA: Information Science Reference.

Hudson, K. (2011). Applied Training in Virtual Environments. In I. Management Association (Ed.), Gaming and Simulations: Concepts, Methodologies, Tools and Applications (pp. 928-940). Hershey, PA: Information Science Reference. doi:10.4018/978-1-60960-195-9.ch402

Huhtinen, A. (2012). From Military Threats to Everyday Fear: Computer Games as the Representation of Military Information Operations. [IJCWT]. *International Journal of Cyber Warfare & Terrorism*, *2*(2), 1–10. doi:10.4018/ijcwt.2012040101

Hyatt, K. J., Barron, J. L., & Noakes, M. A. (2013). Video Gaming for STEM Education. In H. Yang, & S. Wang (Eds.), *Cases on E-Learning Management: Development and Implementation* (pp. 103–117). Hershey, PA: Information Science Reference.

Iacovides, I., Aczel, J., Scanlon, E., Taylor, J., & Woods, W. (2011). Motivation, Engagement and Learning through Digital Games. [IJVPLE]. *International Journal of Virtual and Personal Learning Environments*, *2*(2), 1–16. doi:10.4018/jvple.2011040101

Iacovides, I., Aczel, J., Scanlon, E., Taylor, J., & Woods, W. (2013). Motivation, Engagement and Learning through Digital Games. In M. Thomas (Ed.), *Technologies, Innovation, and Change in Personal and Virtual Learning Environments* (pp. 125–140). Hershey, PA: Information Science Reference.

Ibrahim, A., Vela, F. L., Rodríguez, P. P., Sánchez, J. L., & Zea, N. P. (2012). Playability Guidelines for Educational Video Games: A Comprehensive and Integrated Literature Review. [IJGBL]. *International Journal of Game-Based Learning*, *2*(4), 18–40. doi:10.4018/ijgbl.2012100102

Jackson, R., Robinson, W., & Simon, B. (2014). Gleaning Strategies for Knowledge Sharing and Collective Assessment in the Art Classroom from the Videogame, "Little Big Planet's Creator Spotlights. In V. Venkatesh, J. Wallin, J. Castro, & J. Lewis (Eds.), *Educational, Psychological, and Behavioral Considerations in Niche Online Communities* (pp. 14–32). Hershey, PA: Information Science Reference.

Jacobs, M. (2012). Playing "Nice": What Online Gaming Can Teach Us about Multiculturalism. In K. St.Amant, & S. Kelsey (Eds.), *Computer-Mediated Communication across Cultures: International Interactions in Online Environments* (pp. 32–44). Hershey, PA: Information Science Reference.

Jacobs, M. (2014). Playing "Nice": What Online Gaming Can Teach Us about Multiculturalism. In I. Management Association (Ed.), Cross-Cultural Interaction: Concepts, Methodologies, Tools and Applications (pp. 963-975). Hershey, PA: Information Science Reference. doi:10.4018/978-1-4666-4979-8.ch054

Jacobson, J. (2011). Digital Dome versus Desktop Display in an Educational Game: Gates of Horus. [IJGCMS]. *International Journal of Gaming and Computer-Mediated Simulations*, *3*(1), 13–32. doi:10.4018/jgcms.2011010102

Jamaludin, A., & San Chee, Y. (2011). Investigating Youth's Life Online Phenomena: Subverting Dichotomies through Negotiation of Offline and Online Identities. [IJGCMS]. *International Journal of Gaming and Computer-Mediated Simulations*, *3*(4), 1–18. doi:10.4018/jgcms.2011100101

James, C. L., & Wright, V. H. (2011). Teacher Gamers vs. Teacher Non-Gamers. In I. Management Association (Ed.), Instructional Design: Concepts, Methodologies, Tools and Applications (pp. 1085-1103). Hershey, PA: Information Science Reference. doi:10.4018/978-1-60960-503-2.ch421

Jin, P. (2011). Methodological Considerations in Educational Research Using Serious Games. In I. Management Association (Ed.), Gaming and Simulations: Concepts, Methodologies, Tools and Applications (pp. 1078-1107). Hershey, PA: Information Science Reference. doi:10.4018/978-1-60960-195-9.ch412

Jin, S., DaCosta, B., & Seok, S. (2014). Social Skills Development for Children with Autism Spectrum Disorders through the Use of Interactive Storytelling Games. In B. DaCosta, & S. Seok (Eds.), *Assistive Technology Research, Practice, and Theory* (pp. 144–159). Hershey, PA: Medical Information Science Reference.

Jonathan, T., Bruno, B., & Abdenour, B. (2012). Understanding and Implementing Adaptive Difficulty Adjustment in Video Games. In A. Kumar, J. Etheredge, & A. Boudreaux (Eds.), *Algorithmic and Architectural Gaming Design: Implementation and Development* (pp. 82–106). Hershey, PA: Information Science Reference.

Jones, D. A., & Chang, M. (2012). Multiplayer Online Role Playing Game for Teaching Youth Finance in Canada. [IJOPCD]. *International Journal of Online Pedagogy and Course Design*, *2*(2), 44–59. doi:10.4018/ijopcd.2012040104

Jones, G., & Warren, S. J. (2011). Issues and Concerns of K-12 Educators on 3-D Multi-User Virtual Environments in Formal Classroom Settings. [IJGCMS]. *International Journal of Gaming and Computer-Mediated Simulations*, *3*(1), 1–12. doi:10.4018/jgcms.2011010101

Kafai, Y. B., Fields, D., & Searle, K. A. (2012). Multi-Modal Investigations of Relationship Play in Virtual Worlds. In R. Ferdig, & S. de Freitas (Eds.), *Interdisciplinary Advancements in Gaming, Simulations and Virtual Environments: Emerging Trends* (pp. 41–50). Hershey, PA: Information Science Reference.

Kailas, A., & Chong, C. (2012). Capturing Basic Movements for Mobile Gaming Platforms Embedded with Motion Sensors. [IJEHMC]. *International Journal of E-Health and Medical Communications*, *3*(4), 1–14. doi:10.4018/jehmc.2012100101

Kasapakis, V., & Gavalas, D. (2014). Design Aspects and Context Awareness in Pervasive Games. In B. Guo, D. Riboni, & P. Hu (Eds.), *Creating Personal, Social, and Urban Awareness through Pervasive Computing* (pp. 131–156). Hershey, PA: Information Science Reference.

Kasimati, A., Mysirlaki, S., Bouta, H., & Paraskeva, F. (2014). Ubiquitous Game-Based Learning in Higher Education: A Framework towards the Effective Integration of Game-Based Learning in Higher Education using Emerging Ubiquitous Technologies. In J. Pelet, & P. Papadopoulou (Eds.), *User Behavior in Ubiquitous Online Environments* (pp. 106–129). Hershey, PA: Information Science Reference.

Kaufman, G. F., & Flanagan, M. (2013). Lost in Translation: Comparing the Impact of an Analog and Digital Version of a Public Health Game on Players' Perceptions, Attitudes, and Cognitions. [IJGCMS]. *International Journal of Gaming and Computer-Mediated Simulations*, *5*(3), 1–9. doi:10.4018/jgcms.2013070101

Kazimoglu, C., Kiernan, M., Bacon, L., & MacKinnon, L. (2011). Understanding Computational Thinking before Programming: Developing Guidelines for the Design of Games to Learn Introductory Programming through Game-Play. [IJGBL]. *International Journal of Game-Based Learning*, *1*(3), 30–52. doi:10.4018/ijgbl.2011070103

Ke, F. (2011). A Qualitative Meta-Analysis of Computer Games as Learning Tools. In I. Management Association (Ed.), Gaming and Simulations: Concepts, Methodologies, Tools and Applications (pp. 1619-1665). Hershey, PA: Information Science Reference. doi:10.4018/978-1-60960-195-9.ch701

Ke, F., Yildirim, N., & Enfield, J. (2012). Exploring the Design of Game Enjoyment Through the Perspectives of Novice Game Developers. [IJGCMS]. *International Journal of Gaming and Computer-Mediated Simulations*, *4*(4), 45–63. doi:10.4018/jgcms.2012100104

Kelle, S., Sigurðarson, S. E., Westera, W., & Specht, M. (2011). Game Based Lifelong Learning. In G. Magoulas (Ed.), *E-Infrastructures and Technologies for Lifelong Learning: Next Generation Environments* (pp. 337–349). Hershey, PA: Information Science Reference.

Kennedy-Clark, S., & Thompson, K. (2014). A MUVEing Success: Design Strategies for Professional Development in the Use of Multi-User Virtual Environments and Educational Games in Science Education. In I. Management Association (Ed.), K-12 Education: Concepts, Methodologies, Tools, and Applications (pp. 614-638). Hershey, PA: Information Science Reference. doi:10.4018/978-1-4666-4502-8.ch036

Khan, M. M., & Reed, J. (2011). An Evaluation of Neurogames®: A Collection of Computer Games Designed to Improve Literacy and Numeracy. [IJVPLE]. *International Journal of Virtual and Personal Learning Environments*, *2*(2), 17–29. doi:10.4018/jvple.2011040102

Kickmeier-Rust, M. D., Mattheiss, E., Steiner, C., & Albert, D. (2011). A Psycho-Pedagogical Framework for Multi-Adaptive Educational Games. [IJGBL]. *International Journal of Game-Based Learning*, *1*(1), 45–58. doi:10.4018/ijgbl.2011010104

Kickmeier-Rust, M. D., Mattheiss, E., Steiner, C., & Albert, D. (2013). A Psycho-Pedagogical Framework for Multi-Adaptive Educational Games. In P. Felicia (Ed.), *Developments in Current Game-Based Learning Design and Deployment* (pp. 103–117). Hershey, PA: Information Science Reference.

Kim, D. K., Dinu, L. F., & Chung, W. (2013). Online Games as a Component of School Textbooks: A Test Predicting the Diffusion of Interactive Online Games Designed for the Textbook Reformation in South Korea. [IJICTE]. *International Journal of Information and Communication Technology Education*, *9*(2), 52–65. doi:10.4018/jicte.2013040105

King, E. (2011). Possibility Spaces: Using The Sims 2 as a Sandbox to Explore Possible Selves with At-Risk Teenage Males. [IJGBL]. *International Journal of Game-Based Learning*, *1*(2), 34–51. doi:10.4018/ijgbl.2011040103

King, E. (2013). Possibility Spaces: Using The Sims 2 as a Sandbox to Explore Possible Selves with At-Risk Teenage Males. In P. Felicia (Ed.), *Developments in Current Game-Based Learning Design and Deployment* (pp. 169–187). Hershey, PA: Information Science Reference.

Kinsell, C., DaCosta, B., & Nasah, A. (2014). Simulation Games as Interventions in the Promotion of Social Skills Development among Children with Autism Spectrum Disorders. In B. DaCosta, & S. Seok (Eds.), *Assistive Technology Research, Practice, and Theory* (pp. 160–180). Hershey, PA: Medical Information Science Reference.

Klein, A. Z., Freitas, A., Machado, L., Junior, J. C., Graziola, P. G., & Schlemmer, E. (2014). The Use of Virtual Worlds in Management Education: An Investigation of Current Practices in Second Life. [IJICTE]. *International Journal of Information and Communication Technology Education*, *10*(1), 61–78. doi:10.4018/ijicte.2014010106

Knight, J. F. (2013). Acceptability of Video Games Technology for Medical Emergency Training. [IJGCMS]. *International Journal of Gaming and Computer-Mediated Simulations*, *5*(4), 86–99. doi:10.4018/ijgcms.2013100105

Konert, J., Wendel, V., Richter, K., & Göbel, S. (2013). Collaborative Learning and Game Mastering in Multiplayer Games. In K. Bredl, & W. Bösche (Eds.), *Serious Games and Virtual Worlds in Education, Professional Development, and Healthcare* (pp. 85–104). Hershey, PA: Information Science Reference.

Kongmee, I., Strachan, R., Pickard, A., & Montgomery, C. (2012). A Case Study of Using Online Communities and Virtual Environment in Massively Multiplayer Role Playing Games (MMORPGs) as a Learning and Teaching Tool for Second Language Learners. [IJVPLE]. *International Journal of Virtual and Personal Learning Environments*, *3*(4), 1–15. doi:10.4018/jvple.2012100101

Kranz, M., Murmann, L., & Michahelles, F. (2013). Research in the Large: Challenges for Large-Scale Mobile Application Research- A Case Study about NFC Adoption using Gamification via an App Store. [IJMHCI]. *International Journal of Mobile Human Computer Interaction*, *5*(1), 45–61. doi:10.4018/jmhci.2013010103

Related References

Kristiansen, E. (2013). Design Games for In-Situ Design. [IJMHCI]. *International Journal of Mobile Human Computer Interaction*, *5*(3), 1–22. doi:10.4018/jmhci.2013070101

Kulman, R., Stoner, G., Ruffolo, L., Marshall, S., Slater, J., Dyl, A., & Cheng, A. (2014). Teaching Executive Functions, Self-Management, and Ethical Decision-Making through Popular Videogame Play. In I. Management Association (Ed.),Assistive Technologies: Concepts, Methodologies, Tools, and Applications (pp. 771-785). Hershey, PA: Information Science Reference. doi:10.4018/978-1-4666-4422-9.ch039

Kwok, N. W., & Khoo, A. (2011). Gamers' Motivations and Problematic Gaming: An Exploratory Study of Gamers in World of Warcraft. [IJCBPL]. *International Journal of Cyber Behavior, Psychology and Learning*, *1*(3), 34–49. doi:10.4018/ijcbpl.2011070103

Kwok, N. W., & Khoo, A. (2013). Gamers' Motivations and Problematic Gaming: An Exploratory Study of Gamers in World of Warcraft. In R. Zheng (Ed.), *Evolving Psychological and Educational Perspectives on Cyber Behavior* (pp. 64–81). Hershey, PA: Information Science Reference.

Kyzy, J. T. (2011). "World of Uncertainty" Game for Decision-Makers. [IJVPLE]. *International Journal of Virtual and Personal Learning Environments*, *2*(2), 40–45. doi:10.4018/jvple.2011040104

Lai, C., Chu, C., Liu, H., Yang, S., & Chen, W. (2013). An Examination of Game-Based Learning from Theories of Flow Experience and Cognitive Load. [IJDET]. *International Journal of Distance Education Technologies*, *11*(4), 17–29. doi:10.4018/ijdet.2013100102

Lee, L. H., & San Chee, Y. (2013). Gestural Articulations of Embodied Spatiality: What Gestures Reveal about Students' Sense-Making of Charged Particle Dynamics in a 3D Game World. [IJGCMS]. *International Journal of Gaming and Computer-Mediated Simulations*, *5*(4), 17–47. doi:10.4018/ijgcms.2013100102

Lee, Y. (2013). Are Good Games Also Good Problems?: Content Analysis of Problem Types and Learning Principles in Environmental Education Games. [IJGBL]. *International Journal of Game-Based Learning*, *3*(4), 47–61. doi:10.4018/ijgbl.2013100104

Lei, L., Wanqiang, S., & Tat, E. T. (2012). Online Games for Children. In R. Sharma, M. Tan, & F. Pereira (Eds.), *Understanding the Interactive Digital Media Marketplace: Frameworks, Platforms, Communities and Issues* (pp. 300–315). Hershey, PA: Information Science Reference.

Lei, L., Wanqiang, S., & Tat, E. T. (2014). Online Games for Children. In I. Management Association (Ed.), K-12 Education: Concepts, Methodologies, Tools, and Applications (pp. 598-613). Hershey, PA: Information Science Reference. doi:10.4018/978-1-4666-4502-8.ch035

Lelardeux, C., Alvarez, J., Montaut, T., Galaup, M., & Lagarrigue, P. (2013). Healthcare Games and the Metaphoric Approach. In S. Arnab, I. Dunwell, & K. Debattista (Eds.), *Serious Games for Healthcare: Applications and Implications* (pp. 24–49). Hershey, PA: Medical Information Science Reference.

Lettieri, N., Fabiani, E., Polcini, A. T., De Chiara, R., & Scarano, V. (2011). Emerging Paradigms in Legal Education: A Learning Environment to Teach Law through Online Role Playing Games. In P. Felicia (Ed.), *Handbook of Research on Improving Learning and Motivation through Educational Games: Multidisciplinary Approaches* (pp. 1019–1035). Hershey, PA: Information Science Reference.

Li, K., Chen, C., Wang, H., & Heh, J. (2012). Game-Based Pilot System for English Learning. [IJOPCD]. *International Journal of Online Pedagogy and Course Design*, *2*(2), 86–99. doi:10.4018/ijopcd.2012040106

Li, Z., Liu, F., & Boyer, J. (2011). Amusing Minds for Joyful Learning through E-Gaming. In I. Management Association (Ed.), Gaming and Simulations: Concepts, Methodologies, Tools and Applications (pp. 1280-1297). Hershey, PA: Information Science Reference. doi:10.4018/978-1-60960-195-9.ch503

Lim, T., Louchart, S., Suttie, N., Ritchie, J., Aylett, R., & Stanescu, I. A. et al. (2013). Strategies for Effective Digital Games Development and Implementation. In Y. Baek, & N. Whitton (Eds.), *Cases on Digital Game-Based Learning: Methods, Models, and Strategies* (pp. 168–198). Hershey, PA: Information Science Reference.

Lin, K., Wei, Y. C., & Hung, J. C. (2012). The Effects of Online Interactive Games on High School Students' Achievement and Motivation in History Learning. [IJDET]. *International Journal of Distance Education Technologies*, *10*(4), 96–105. doi:10.4018/jdet.2012100108

Lin, T., Wu, Z., Tang, N., & Wu, S. (2013). Exploring the Effects of Display Characteristics on Presence and Emotional Responses of Game Players. [IJTHI]. *International Journal of Technology and Human Interaction*, *9*(1), 50–63. doi:10.4018/jthi.2013010104

Lin, Y. H. (2012). Playing as Producing: Convergence Culture and Localization of EA Digital Games in Taiwan. In I. Management Association (Ed.), Computer Engineering: Concepts, Methodologies, Tools and Applications (pp. 1646-1659). Hershey, PA: Engineering Science Reference. doi:10.4018/978-1-61350-456-7.ch702

Linares, K., Subrahmanyam, K., Cheng, R., & Guan, S. A. (2011). A Second Life Within Second Life: Are Virtual World Users Creating New Selves and New Lives? [IJCBPL]. *International Journal of Cyber Behavior, Psychology and Learning*, *1*(3), 50–71. doi:10.4018/ijcbpl.2011070104

Linehan, C., Lawson, S., Doughty, M., Kirman, B., Haferkamp, N., & Krämer, N. C. et al. (2012). Teaching Group Decision Making Skills to Emergency Managers via Digital Games. In A. Lugmayr, H. Franssila, P. Näränen, O. Sotamaa, J. Vanhala, & Z. Yu (Eds.), *Media in the Ubiquitous Era: Ambient, Social and Gaming Media* (pp. 111–129). Hershey, PA: Information Science Reference.

Linehan, C., Lawson, S., Doughty, M., Kirman, B., Haferkamp, N., Krämer, N. C., et al. (2014). Teaching Group Decision Making Skills to Emergency Managers via Digital Games. In I. Management Association (Ed.), Crisis Management: Concepts, Methodologies, Tools and Applications (pp. 667-686). Hershey, PA: Information Science Reference. doi:10.4018/978-1-4666-4707-7.ch032

Linek, S. B. (2011). As You Like It: What Media Psychology Can Tell Us About Educational Game Design. In P. Felicia (Ed.), *Handbook of Research on Improving Learning and Motivation through Educational Games: Multidisciplinary Approaches* (pp. 606–632). Hershey, PA: Information Science Reference.

Linek, S. B., Marte, B., & Albert, D. (2011). Background Music in Educational Games: Motivational Appeal and Cognitive Impact. [IJGBL]. *International Journal of Game-Based Learning*, *1*(3), 53–64. doi:10.4018/ijgbl.2011070104

Litinski, V. (2013). Games for Health: Building the Case. [IJGCMS]. *International Journal of Gaming and Computer-Mediated Simulations*, *5*(3), 108–115. doi:10.4018/jgcms.2013070108

Loh, C. S. (2013). Improving the Impact and Return of Investment of Game-Based Learning. [IJVPLE]. *International Journal of Virtual and Personal Learning Environments*, *4*(1), 1–15. doi:10.4018/jvple.2013010101

Lorentz, P. (2012). Is there a Virtual Socialization by Acting Virtual Identities?: Case Study: The Sims. In N. Zagalo, L. Morgado, & A. Boa-Ventura (Eds.), *Virtual Worlds and Metaverse Platforms: New Communication and Identity Paradigms* (pp. 206–218). Hershey, PA: Information Science Reference.

Lowe, R. (2012). Implementing Computer Gaming Technology in Architectural Design Curricula: Testing Architecture with the Rich Intertwining of Real-Time Spatial, Material, Lighting and Physical Simulations. In N. Gu, & X. Wang (Eds.), *Computational Design Methods and Technologies: Applications in CAD, CAM and CAE Education* (pp. 199–224). Hershey, PA: Information Science Reference.

Lund, K., Lochrie, M., & Coulton, P. (2012). Designing Scalable Location Based Games that Encourage Emergent Behaviour. [IJACI]. *International Journal of Ambient Computing and Intelligence*, *4*(4), 1–20. doi:10.4018/jaci.2012100101

Ma, Y., Williams, D., & Prejean, L. (2012). Understanding the Relationships Among Various Design Components in a Game-Based Learning Environment. [IJGCMS]. *International Journal of Gaming and Computer-Mediated Simulations*, *4*(1), 68–85. doi:10.4018/jgcms.2012010104

Maciuszek, D., Ladhoff, S., & Martens, A. (2011). Content Design Patterns for Game-Based Learning. [IJGBL]. *International Journal of Game-Based Learning*, *1*(3), 65–82. doi:10.4018/ijgbl.2011070105

Management Association. I. (2011). Development of Game-Based Training Systems. In Instructional Design: Concepts, Methodologies, Tools and Applications (pp. 431-463). Hershey, PA: Information Science Reference. doi:10.4018/978-1-60960-503-2.ch215

Mancuso, V. P., Hamilton, K., Tesler, R., Mohammed, S., & McNeese, M. (2013). An Experimental Evaluation of the Effectiveness of Endogenous and Exogenous Fantasy in Computer-Based Simulation Training. [IJGCMS]. *International Journal of Gaming and Computer-Mediated Simulations*, *5*(1), 50–65. doi:10.4018/jgcms.2013010104

Manessis, D. (2014). The Importance of Future Kindergarten Teachers' Beliefs about the Usefulness of Games Based Learning. [IJGBL]. *International Journal of Game-Based Learning*, *4*(1), 78–90. doi:10.4018/IJGBL.2014010105

Marache-Francisco, C., & Brangier, E. (2014). The Gamification Experience: UXD with a Gamification Background. In K. Blashki, & P. Isaias (Eds.), *Emerging Research and Trends in Interactivity and the Human-Computer Interface* (pp. 205–223). Hershey, PA: Information Science Reference.

Marklund, B. B., Backlund, P., Dahlin, C., Engström, H., & Wilhelmsson, U. (2014). A Game Based Approach to Support Social Presence and Awareness in Distributed Project-Based Learning. [IJGBL]. *International Journal of Game-Based Learning*, *4*(1), 1–20. doi:10.4018/IJGBL.2014010101

Marlow, C. M. (2012). Making Games for Environmental Design Education: Revealing Landscape Architecture. [IJGCMS]. *International Journal of Gaming and Computer-Mediated Simulations*, *4*(2), 60–83. doi:10.4018/jgcms.2012040104

Marques, L. B., & das Graças de Souza, D. (2013). Behavioral Evaluation of Preference for Game-Based Teaching Procedures. [IJGBL]. *International Journal of Game-Based Learning*, *3*(1), 51–62. doi:10.4018/ijgbl.2013010104

Marston, H. R., & McClenaghan, P. A. (2013). Play Yourself Fit: Exercise + Videogames = Exergames. In K. Bredl, & W. Bösche (Eds.), *Serious Games and Virtual Worlds in Education, Professional Development, and Healthcare* (pp. 241–257). Hershey, PA: Information Science Reference.

Martinez-Garza, M. M., & Clark, D. (2013). Teachers and Teaching in Game-Based Learning Theory and Practice. In M. Khine, & I. Saleh (Eds.), *Approaches and Strategies in Next Generation Science Learning* (pp. 147–163). Hershey, PA: Information Science Reference.

Martinez-Garza, M. M., Clark, D., & Nelson, B. (2013). Advances in Assessment of Students' Intuitive Understanding of Physics through Gameplay Data. [IJGCMS]. *International Journal of Gaming and Computer-Mediated Simulations*, *5*(4), 1–16. doi:10.4018/ijgcms.2013100101

Martins, H. F. (2012). The Use of a Business Simulation Game in a Management Course. In M. Cruz-Cunha (Ed.), *Handbook of Research on Serious Games as Educational, Business and Research Tools* (pp. 693–707). Hershey, PA: Information Science Reference.

Martins, T., Carvalho, V., & Soares, F. (2012). An Overview on the Use of Serious Games in Physical Therapy and Rehabilitation. In M. Cruz-Cunha (Ed.), *Handbook of Research on Serious Games as Educational, Business and Research Tools* (pp. 1175–1187). Hershey, PA: Information Science Reference.

Martins, T., Carvalho, V., & Soares, F. (2014). An Overview on the Use of Serious Games in Physical Therapy and Rehabilitation. In I. Management Association (Ed.),Assistive Technologies: Concepts, Methodologies, Tools, and Applications (pp. 758-770). Hershey, PA: Information Science Reference. doi:10.4018/978-1-4666-4422-9.ch038

Mayer, I., Bekebrede, G., Warmelink, H., & Zhou, Q. (2014). A Brief Methodology for Researching and Evaluating Serious Games and Game-Based Learning. In T. Connolly, T. Hainey, E. Boyle, G. Baxter, & P. Moreno-Ger (Eds.), *Psychology, Pedagogy, and Assessment in Serious Games* (pp. 357–393). Hershey, PA: Information Science Reference.

McCall, R., Koenig, V., & Kracheel, M. (2013). Using Gamification and Metaphor to Design a Mobility Platform for Commuters. [IJMHCI]. *International Journal of Mobile Human Computer Interaction*, *5*(1), 1–15. doi:10.4018/jmhci.2013010101

McDaniel, R., & Fiore, S. M. (2012). Best Practices for the Design and Development of Ethical Learning Video Games. [IJCEE]. *International Journal of Cyber Ethics in Education*, *2*(4), 1–23. doi:10.4018/ijcee.2012100101

McDaniel, R., & Kenny, R. (2013). Evaluating the Relationship between Cognitive Style and Pre-Service Teachers' Preconceived Notions about Adopting Console Video Games for Use in Future Classrooms. [IJGBL]. *International Journal of Game-Based Learning*, *3*(2), 55–76. doi:10.4018/ijgbl.2013040104

McLean, L., & Griffiths, M. D. (2013). Female Gamers: A Thematic Analysis of Their Gaming Experience. [IJGBL]. *International Journal of Game-Based Learning*, *3*(3), 54–71. doi:10.4018/ijgbl.2013070105

McLean, L., & Griffiths, M. D. (2013). Gamers' Attitudes towards Victims of Crime: An Interview Study Using Vignettes. [IJCBPL]. *International Journal of Cyber Behavior, Psychology and Learning*, *3*(2), 13–33. doi:10.4018/ijcbpl.2013040102

McLean, L., & Griffiths, M. D. (2013). Violent Video Games and Attitudes Towards Victims of Crime: An Empirical Study Among Youth. [IJCBPL]. *International Journal of Cyber Behavior, Psychology and Learning*, *3*(3), 1–16. doi:10.4018/ijcbpl.2013070101

McQuire, S., Pedell, S., Gibbs, M., Vetere, F., Papastergiadis, N., & Downs, J. (2012). Public Screens: From Display to Interaction. [IJEPR]. *International Journal of E-Planning Research*, *1*(2), 23–43. doi:10.4018/ijepr.2012040102

Mehl-Schneider, T., & Steinmetz, S. (2014). Video Games as a Form of Therapeutic Intervention for Children with Autism Spectrum Disorders. In N. Silton (Ed.), *Innovative Technologies to Benefit Children on the Autism Spectrum* (pp. 197–211). Hershey, PA: Medical Information Science Reference.

Mehm, F., Göbel, S., & Steinmetz, R. (2013). An Authoring Tool for Educational Adventure Games: Concept, Game Models and Authoring Processes. [IJGBL]. *International Journal of Game-Based Learning*, *3*(1), 63–79. doi:10.4018/ijgbl.2013010105

Meletiou-Mavrotheris, M. (2013). Integrating Game-Enhanced Mathematics Learning into the Pre-Service Training of Teachers. In S. de Freitas, M. Ott, M. Popescu, & I. Stanescu (Eds.), *New Pedagogical Approaches in Game Enhanced Learning: Curriculum Integration* (pp. 159–179). Hershey, PA: Information Science Reference.

Meletiou-Mavrotheris, M. (2014). Integrating Game-Enhanced Mathematics Learning into the Pre-Service Training of Teachers. In I. Management Association (Ed.), K-12 Education: Concepts, Methodologies, Tools, and Applications (pp. 1555-1575). Hershey, PA: Information Science Reference. doi:10.4018/978-1-4666-4502-8.ch090

Mena, R. J. (2014). The Quest for a Massively Multiplayer Online Game that Teaches Physics. In T. Connolly, T. Hainey, E. Boyle, G. Baxter, & P. Moreno-Ger (Eds.), *Psychology, Pedagogy, and Assessment in Serious Games* (pp. 292–316). Hershey, PA: Information Science Reference.

Méndez, M. R., & Martínez, E. C. (2013). GAMESTAR(T): An AR-SGAMES Project. In C. Gonzalez (Ed.), *Student Usability in Educational Software and Games: Improving Experiences* (pp. 336–371). Hershey, PA: Information Science Reference.

Metcalf, D., Graffeo, C., & Read, L. (2011). Fundamental Design Elements of Pervasive Games for Blended Learning. In A. Kitchenham (Ed.), *Blended Learning across Disciplines: Models for Implementation* (pp. 148–172). Hershey, PA: Information Science Reference.

Metcalf, S., Kamarainen, A., Tutwiler, M. S., Grotzer, T., & Dede, C. (2011). Ecosystem Science Learning via Multi-User Virtual Environments. [IJGCMS]. *International Journal of Gaming and Computer-Mediated Simulations, 3*(1), 86–90. doi:10.4018/jgcms.2011010107

Miller, K. B. (2013). Gaming as a Woman: Gender Difference Issues in Video Games and Learning. In S. D'Agustino (Ed.), *Immersive Environments, Augmented Realities, and Virtual Worlds: Assessing Future Trends in Education* (pp. 106–122). Hershey, PA: Information Science Reference.

Milolidakis, G., Kimble, C., & Grenier, C. (2011). A Practice-Based Analysis of Social Interaction in a Massively Multiplayer Online Gaming Environment. In M. Cruz-Cunha, V. Varvalho, & P. Tavares (Eds.), *Business, Technological, and Social Dimensions of Computer Games: Multidisciplinary Developments* (pp. 32–48). Hershey, PA: Information Science Reference.

Minovic, M., Milovanovic, M., Minovic, J., & Starcevic, D. (2012). Integrating an Educational Game in Moodle LMS. [IJDET]. *International Journal of Distance Education Technologies, 10*(4), 17–25. doi:10.4018/jdet.2012100102

Minovic, M., Milovanovic, M., & Starcevic, D. (2011). Delivering Educational Games to Mobile Devices. [IJKSR]. *International Journal of Knowledge Society Research, 2*(2), 47–58. doi:10.4018/jksr.2011040105

Mitgutsch, K. (2011). Playful Learning Experiences: Meaningful Learning Patterns in Players' Biographies. [IJGCMS]. *International Journal of Gaming and Computer-Mediated Simulations, 3*(3), 54–68. doi:10.4018/jgcms.2011070104

Monjelat, N., Checa, M., Varela, A. B., Del Castillo, H., & Herrero, D. (2013). Using the Sims 3 for Narrative Construction in Secondary Education: A Multimedia Experience in Language Classes. In S. de Freitas, M. Ott, M. Popescu, & I. Stanescu (Eds.), *New Pedagogical Approaches in Game Enhanced Learning: Curriculum Integration* (pp. 180–213). Hershey, PA: Information Science Reference.

Moore, D. R., & Hsiao, E. (2012). Concept Learning and the Limitations of Arcade-Style Games. [IJGBL]. *International Journal of Game-Based Learning, 2*(3), 1–10. doi:10.4018/ijgbl.2012070101

Möring, S. (2013). The Metaphor-Simulation Paradox in the Study of Computer Games. [IJGCMS]. *International Journal of Gaming and Computer-Mediated Simulations*, *5*(4), 48–74. doi:10.4018/ijgcms.2013100103

Moseley, A. (2012). An Alternate Reality for Education?: Lessons to be Learned from Online Immersive Games. [IJGBL]. *International Journal of Game-Based Learning*, *2*(3), 32–50. doi:10.4018/ijgbl.2012070103

Moseley, A. (2014). A Case for Integration: Assessment and Games. In T. Connolly, T. Hainey, E. Boyle, G. Baxter, & P. Moreno-Ger (Eds.), *Psychology, Pedagogy, and Assessment in Serious Games* (pp. 342–356). Hershey, PA: Information Science Reference.

Mourato, F., Birra, F., & Próspero dos Santos, M. (2013). Using Graph-Based Analysis to Enhance Automatic Level Generation for Platform Videogames. [IJCICG]. *International Journal of Creative Interfaces and Computer Graphics*, *4*(1), 49–70. doi:10.4018/ijcicg.2013010104

Murphy, J., & Zagal, J. (2011). Videogames and the Ethics of Care. [IJGCMS]. *International Journal of Gaming and Computer-Mediated Simulations*, *3*(3), 69–81. doi:10.4018/jgcms.2011070105

Nacke, L. E., & Grimshaw, M. (2011). Player-Game Interaction Through Affective Sound. In M. Grimshaw (Ed.), *Game Sound Technology and Player Interaction: Concepts and Developments* (pp. 264–285). Hershey, PA: Information Science Reference.

Nahl, D., & James, L. (2013). Gamification in Instruction and the Management of Intersubjectivity in Online University Courses. [IJWP]. *International Journal of Web Portals*, *5*(2), 48–62. doi:10.4018/jwp.2013040104

Nap, H. H., & Diaz-Orueta, U. (2013). Rehabilitation Gaming. In S. Arnab, I. Dunwell, & K. Debattista (Eds.), *Serious Games for Healthcare: Applications and Implications* (pp. 50–75). Hershey, PA: Medical Information Science Reference.

Nap, H. H., & Diaz-Orueta, U. (2014). Rehabilitation Gaming. In J. Bishop (Ed.), *Gamification for Human Factors Integration: Social, Education, and Psychological Issues* (pp. 122–147). Hershey, PA: Information Science Reference.

Neshati, R., & Daim, T. (2012). Multidimensional Assessment of Emerging Technologies: Case of Next Generation Internet and Online Gaming Application. In J. Wang (Ed.), *Societal Impacts on Information Systems Development and Applications* (pp. 1–23). Hershey, PA: Information Science Reference.

Neto, J., & Mendes, P. (2012). Game4Manager: More Than Virtual Managers. In M. Cruz-Cunha (Ed.), *Handbook of Research on Serious Games as Educational, Business and Research Tools* (pp. 108–134). Hershey, PA: Information Science Reference.

Ng, E. M. (2011). Exploring the Gender Differences of Student Teachers when using an Educational Game to Learn Programming Concepts. In P. Felicia (Ed.), *Handbook of Research on Improving Learning and Motivation through Educational Games: Multidisciplinary Approaches* (pp. 550–566). Hershey, PA: Information Science Reference.

Ninaus, M., Witte, M., Kober, S. E., Friedrich, E. V., Kurzmann, J., & Hartsuiker, E. et al. (2014). Neurofeedback and Serious Games. In T. Connolly, T. Hainey, E. Boyle, G. Baxter, & P. Moreno-Ger (Eds.), *Psychology, Pedagogy, and Assessment in Serious Games* (pp. 82–110). Hershey, PA: Information Science Reference.

Nygren, E., Sutinen, E., Blignaut, A. S., Laine, T. H., & Els, C. J. (2012). Motivations for Play in the UFractions Mobile Game in Three Countries. [IJMBL]. *International Journal of Mobile and Blended Learning*, *4*(2), 30–48. doi:10.4018/jmbl.2012040103

Oksanen, K., & Hämäläinen, R. (2013). Perceived Sociability and Social Presence in a Collaborative Serious Game. [IJGBL]. *International Journal of Game-Based Learning*, *3*(1), 34–50. doi:10.4018/ijgbl.2013010103

Olla, V. (2012). Case Study of Game-Based Learning in a Citizenship Education K-12 Classroom: Opportunities and Challenges. In I. Chen, & D. McPheeters (Eds.), *Cases on Educational Technology Integration in Urban Schools* (pp. 154–169). Hershey, PA: Information Science Reference.

Olla, V. (2014). Case Study of Game-Based Learning in a Citizenship Education K-12 Classroom: Opportunities and Challenges. In I. Management Association (Ed.), K-12 Education: Concepts, Methodologies, Tools, and Applications (pp. 881-891). Hershey, PA: Information Science Reference. doi:10.4018/978-1-4666-4502-8.ch052

Orr, K., & McGuinness, C. (2014). What is the "Learning" in Games-Based Learning? In T. Connolly, T. Hainey, E. Boyle, G. Baxter, & P. Moreno-Ger (Eds.) Psychology, Pedagogy, and Assessment in Serious Games (pp. 221-242). Hershey, PA: Information Science Reference. doi:10.4018/978-1-4666-4773-2.ch011

Ortiz, J. A. (2011). Knowing the Game: A Review of Videogames and Entertainment Software in the United States - Trends and Future Research Opportunities. In M. Cruz-Cunha, V. Varvalho, & P. Tavares (Eds.), *Business, Technological, and Social Dimensions of Computer Games: Multidisciplinary Developments* (pp. 293–311). Hershey, PA: Information Science Reference.

Ortiz de Gortari, A. B., Aronsson, K., & Griffiths, M. (2011). Game Transfer Phenomena in Video Game Playing: A Qualitative Interview Study. [IJCBPL]. *International Journal of Cyber Behavior, Psychology and Learning*, *1*(3), 15–33. doi:10.4018/ijcbpl.2011070102

Ortiz de Gortari, A. B., Aronsson, K., & Griffiths, M. (2013). Game Transfer Phenomena in Video Game Playing: A Qualitative Interview Study. In R. Zheng (Ed.), *Evolving Psychological and Educational Perspectives on Cyber Behavior* (pp. 170–189). Hershey, PA: Information Science Reference.

Ott, M., Popescu, M. M., Stanescu, I. A., & de Freitas, S. (2013). Game-Enhanced Learning: Preliminary Thoughts on Curriculum Integration. In S. de Freitas, M. Ott, M. Popescu, & I. Stanescu (Eds.), *New Pedagogical Approaches in Game Enhanced Learning: Curriculum Integration* (pp. 1–19). Hershey, PA: Information Science Reference.

Özkan-Czerkawski, B. (2012). Digital Games: Are They the Future of E-Learning Environments? In H. Wang (Ed.), *Interactivity in E-Learning: Case Studies and Frameworks* (pp. 166–178). Hershey, PA: Information Science Reference.

Paavilainen, J., Korhonen, H., & Saarenpää, H. (2012). Comparing Two Playability Heuristic Sets with Expert Review Method: A Case Study of Mobile Game Evaluation. In A. Lugmayr, H. Franssila, P. Näränen, O. Sotamaa, J. Vanhala, & Z. Yu (Eds.), *Media in the Ubiquitous Era: Ambient, Social and Gaming Media* (pp. 29–52). Hershey, PA: Information Science Reference.

Pallot, M., Le Marc, C., Richir, S., Schmidt, C., & Mathieu, J. (2012). Innovation Gaming: An Immersive Experience Environment Enabling Co-creation. In M. Cruz-Cunha (Ed.), *Handbook of Research on Serious Games as Educational, Business and Research Tools* (pp. 1–24). Hershey, PA: Information Science Reference.

Paraskeva, F., Mysirlaki, S., & Vasileiou, V. N. (2012). Developing Self-Regulation Skills in Virtual Worlds: An Educational Scenario Applied in Second Life. [IJOPCD]. *International Journal of Online Pedagogy and Course Design*, *2*(2), 25–43. doi:10.4018/ijopcd.2012040103

Parisi, D. (2011). Game Interfaces as Bodily Techniques. In I. Management Association (Ed.), Gaming and Simulations: Concepts, Methodologies, Tools and Applications (pp. 1033-1047). Hershey, PA: Information Science Reference. doi:10.4018/978-1-60960-195-9.ch409

Park, H. (2013). Effect of Self-Directed Learning Readiness by Learner's Interaction on Social Network Games. [JCIT]. *Journal of Cases on Information Technology*, *15*(3), 47–60. doi:10.4018/jcit.2013070104

Park, H. (2013). Interaction to Facilitate Learning in Social Network Gaming. In H. Yang, & S. Wang (Eds.), *Cases on E-Learning Management: Development and Implementation* (pp. 145–161). Hershey, PA: Information Science Reference.

Parker, J. R., & Becker, K. (2013). The Simulation-Game Controversy: What is a Ludic Simulation? [IJGCMS]. *International Journal of Gaming and Computer-Mediated Simulations*, *5*(1), 1–12. doi:10.4018/jgcms.2013010101

Pearcy, M. (2012). America's Army: "Playful Hatred" in the Social Studies Classroom. [IJGCMS]. *International Journal of Gaming and Computer-Mediated Simulations*, *4*(2), 19–36. doi:10.4018/jgcms.2012040102

Pendegraft, N. (2011). Simulation, Games, and Virtual Environments in IT Education. In I. Management Association (Ed.), Gaming and Simulations: Concepts, Methodologies, Tools and Applications (pp. 1383-1390). Hershey, PA: Information Science Reference. doi:10.4018/978-1-60960-195-9.ch509

Pereira, J. (2013). Beyond Hidden Bodies and Lost Pigs: Student Perceptions of Foreign Language Learning with Interactive Fiction. In Y. Baek, & N. Whitton (Eds.), *Cases on Digital Game-Based Learning: Methods, Models, and Strategies* (pp. 50–80). Hershey, PA: Information Science Reference.

Perry, J. C., Andureu, J., Cavallaro, F. I., Veneman, J., Carmien, S., & Keller, T. (2011). Effective Game use in Neurorehabilitation: User-Centered Perspectives. In P. Felicia (Ed.), *Handbook of Research on Improving Learning and Motivation through Educational Games: Multidisciplinary Approaches* (pp. 683–725). Hershey, PA: Information Science Reference.

Petley, R., Attewell, J., & Savill-Smith, C. (2011). Not Just Playing Around: The MoLeNET Experience of Using Games Technologies to Support Teaching and Learning. [IJVPLE]. *International Journal of Virtual and Personal Learning Environments*, *2*(2), 59–72. doi:10.4018/jvple.2011040106

Petley, R., Parker, G., & Attewell, J. (2011). The Mobile Learning Network: Getting Serious about Games Technologies for Learning. [IJGBL]. *International Journal of Game-Based Learning, 1*(4), 37–48. doi:10.4018/ijgbl.2011100104

Petley, R., Parker, G., & Attewell, J. (2013). The Mobile Learning Network: Getting Serious about Games Technologies for Learning. In P. Felicia (Ed.), *Developments in Current Game-Based Learning Design and Deployment* (pp. 91–102). Hershey, PA: Information Science Reference.

Pivec, P., & Pivec, M. (2011). Digital Games: Changing Education, One Raid at a Time. [IJGBL]. *International Journal of Game-Based Learning, 1*(1), 1–18. doi:10.4018/ijgbl.2011010101

Pragnell, C., & Gatzidis, C. (2012). Addiction in World of Warcraft: A Virtual Ethnography Study. In H. Yang, & S. Yuen (Eds.), *Handbook of Research on Practices and Outcomes in Virtual Worlds and Environments* (pp. 54–74). Hershey, PA: Information Science Reference.

Prayaga, L., Coffey, J. W., & Rasmussen, K. (2011). Strategies to Teach Game Development Across Age Groups. [IJGCMS]. *International Journal of Gaming and Computer-Mediated Simulations, 3*(2), 28–43. doi:10.4018/jgcms.2011040102

Prayaga, L., Coffey, J. W., & Rasmussen, K. (2013). Strategies to Teach Game Development Across Age Groups. In R. Ferdig (Ed.), *Design, Utilization, and Analysis of Simulations and Game-Based Educational Worlds* (pp. 95–110). Hershey, PA: Information Science Reference.

Prescott, J., & Bogg, J. (2013). The Computer Games Industry: New Industry, Same Old Issues. In Gendered Occupational Differences in Science, Engineering, and Technology Careers (pp. 64-77). Hershey, PA: Information Science Reference. doi:10.4018/978-1-4666-2107-7.ch003

Prescott, J., & Bogg, J. (2014). Games and Society: Can Games Make a Better World? In Gender Divide and the Computer Game Industry (pp. 51-72). Hershey, PA: Information Science Reference. doi:10.4018/978-1-4666-4534-9.ch003

Prescott, J., & Bogg, J. (2014). Game Workers and the Gender Divide in the Production of Computer Games. In *Gender Divide and the Computer Game Industry* (pp. 123–146). Hershey, PA: Information Science Reference.

Prescott, J., & Bogg, J. (2014). Introduction: Why the Gender Divide in Computer Games is an Important and Timely Issue. In Gender Divide and the Computer Game Industry (pp. 1-27). Hershey, PA: Information Science Reference. doi:10.4018/978-1-4666-4534-9.ch001

Prescott, J., & Bogg, J. (2014). Play Preferences and the Gendering of Gaming. In *Gender Divide and the Computer Game Industry* (pp. 73–96). Hershey, PA: Information Science Reference.

Prescott, J., & Bogg, J. (2014). The Computer Game Industry, Market, and Culture. In *Gender Divide and the Computer Game Industry* (pp. 28–50). Hershey, PA: Information Science Reference.

Prescott, J., & Bogg, J. (2014). The Experience of Women Game Developers. In *Gender Divide and the Computer Game Industry* (pp. 147–169). Hershey, PA: Information Science Reference.

Preston, J. A., Chastine, J., O'Donnell, C., Tseng, T., & MacIntyre, B. (2012). Game Jams: Community, Motivations, and Learning among Jammers. [IJGBL]. *International Journal of Game-Based Learning*, *2*(3), 51–70. doi:10.4018/ijgbl.2012070104

Qian, Y. (2011). An Investigation of Current Online Educational Games. In I. Management Association (Ed.), Gaming and Simulations: Concepts, Methodologies, Tools and Applications (pp. 1666-1681). Hershey, PA: Information Science Reference. doi:10.4018/978-1-60960-195-9.ch702

Quick, J. M., Atkinson, R. K., & Lin, L. (2012). Empirical Taxonomies of Gameplay Enjoyment: Personality and Video Game Preference. [IJGBL]. *International Journal of Game-Based Learning*, *2*(3), 11–31. doi:10.4018/ijgbl.2012070102

Quick, J. M., Atkinson, R. K., & Lin, L. (2012). The Gameplay Enjoyment Model. [IJGCMS]. *International Journal of Gaming and Computer-Mediated Simulations*, *4*(4), 64–80. doi:10.4018/jgcms.2012100105

Quiroga, M. Á., Román, F. J., Catalán, A., Rodríguez, H., Ruiz, J., & Herranz, M. et al. (2011). Videogame Performance (Not Always) Requires Intelligence. [IJOPCD]. *International Journal of Online Pedagogy and Course Design*, *1*(3), 18–32. doi:10.4018/ijopcd.2011070102

Rai, D., & Beck, J. E. (2012). Math Learning Environment with Game-Like Elements: An Experimental Framework. [IJGBL]. *International Journal of Game-Based Learning*, *2*(2), 90–110. doi:10.4018/ijgbl.2012040106

Raisamo, R., Walldén, S., Suhonen, K., Myllymaa, K., Raisamo, S., & Vänni, K. (2012). Design and Evaluation of Tamhattan: A Multimodal Game Promoting Awareness of Health in a Social and Positive Way. In M. Cruz-Cunha (Ed.), *Handbook of Research on Serious Games as Educational, Business and Research Tools* (pp. 90–107). Hershey, PA: Information Science Reference.

Ray, B. B., Hocutt, M. M., & Hooley, D. (2014). Use of an Online Simulation to Promote Content Learning. [IJOPCD]. *International Journal of Online Pedagogy and Course Design*, *4*(1), 43–57. doi:10.4018/ijopcd.2014010104

Raybourn, E. M. (2011). Honing Emotional Intelligence with Game-Based Crucible Experiences. [IJGBL]. *International Journal of Game-Based Learning*, *1*(1), 32–44. doi:10.4018/ijgbl.2011010103

Razak, A. A., Connolly, T., & Hainey, T. (2012). Teachers' Views on the Approach of Digital Games-Based Learning within the Curriculum for Excellence. [IJGBL]. *International Journal of Game-Based Learning*, *2*(1), 33–51. doi:10.4018/ijgbl.2012010103

Regazzi, J. J. (2014). Learning and Playing get Personal. In Infonomics and the Business of Free: Modern Value Creation for Information Services (pp. 129-149). Hershey, PA: Business Science Reference. doi:10.4018/978-1-4666-4454-0.ch007

Rego, P. A., Moreira, P. M., & Reis, L. P. (2012). New Forms of Interaction in Serious Games for Rehabilitation. In M. Cruz-Cunha (Ed.), *Handbook of Research on Serious Games as Educational, Business and Research Tools* (pp. 1188–1211). Hershey, PA: Information Science Reference.

Remmele, B., & Whitton, N. (2014). Disrupting the Magic Circle: The Impact of Negative Social Gaming Behaviours. In T. Connolly, T. Hainey, E. Boyle, G. Baxter, & P. Moreno-Ger (Eds.), *Psychology, Pedagogy, and Assessment in Serious Games* (pp. 111–126). Hershey, PA: Information Science Reference.

Reymers, K. (2011). Chicken Killers or Bandwidth Patriots?: A Case Study of Ethics in Virtual Reality. [IJT]. *International Journal of Technoethics*, *2*(3), 1–22. doi:10.4018/jte.2011070101

Ribeiro, C., Monteiro, M., Corredoura, S., Candeias, F., & Pereira, J. (2013). Games in Higher Education: Opportunities, Expectations, Challenges, and Results in Medical Education. In S. de Freitas, M. Ott, M. Popescu, & I. Stanescu (Eds.), *New Pedagogical Approaches in Game Enhanced Learning: Curriculum Integration* (pp. 228–247). Hershey, PA: Information Science Reference.

Ribeiro, C., Pereira, J., Calado, C., & Ferreira, C. (2013). Challenges of Introducting Serious Games and Virtual Worlds in Educational Curriculum. In Y. Baek, & N. Whitton (Eds.), *Cases on Digital Game-Based Learning: Methods, Models, and Strategies* (pp. 425–450). Hershey, PA: Information Science Reference.

Rice, J. W. (2012). The Gamification of Learning and Instruction: Game-Based Methods and Strategies for Training and Education. [IJGCMS]. *International Journal of Gaming and Computer-Mediated Simulations*, *4*(4), 81–83. doi:10.4018/jgcms.2012100106

Richard, G. T. (2014). Supporting Visibility and Resilience in Play: Gender-Supportive Online Gaming Communities as a Model of Identity and Confidence Building in Play and Learning. In D. Hickey, & J. Essid (Eds.), *Identity and Leadership in Virtual Communities: Establishing Credibility and Influence* (pp. 170–186). Hershey, PA: Information Science Reference.

Ritzhaupt, A. D., Frey, C., Poling, N., & Johnson, M. C. (2012). Playing Games in School: Video Games and Simulations for Primary and Secondary Education. [IJGCMS]. *International Journal of Gaming and Computer-Mediated Simulations*, *4*(2), 84–88. doi:10.4018/jgcms.2012040105

Rodrigues, R. G., Pinheiro, P. G., & Barbosa, J. (2012). Online Playability: The Social Dimension to the Virtual World. In M. Cruz-Cunha (Ed.), *Handbook of Research on Serious Games as Educational, Business and Research Tools* (pp. 391–421). Hershey, PA: Information Science Reference.

Rodríguez-Hoyos, C., & Gomes, M. J. (2012). Exploring the Educational Power of Serious Games: A Review of Literature. In M. Cruz-Cunha (Ed.), *Handbook of Research on Serious Games as Educational, Business and Research Tools* (pp. 928–946). Hershey, PA: Information Science Reference.

Romero, M., & Usart, M. (2013). Learning with the Support of a Digital Game in the Introduction to Finance Class: Analysis of the Students' Perception of the Game's Ease of Use and Usefulness. In Y. Baek, & N. Whitton (Eds.), *Cases on Digital Game-Based Learning: Methods, Models, and Strategies* (pp. 495–508). Hershey, PA: Information Science Reference.

Rooney, P. (2012). A Theoretical Framework for Serious Game Design: Exploring Pedagogy, Play and Fidelity and their Implications for the Design Process. [IJGBL]. *International Journal of Game-Based Learning*, 2(4), 41–60. doi:10.4018/ijgbl.2012100103

Rosas, O. V., & Dhen, G. (2012). One Self to Rule Them All: A Critical Discourse Analysis of French-Speaking Players' Identity Construction in World of Warcraft. In N. Zagalo, L. Morgado, & A. Boa-Ventura (Eds.), *Virtual Worlds and Metaverse Platforms: New Communication and Identity Paradigms* (pp. 337–366). Hershey, PA: Information Science Reference.

Ruggiero, D. (2013). Persuasive Games as Social Action Agents: Challenges and Implications in Learning and Society. [IJGCMS]. *International Journal of Gaming and Computer-Mediated Simulations*, *5*(4), 75–85. doi:10.4018/ijgcms.2013100104

San Millán, E. D., & Priego, R. G. (2014). Learning by Playing: Is Gamification a Keyword in the New Education Paradigm? In F. García-Peñalvo, & A. Seoane Pardo (Eds.), *Online Tutor 2.0: Methodologies and Case Studies for Successful Learning* (pp. 16–69). Hershey, PA: Information Science Reference.

Sandaruwan, D., Kodikara, N., Keppitiyagama, C., Rosa, R., Dias, K., Senadheera, R., & Manamperi, K. (2012). Low Cost Immersive VR Solutions for Serious Gaming. In A. Kumar, J. Etheredge, & A. Boudreaux (Eds.), *Algorithmic and Architectural Gaming Design: Implementation and Development* (pp. 407–429). Hershey, PA: Information Science Reference.

Santo, A. E., Rijo, R., Monteiro, J., Henriques, I., Matos, A., & Rito, C. et al. (2012). Games Improving Disorders of Attention Deficit and Hyperactivity. In M. Cruz-Cunha (Ed.), *Handbook of Research on Serious Games as Educational, Business and Research Tools* (pp. 1160–1174). Hershey, PA: Information Science Reference.

Santo, A. E., Rijo, R., Monteiro, J., Henriques, I., Matos, A., Rito, C., et al. (2014). Games Improving Disorders of Attention Deficit and Hyperactivity. In I. Management Association (Ed.), K-12 Education: Concepts, Methodologies, Tools, and Applications (pp. 563-577). Hershey, PA: Information Science Reference. doi:10.4018/978-1-4666-4502-8.ch033

Sardone, N., & Devlin-Scherer, R. (2011). Multi-User Virtual Learning Environments in Education. In I. Management Association (Ed.), Gaming and Simulations: Concepts, Methodologies, Tools and Applications (pp. 1967-1980). Hershey, PA: Information Science Reference. doi:10.4018/978-1-60960-195-9.ch805

Saridaki, M., & Mourlas, C. (2011). Motivational Aspects of Gaming for Students with Intellectual Disabilities. [IJGBL]. *International Journal of Game-Based Learning*, *1*(4), 49–59. doi:10.4018/ijgbl.2011100105

Saridaki, M., & Mourlas, C. (2011). Motivating the Demotivated Classroom: Gaming as a Motivational Medium for Students with Intellectual Disability and their Educators. In P. Felicia (Ed.), *Handbook of Research on Improving Learning and Motivation through Educational Games: Multidisciplinary Approaches* (pp. 759–775). Hershey, PA: Information Science Reference.

Saridaki, M., & Mourlas, C. (2013). Integrating Serious Games in the Educational Experience of Students with Intellectual Disabilities: Towards a Playful and Integrative Model. [IJGBL]. *International Journal of Game-Based Learning*, *3*(3), 10–20. doi:10.4018/ijgbl.2013070102

Saridaki, M., & Mourlas, C. (2013). Motivational Aspects of Gaming for Students with Intellectual Disabilities. In P. Felicia (Ed.), *Developments in Current Game-Based Learning Design and Deployment* (pp. 144–154). Hershey, PA: Information Science Reference.

Sauvé, L., Renaud, L., & Kaufman, D. (2011). Games, Simulations, and Simulation Games for Learning: Definitions and Distinctions. In I. Management Association (Ed.),Gaming and Simulations: Concepts, Methodologies, Tools and Applications (pp. 168-193). Hershey, PA: Information Science Reference. doi:10.4018/978-1-60960-195-9.ch110

Sawyer, B. (2011). Research Essay: What Will Serious Games of the Future Look Like? [IJGCMS]. *International Journal of Gaming and Computer-Mediated Simulations*, *3*(3), 82–90. doi:10.4018/jgcms.2011070106

Schiller, E., Schultes, M., Strohmeier, D., & Spiel, C. (2011). Gaming and Aggression: The Importance of Age-Appropriateness in Violent Video Games. In E. Dunkels, G. Franberg, & C. Hallgren (Eds.), *Youth Culture and Net Culture: Online Social Practices* (pp. 316–337). Hershey, PA: Information Science Reference.

Schmitz, B., Klemke, R., & Specht, M. (2013). A Learning Outcome-Oriented Approach towards Classifying Pervasive Games for Learning using Game Design Patterns and Contextual Information. [IJMBL]. *International Journal of Mobile and Blended Learning*, *5*(4), 59–71. doi:10.4018/ijmbl.2013100104

Schott, G., & Selwyn, N. (2013). Game Literacy: Assessing its Value for Both Classification and Public Perceptions of Games in a New Zealand Context. In I. Management Association (Ed.), Digital Literacy: Concepts, Methodologies, Tools, and Applications (pp. 347-363). Hershey, PA: Information Science Reference. doi:10.4018/978-1-4666-1852-7.ch018

Schultheiss, D., & Helm, M. (2013). Gaming in School: Factors Influencing the Use of Serious Games in Public Schools in Middle Germany. In K. Bredl, & W. Bösche (Eds.), *Serious Games and Virtual Worlds in Education, Professional Development, and Healthcare* (pp. 145–158). Hershey, PA: Information Science Reference.

Schulzke, M. (2013). The Bioethics of Digital Dystopias. [IJT]. *International Journal of Technoethics, 4*(2), 46–57. doi:10.4018/jte.2013070104

Seitz, D. S., & Uram, C. (2011). Gaming and Simulation. In I. Management Association (Ed.), Instructional Design: Concepts, Methodologies, Tools and Applications (pp. 1006-1022). Hershey, PA: Information Science Reference. doi:10.4018/978-1-60960-503-2.ch417

Seo, K. K., & Johnson, C. (2014). Exploring Massively Multiplayer Online Gaming as an Emerging Trend in Distance Education. In T. Yuzer, & G. Eby (Eds.), *Handbook of Research on Emerging Priorities and Trends in Distance Education: Communication, Pedagogy, and Technology* (pp. 91–102). Hershey, PA: Information Science Reference.

Sharritt, M., & Suthers, D. D. (2011). Levels of Failure and Learning in Games. [IJGCMS]. *International Journal of Gaming and Computer-Mediated Simulations, 3*(4), 54–69. doi:10.4018/jgcms.2011100104

Shelton, A. K. (2012). Two Tickets for Paradise: Gaming and Tourism. In P. Ordóñez de Pablos, R. Tennyson, & J. Zhao (Eds.), *Global Hospitality and Tourism Management Technologies* (pp. 91–101). Hershey, PA: Business Science Reference.

Shelton, B. E., Satwicz, T., & Caswell, T. (2011). Historical Perspectives on Games and Education from the Learning Sciences. [IJGBL]. *International Journal of Game-Based Learning, 1*(3), 83–106. doi:10.4018/ijgbl.2011070106

Shelton, B. E., Satwicz, T., & Caswell, T. (2013). Historical Perspectives on Games and Education from the Learning Sciences. In P. Felicia (Ed.), *Developments in Current Game-Based Learning Design and Deployment* (pp. 339–364). Hershey, PA: Information Science Reference.

Shin, N., Norris, C., & Soloway, E. (2011). Mobile Gaming Environment: Learning and Motivational Effects. In P. Felicia (Ed.), *Handbook of Research on Improving Learning and Motivation through Educational Games: Multidisciplinary Approaches* (pp. 467–481). Hershey, PA: Information Science Reference.

Sierra, W., & Eyman, D. (2013). "I Rolled the Dice with Trade Chat and This is What I Got": Demonstrating Context-Dependent Credibility in Virtual Worlds. In M. Folk, & S. Apostel (Eds.), *Online Credibility and Digital Ethos: Evaluating Computer-Mediated Communication* (pp. 332–352). Hershey, PA: Information Science Reference.

Simão de Vasconcellos, M., & Soares de Araújo, I. (2013). Massively Multiplayer Online Role Playing Games for Health Communication in Brazil. In K. Bredl, & W. Bösche (Eds.), *Serious Games and Virtual Worlds in Education, Professional Development, and Healthcare* (pp. 294–312). Hershey, PA: Information Science Reference.

Šimko, J., Tvarožek, M., & Bieliková, M. (2011). Semantics Discovery via Human Computation Games. [IJSWIS]. *International Journal on Semantic Web and Information Systems*, *7*(3), 23–45. doi:10.4018/jswis.2011070102

Simões, P. D., & Ferreira, C. G. (2012). The Use of Digital Games to Stimulate Behaviors. In M. Cruz-Cunha (Ed.), *Handbook of Research on Serious Games as Educational, Business and Research Tools* (pp. 1145–1159). Hershey, PA: Information Science Reference.

Sintoris, C., Stoica, A., Papadimitriou, I., Yiannoutsou, N., Komis, V., & Avouris, N. (2012). MuseumScrabble: Design of a Mobile Game for Children's Interaction with a Digitally Augmented Cultural Space. In J. Lumsden (Ed.), *Social and Organizational Impacts of Emerging Mobile Devices: Evaluating Use* (pp. 124–142). Hershey, PA: Information Science Reference.

Steinkuehler, C., Alagoz, E., King, E., & Martin, C. (2012). A Cross Case Analysis of Two Out-of-School Programs Based on Virtual Worlds. [IJGCMS]. *International Journal of Gaming and Computer-Mediated Simulations*, *4*(1), 25–54. doi:10.4018/jgcms.2012010102

Svingby, G., & Nilsson, E. M. (2011). Research Review: Empirical Studies on Computer Game Play in Science Education. In P. Felicia (Ed.), *Handbook of Research on Improving Learning and Motivation through Educational Games: Multidisciplinary Approaches* (pp. 1–28). Hershey, PA: Information Science Reference.

Szilas, N., & Acosta, M. (2011). A Theoretical Background for Educational Video Games: Games, Signs, Knowledge. In P. Felicia (Ed.), *Handbook of Research on Improving Learning and Motivation through Educational Games: Multidisciplinary Approaches* (pp. 215–238). Hershey, PA: Information Science Reference.

Tai, Z., & Zeng, H. (2011). Mobile Games in China: Formation, Ferment, and Future. In D. Jin (Ed.), *Global Media Convergence and Cultural Transformation: Emerging Social Patterns and Characteristics* (pp. 276–295). Hershey, PA: Information Science Reference.

Talab, R. S., & S., H. R. (2011). Constructivist and Constructionist Approaches to Graduate Teaching in Second Life: Ethical Considerations and Legal Implications. [IJCEE]. *International Journal of Cyber Ethics in Education*, *1*(1), 36–57. doi:10.4018/ijcee.2011010104

Talbot, T. B. (2013). Playing with Biology: Making Medical Games that Appear Lifelike. [IJGCMS]. *International Journal of Gaming and Computer-Mediated Simulations*, *5*(3), 83–96. doi:10.4018/jgcms.2013070106

Talbot, T. B., Sagae, K., John, B., & Rizzo, A. A. (2012). Sorting Out the Virtual Patient: How to Exploit Artificial Intelligence, Game Technology and Sound Educational Practices to Create Engaging Role-Playing Simulations. [IJGCMS]. *International Journal of Gaming and Computer-Mediated Simulations*, *4*(3), 1–19. doi:10.4018/jgcms.2012070101

Tan, S., Baxa, J., & Spackman, M. P. (2012). Effects of Built-in Audio versus Unrelated Background Music on Performance in an Adventure Role-Playing Game. In R. Ferdig, & S. de Freitas (Eds.), *Interdisciplinary Advancements in Gaming, Simulations and Virtual Environments: Emerging Trends* (pp. 142–164). Hershey, PA: Information Science Reference.

Tan, W. H. (2013). Game Coaching System Design and Development: A Retrospective Case Study of FPS Trainer. [IJGBL]. *International Journal of Game-Based Learning*, *3*(2), 77–90. doi:10.4018/ijgbl.2013040105

Tan, W. H., Neill, S., & Johnston-Wilder, S. (2012). How do Professionals' Attitudes Differ between what Game-Based Learning could Ideally Achieve and what is Usually Achieved. [IJGBL]. *International Journal of Game-Based Learning*, *2*(1), 1–15. doi:10.4018/ijgbl.2012010101

Tang, S., & Hanneghan, M. (2014). Designing Educational Games: A Pedagogical Approach. In J. Bishop (Ed.), *Gamification for Human Factors Integration: Social, Education, and Psychological Issues* (pp. 181–198). Hershey, PA: Information Science Reference.

Teng, C., Jeng, S., Chang, H. K., & Wu, S. (2012). Who Plays Games Online?: The Relationship Between Gamer Personality and Online Game Use. [IJEBR]. *International Journal of E-Business Research, 8*(4), 1–14. doi:10.4018/jebr.2012100101

Tennyson, R. D., & Jorczak, R. L. (2011). Electronic Games Improve Adult Learning in Diverse Populations. In I. Management Association (Ed.), Gaming and Simulations: Concepts, Methodologies, Tools and Applications (pp. 1495-1512). Hershey, PA: Information Science Reference. doi:10.4018/978-1-60960-195-9.ch602

Thacker, S., & Griffiths, M. D. (2012). An Exploratory Study of Trolling in Online Video Gaming. [IJCBPL]. *International Journal of Cyber Behavior, Psychology and Learning, 2*(4), 17–33. doi:10.4018/ijcbpl.2012100102

Thaler, S., Simperl, E., Siorpaes, K., & Wölger, S. (2012). SpotTheLink: A Game-Based Approach to the Alignment of Ontologies. In S. Brüggemann, & C. d'Amato (Eds.), *Collaboration and the Semantic Web: Social Networks, Knowledge Networks, and Knowledge Resources* (pp. 40–63). Hershey, PA: Information Science Reference.

Tiong, T. K., Tianyi, G., Sopra, R., & Sharma, R. S. (2012). The Issue of Fragmentation on Mobile Games Platforms. In R. Sharma, M. Tan, & F. Pereira (Eds.), *Understanding the Interactive Digital Media Marketplace: Frameworks, Platforms, Communities and Issues* (pp. 89–96). Hershey, PA: Information Science Reference.

Tobias, S., Rudy, D., & Ispa, J. (2011). Relations Between Videogame Play and 8th-Graders' Mathematics Achievement. [IJGCMS]. *International Journal of Gaming and Computer-Mediated Simulations, 3*(4), 33–53. doi:10.4018/jgcms.2011100103

Tootell, H., & Freeman, A. (2014). The Applicability of Gaming Elements to Early Childhood Education. In J. Bishop (Ed.), *Gamification for Human Factors Integration: Social, Education, and Psychological Issues* (pp. 225–241). Hershey, PA: Information Science Reference.

Tran, B. (2014). Rhetoric of Play: Utilizing the Gamer Factor in Selecting and Training Employees. In T. Connolly, T. Hainey, E. Boyle, G. Baxter, & P. Moreno-Ger (Eds.), *Psychology, Pedagogy, and Assessment in Serious Games* (pp. 175–203). Hershey, PA: Information Science Reference.

Trepte, S., Reinecke, L., & Behr, K. (2011). Playing Myself or Playing to Win?: Gamers' Strategies of Avatar Creation in Terms of Gender and Sex. In R. Ferdig (Ed.), *Discoveries in Gaming and Computer-Mediated Simulations: New Interdisciplinary Applications* (pp. 329–352). Hershey, PA: Information Science Reference.

Trespalacios, J. H., & Chamberlin, B. (2012). 21st Century Learning: The Role of Serious Games. In M. Cruz-Cunha (Ed.), *Handbook of Research on Serious Games as Educational, Business and Research Tools* (pp. 782–799). Hershey, PA: Information Science Reference.

Tsai, C., Hsu, P., & Tseng, H. (2013). Exploring the Effects of Web-Mediated Game-Based Learning and Self-Regulated Learning on Students' Learning. [IJICTE]. *International Journal of Information and Communication Technology Education*, *9*(2), 39–51. doi:10.4018/jicte.2013040104

Tseng, F., & Teng, C. (2011). An Empirical Investigation into the Sources of Customer Dissatisfaction with Online Games. [IJEBR]. *International Journal of E-Business Research*, *7*(4), 17–30. doi:10.4018/jebr.2011100102

Tseng, F., & Teng, C. (2013). An Empirical Investigation into the Sources of Customer Dissatisfaction with Online Games. In I. Lee (Ed.), *Mobile Applications and Knowledge Advancements in E-Business* (pp. 273–286). Hershey, PA: Business Science Reference.

Tzemopoulos, A. (2014). The Online Community of Second Life and the Residents of Virtual Ability Island. In V. Venkatesh, J. Wallin, J. Castro, & J. Lewis (Eds.), *Educational, Psychological, and Behavioral Considerations in Niche Online Communities* (pp. 275–296). Hershey, PA: Information Science Reference.

Umarov, I., & Mozgovoy, M. (2012). Believable and Effective AI Agents in Virtual Worlds: Current State and Future Perspectives. [IJGCMS]. *International Journal of Gaming and Computer-Mediated Simulations*, *4*(2), 37–59. doi:10.4018/jgcms.2012040103

Ursyn, A. (2014). Challenges in Game Design. In Computational Solutions for Knowledge, Art, and Entertainment: Information Exchange Beyond Text (pp. 413-428). Hershey, PA: Information Science Reference. doi:10.4018/978-1-4666-4627-8.ch021

Usart, M., & Romero, M. (2014). Time Factor Assessment in Game-Based Learning: Time Perspective and Time-on-Task as Individual Differences between Players. In T. Connolly, T. Hainey, E. Boyle, G. Baxter, & P. Moreno-Ger (Eds.), *Psychology, Pedagogy, and Assessment in Serious Games* (pp. 62–81). Hershey, PA: Information Science Reference.

Uzun, L., Ekin, M. T., & Kartal, E. (2013). The Opinions and Attitudes of the Foreign Language Learners and Teachers Related to the Traditional and Digital Games: Age and Gender Differences. [IJGBL]. *International Journal of Game-Based Learning*, *3*(2), 91–111. doi:10.4018/ijgbl.2013040106

van de Laar, B., Reuderink, B., Bos, D. P., & Heylen, D. (2012). Evaluating User Experience of Actual and Imagined Movement in BCI Gaming. In R. Ferdig, & S. de Freitas (Eds.), *Interdisciplinary Advancements in Gaming, Simulations and Virtual Environments: Emerging Trends* (pp. 266–280). Hershey, PA: Information Science Reference.

van Rosmalen, P., Wilson, A., & Hummel, H. G. (2014). Games for and by Teachers and Learners. In T. Connolly, T. Hainey, E. Boyle, G. Baxter, & P. Moreno-Ger (Eds.), *Psychology, Pedagogy, and Assessment in Serious Games* (pp. 243–269). Hershey, PA: Information Science Reference.

van Staalduinen, J. (2011). A First Step towards Integrating Educational Theory and Game Design. In P. Felicia (Ed.), *Handbook of Research on Improving Learning and Motivation through Educational Games: Multidisciplinary Approaches* (pp. 98–117). Hershey, PA: Information Science Reference.

Vandercruysse, S., Vandewaetere, M., & Clarebout, G. (2012). Game-Based Learning: A Review on the Effectiveness of Educational Games. In M. Cruz-Cunha (Ed.), *Handbook of Research on Serious Games as Educational, Business and Research Tools* (pp. 628–647). Hershey, PA: Information Science Reference.

Vitolo, T. M. (2012). Crossing the Chasm: Hurdles to Acceptance and Success of Serious Games. In M. Cruz-Cunha (Ed.), *Handbook of Research on Serious Games as Educational, Business and Research Tools* (pp. 280–289). Hershey, PA: Information Science Reference.

Voulgari, I., & Komis, V. (2011). Collaborative Learning in Massively Multiplayer Online Games: A Review of Social, Cognitive and Motivational Perspectives. In P. Felicia (Ed.), *Handbook of Research on Improving Learning and Motivation through Educational Games: Multidisciplinary Approaches* (pp. 370–394). Hershey, PA: Information Science Reference.

Wallner, G., Kriglstein, S., & Biba, J. (2014). Evaluating Games in Classrooms: A Case Study with DOGeometry. In I. Management Association (Ed.), K-12 Education: Concepts, Methodologies, Tools, and Applications (pp. 1220-1234). Hershey, PA: Information Science Reference. doi:10.4018/978-1-4666-4502-8.ch071

Wang, S. (2012). Effects of Playing a History-Simulation Game: Romance of Three Kingdoms. In R. Ferdig, & S. de Freitas (Eds.), *Interdisciplinary Advancements in Gaming, Simulations and Virtual Environments: Emerging Trends* (pp. 97–119). Hershey, PA: Information Science Reference.

Wang, V., Tucker, J. V., & Haines, K. (2013). Viewing Cybercommunities through the Lens of Modernity: The Case of Second Life. [IJVCSN]. *International Journal of Virtual Communities and Social Networking*, *5*(1), 75–90. doi:10.4018/jvcsn.2013010105

Wanless-Sobel, C. (2011). Educational Gaming Avatars. In I. Management Association (Ed.), Gaming and Simulations: Concepts, Methodologies, Tools and Applications (pp. 1023-1032). Hershey, PA: Information Science Reference. doi:10.4018/978-1-60960-195-9.ch408

Warren, S. J., & Gratch, J. S. (2013). Employing a Critical Lens on Instructor Perceptions of Learning Games: Introduction to a Method. [IJVPLE]. *International Journal of Virtual and Personal Learning Environments*, *4*(3), 1–17. doi:10.4018/jvple.2013070101

Warren, S. J., Jones, G., Dolliver, B., & Stein, R. A. (2012). Investigating Games and Simulations in Educational Research and Theory: Enhancing Academic Communication and Scholarship with a Common Language. [IJGCMS]. *International Journal of Gaming and Computer-Mediated Simulations*, *4*(4), 1–18. doi:10.4018/jgcms.2012100101

Warren, S. J., & Lin, L. (2012). Ethical Considerations for Learning Game, Simulation, and Virtual World Design and Development. In H. Yang, & S. Yuen (Eds.), *Handbook of Research on Practices and Outcomes in Virtual Worlds and Environments* (pp. 1–18). Hershey, PA: Information Science Reference.

Warren, S. J., & Lin, L. (2014). Ethical Considerations for Learning Game, Simulation, and Virtual World Design and Development. In I. Management Association (Ed.), K-12 Education: Concepts, Methodologies, Tools, and Applications (pp. 292-309). Hershey, PA: Information Science Reference. doi:10.4018/978-1-4666-4502-8.ch017

Watson, W. R., & Fang, J. (2012). PBL as a Framework for Implementing Video Games in the Classroom. [IJGBL]. *International Journal of Game-Based Learning*, *2*(1), 77–89. doi:10.4018/ijgbl.2012010105

Weiss, A., & Tettegah, S. (2012). World of Race War: Race and Learning in World of Warcraft. [IJGCMS]. *International Journal of Gaming and Computer-Mediated Simulations*, *4*(4), 33–44. doi:10.4018/jgcms.2012100103

White, M. M. (2012). Designing Tutorial Modalities and Strategies for Digital Games: Lessons from Education. [IJGBL]. *International Journal of Game-Based Learning*, *2*(2), 13–34. doi:10.4018/ijgbl.2012040102

Whitton, N. (2011). Encouraging Engagement in Game-Based Learning. [IJGBL]. *International Journal of Game-Based Learning*, *1*(1), 75–84. doi:10.4018/ijgbl.2011010106

Wiemeyer, J., & Hardy, S. (2013). Serious Games and Motor Learning: Concepts, Evidence, Technology. In K. Bredl, & W. Bösche (Eds.), *Serious Games and Virtual Worlds in Education, Professional Development, and Healthcare* (pp. 197–220). Hershey, PA: Information Science Reference.

Wiemeyer, J., & Schneider, P. (2012). Applying Serious Games to Motor Learning in Sport. [IJGBL]. *International Journal of Game-Based Learning*, *2*(4), 61–73. doi:10.4018/ijgbl.2012100104

Williamson, B., & Sandford, R. (2011). Playful Pedagogies: Cultural and Curricular Approaches to Game-Based Learning in the School Classroom. In P. Felicia (Ed.), *Handbook of Research on Improving Learning and Motivation through Educational Games: Multidisciplinary Approaches* (pp. 846–859). Hershey, PA: Information Science Reference.

Winny, S. (2013). Serious Games as an Instrument of Non-Formal Learning: A Review of Web-Based Learning Experiences on the Issue of Renewable Energy. In K. Bredl, & W. Bösche (Eds.), *Serious Games and Virtual Worlds in Education, Professional Development, and Healthcare* (pp. 174–195). Hershey, PA: Information Science Reference.

Wodehouse, A. J., & Ion, W. J. (2012). Computer Gaming Scenarios for Product Development Teams. In R. Ferdig, & S. de Freitas (Eds.), *Interdisciplinary Advancements in Gaming, Simulations and Virtual Environments: Emerging Trends* (pp. 216–233). Hershey, PA: Information Science Reference.

Wu, Y. (2013). Using Educational Computer Games for Science Teaching: Experiences and Perspectives of Elementary Science Teachers in Taiwan. [IJOPCD]. *International Journal of Online Pedagogy and Course Design*, *3*(4), 16–28. doi:10.4018/ijopcd.2013100102

Xu, C., & Xu, D. N. (2012). From a Game Story to a Real 2D Game. In A. Kumar, J. Etheredge, & A. Boudreaux (Eds.), *Algorithmic and Architectural Gaming Design: Implementation and Development* (pp. 359–389). Hershey, PA: Information Science Reference.

Yates, M., & Hurry, J. (2013). The Moderating Role of Video Game Play in the Relationship Between Stress and Externalising Behaviours in Adolescent Males. [IJCBPL]. *International Journal of Cyber Behavior, Psychology and Learning*, *3*(3), 17–33. doi:10.4018/ijcbpl.2013070102

Young, M., Killen, M., Lee-Kim, J., & Park, Y. (2012). Introducing Cool School: Where Peace Rules and Conflict Resolution can be Fun. [IJGBL]. *International Journal of Game-Based Learning*, *2*(4), 74–83. doi:10.4018/ijgbl.2012100105

Yu, C., & Olinzock, A. (2011). Creating Computer Games for Class Instruction. In M. Cruz-Cunha, V. Varvalho, & P. Tavares (Eds.), *Computer Games as Educational and Management Tools: Uses and Approaches* (pp. 151–166). Hershey, PA: Information Science Reference.

Yu, C., Wu, J., & Johnson, A. (2012). Serious Games: Issues and Challenges for Teaching and Training. In M. Cruz-Cunha (Ed.), *Handbook of Research on Serious Games as Educational, Business and Research Tools* (pp. 559–577). Hershey, PA: Information Science Reference.

Yusof, S. A. (2012). Norms, Practices, and Rules of Virtual Community of Online Gamers: Applying the Institutional Theoretical Lens. In M. Cruz-Cunha (Ed.), *Handbook of Research on Serious Games as Educational, Business and Research Tools* (pp. 378–390). Hershey, PA: Information Science Reference.

Zaharias, P., Gatzoulis, C., & Chrysanthou, Y. (2012). Exploring User Experience While Playing Educational Games: Focus on Temporality and Attractiveness. [IJGCMS]. *International Journal of Gaming and Computer-Mediated Simulations*, *4*(4), 19–32. doi:10.4018/jgcms.2012100102

Zhou, S. X., & Leung, L. (2012). Gratification, Loneliness, Leisure Boredom, and Self-Esteem as Predictors of SNS-Game Addiction and Usage Pattern Among Chinese College Students. [IJCBPL]. *International Journal of Cyber Behavior, Psychology and Learning*, *2*(4), 34–48. doi:10.4018/ijcbpl.2012100103

Ziaeehezarjeribi, Y., & Graves, I. (2011). Behind the MASK: Motivation through Avatar Skills and Knowledge. [IJGCMS]. *International Journal of Gaming and Computer-Mediated Simulations*, *3*(4), 19–32. doi:10.4018/jgcms.2011100102

Ziaeehezarjeribi, Y., & Graves, I. (2013). Behind the MASK: Motivation through Avatar Skills and Knowledge. In R. Ferdig (Ed.), *Design, Utilization, and Analysis of Simulations and Game-Based Educational Worlds* (pp. 225–239). Hershey, PA: Information Science Reference.

Compilation of References

Aarseth, E. J. (1997). *Cybertext: Perspectives on Ergodic Literature*. Baltimore, MD: Johns Hopkins University Press.

Abrahamse, W., Steg, L., Vlek, C., & Rothengatter, T. (2005). A review of interventions studies aimed at household energy conservation. *Journal of Environmental Psychology*, *25*(3), 273–291. doi:10.1016/j.jenvp.2005.08.002

Abrams, D., & Hogg, M. A. (1988). Comments on the motivational status of self-esteem in social identity and inter-group discrimination. *European Journal of Social Psychology*, *18*(4), 317–334. doi:10.1002/ejsp.2420180403

Abt, C. C. (1970). *Serious games*. New York, NY: The Viking Press.

Acculturation. (2013). In *Oxford dictionary*. Retrieved July 20, 2013, from http://oxforddictionaries.com/definition/english/acculturate?q=acculturation#acculturate__8

Adams, S. et al. (2005). The effect social desirability and social approval on self-reports of physical activity. *American Journal of Epidemiology*, *161*(4), 389–398. doi:10.1093/aje/kwi054 PMID:15692083

Adobe Systems. (n.d.). *Adobe Flash*. Retrieved January 26, 2014, from http://www.adobe.com/ca/products/flash.html

Agre, P. E. (1998). Designing genres for new media: Social, economic and political contexts. In S. G. Jones (Ed.), *Cybersociety 2.0: Revisting computer-mediated communication and community* (pp. 69–99). London: Sage Publications Inc. doi:10.4135/9781452243689.n3

Albarracin, D., Johnson, B. T., & Zanna, M. P. (2005). *Handbook of attitudes*. Hoboken, NJ: Lawrence Erlbaum Associates.

Alexander, C. (1979). *The timeless way of building*. New York: Oxford University Press.

Althof, W., & Berkowitz, M. W. (2006). Moral education and character education: Their relationship and roles in citizenship education. *Journal of Moral Education*, *35*(4), 495–518. doi:10.1080/03057240601012204

Anderson, C. A., Shibuya, A., Ihori, N., Swing, E. L., Bushman, B. J., & Sakamoto, A. et al. (2010). Violent video game effects on aggression, empathy, and prosocial behavior in Eastern and Western countries: A meta-analytic review. *Psychological Bulletin*, *136*(2), 151–173. doi:10.1037/a0018251 PMID:20192553

Angenot, M. (1989). *1889, Un état du discours social*. Longueuil: Le Préambule.

Angenot, M. (1986). *Le cru et le faisandé: Sexe, discours social et littérature à la Belle époque*. Bruxelles: Éditions Labor.

Annunziato, R. A., Emre, S., Shneider, B., Barton, C., Dugan, C. A., & Shemesh, E. (2007). Adherence and medical outcomes in pediatric liver transplant recipients who transition to adult services. *Pediatric Transplantation, 11*(6), 608–614. doi:10.1111/j.1399-3046.2007.00689.x PMID:17663682

Bakardjieve, M. (2007). Virtual togetherness: An everyday-life perspective. In D. Bell, & B. M. Kennedy (Eds.), *The Cybercultures reader* (pp. 236–253). London: Rouledge.

Banet, A. Jr, & Hayden, C. (1977). A tavistock primer. In J. E. Jones, & J. W. Pfeiffer (Eds.), *The 1977 annual handbook for group facilitators* (6th ed.). Pfeiffer & Company.

Bäng, M., Svahn, M., & Gustafsson, A. (2009). Persuasive design of a mobile energy conservation game with direct feedback and social cues. In *Proceedings of DiGRA 2009 Breaking New Ground: Innovation in Games, Play, Practice and Theory*. DiGRA.

Bång, M., Gustafsson, A., & Katzeff, C. (2007). Promoting new patterns in household energy consumption with pervasive learning games. In *Persuasive Technology* (pp. 55–63). Springer. doi:10.1007/978-3-540-77006-0_7

Bang, M., Torstensson, C., & Katzeff, C. (2006). The Power House: A persuasive computer game designed to raise awareness of domestic energy consumption. In *First International Conference on Persuasive Computing for Well-being* (LNCS) (pp. 123 -132). Berlin: Springer.

Bateson, G. (1956). *The Message "This is Play"*. New York: Macy Foundation.

Baumeister, R. F., & Leary, M. R. (1995). The need to belong: Desire for interpersonal attachments as a fundamental human motivation. *Psychological Bulletin, 117*(3), 497–529. doi:10.1037/0033-2909.117.3.497 PMID:7777651

Beach, W. (1992). *Ethical education in American public schools*. Washington, DC: National Education Association.

Beale, I. L., Kato, P. M., Marin-Bowling, V. M., Guthrie, N., & Cole, S. W. (2007). Improvement in cancer-related knowledge following use of a psychoeducational video game for adolescents and young adults with cancer. *The Journal of Adolescent Health, 41*(3), 263–270. doi:10.1016/j.jadohealth.2007.04.006 PMID:17707296

Bender, T. (1978). *Community and social change in America*. New Brunswick, NJ: Rutgers University Press.

Benedict, R. (1934). *Patterns of culture* (Vol. 8). Boston: Houghton Mifflin Harcourt.

Benford, S., Rowland, D., Flintham, M., Drozd, A., Hull, R., & Reid, J. et al. (2005). Life on the edge: Supporting collaboration in location-based experiences. In *Proceedings of the SIGCHI Conference on Human Factors in Computing Systems*.Portland, OR: ACM. doi:10.1145/1054972.1055072

Bertolo, M., & Mariani, I. (2014). *Game Deisgn. Gioco e giocare tra teoria e progetto*. Milano: Pearson.

Bingham, P. M., Bates, J. H. T., Thompson-Figueroa, J., & Lahiri, T. (2010). A breath biofeedback computer game for children with cystic fibrosis. *Clinical Pediatrics*, *49*(4), 337–342. doi:10.1177/0009922809348022 PMID:20118101

Bion, A. C. (1961). *Experiences groups*. New York: Basic Books. doi:10.4324/9780203359075

Bion, A. C. (1970). *Attention and interpretation: Scientific approach to insight in psychoanalysis and groups*. New York: Basic Books.

Bion, W. R. (1977). Attention and interpretation: Container and contained. In *Seven servants: Four works*. New York: Jason Aronson.

Blackler, F., & McDonald, S. (2000). Power, mastery and organizational learning. *Journal of Management Studies*, *37*(6), 833–851. doi:10.1111/1467-6486.00206

Blasi, A. (1983). Moral cognition and moral action: A theoretical perspective. *Developmental Review*, *3*(2), 178–210. doi:10.1016/0273-2297(83)90029-1

Bloom, B. S., Engelhart, M. D., Furst, E. J., Hill, W. H., & Krathwohl, D. R. (1956). *Taxonomy of educational objectives: Handbook I: Cognitive domain*. New York: David McKay.

Blumberg, F. C., & Ismailer, S. S. (2009). What do children learn from playing digital games? In *Serious games – Mechanisms and effects*. New York: Routledge, Taylor and Francis.

Bodenheimer, T., Chen, E., & Bennett, H. D. (2009). Confronting the growing burden of chronic disease: Can the U.S. health care workforce do the job? *Health Affairs*, *28*(1), 64–74. doi:10.1377/hlthaff.28.1.64 PMID:19124856

Bogost, I. (2008). Fine processing. In *Persuasive technology* (pp. 13–22). Springer. doi:10.1007/978-3-540-68504-3_2

Bogost, I. (2010). *Persuasive games: The expressive power of videogames*. Cambridge, MA: MIT Press.

Boin, R. A., 't Hart, P., Stern, E., & Sundelius, B. (2005). *The politics of crisis management: Public leadership under pressure*. Cambridge, UK: Cambridge University Press. doi:10.1017/CBO9780511490880

Bowie, N. E. (1993). Does it pay to bluff in business? In T. L. Beauchamp, & N. E. Bowie (Eds.), *Ethical theory and business* (pp. 443–448). Englewood Cliffs, NJ: Prentice Hall, Inc.

Bowling, A., Gabriel, Z., Dykes, J., Dowding, L. M., Fleissig, A., Banister, D., & Sutton, S. (2003). Let's ask them: A national survey of definitions of quality of life and its enhancement among people aged 65 and over. *International Journal of Aging & Human Development*, *56*(4), 269–306. doi:10.2190/BF8G-5J8L-YTRF-6404 PMID:14738211

Bozeman, D. P., Perrewe, P. L., Kacmar, K. M., Hochwarter, W. A., & Brymer, R. A. (1996). *An examination of reactions to perceptions of organizational politics*. Paper presented at the 1996 Southern Management Association Meetings. New Orleans, LA.

Bronzwaer, S. (2011, August 11). Extreemrechts wil BBQ, wat nu?. *NRC Handelsblad*.

Brown, S. J., Lieberman, D. A., Gemeny, B. A., Fan, Y. C., Wilson, D. M., & Pasta, D. J. (1997). Educational video game for juvenile diabetes: Results of a controlled trial. *Medical Informatics*, *22*(1), 77–89. doi:10.3109/14639239709089835 PMID:9183781

Burgoyne, J., & Jackson, B. (1997). The arena thesis: Management development as a pluralistic meeting point. In J. Burgoyne & M. Reynolds (Eds.), *Management learning: Integrating perspectives in theory and practice* (pp. 54-70). London: Sage.

Busch, M. C. M., & Schrijvers, C. T. (2010). *Effecten van leefstijlinterventies gericht op lagere sociaaleconomische groepen*. RIVM.

Campaign Trail . (n.d.). Retrieved April 28, 2014, from http://sf0.org/tasks/Campaign-Trail

Carr, A. Z. (1968). Is business bluffing ethical? *Harvard Business Review*.

Carstensen, L. L. (1993). Motivation for social contact across the life span. In JacobJ. (Ed.), *Nebraska Symposium on Motivation: Developmental perspectives on motivation* (pp. 40, 209-254). Lincoln, NE: University of Nebraska Press.

Castells, M. (2004). *The network society: A cross-cultural perspective*. London: Edward Elgar. doi:10.4337/9781845421663

Catalano, R. F., Fagan, A. A., Gavin, L. E., Greenberg, M. T., Irwin, C. E. Jr, Ross, D. A., & Shek, D. T. (2012). Worldwide application of prevention science in adolescent health. *Lancet*, *379*(9826), 1653–1664. doi:10.1016/S0140-6736(12)60238-4 PMID:22538180

Certeau, M. (Ed.). (1990). L'invention du quotidien: Vol. 1. *Arts de faire*. Paris: Éditions Gallimard.

Chaiken, S. (1987). The heuristic model of persuasion. In M. P. Zanna, J. M. Olson, & C. P. Herman (Eds.), *Social influence the Ontario symposium* (5th ed.). Hillsdale, NJ: Lawrence Erlbaum Associates.

Chaiken, S., Lieberman, A., & Eagly, A. (1989). Heuristic and systemic information processing within and beyond the persuasion context. In J. Uleman, & J. Bargh (Eds.), *Unintended thought* (p. 212). New York, NY: Guilford.

Chaiken, S., Pomerantz, E. M., & Giner-Sorolla, R. (1995). Structural consistency and attitude strength. In R. E. Petty, & J. A. Krosnick (Eds.), *Atitude strength, antecedents and consequences* (pp. 387–412). Mahwah, NJ: Lawrence Erlbaum.

Chen, J., & Ringel, M. (2001). *Can advergaming be the future of interactive advertising?* Retrieved from http://www.kpe.com/ourwork/pdf/advergaming.pdf

Chen, S., & Chaiken, S. (1999). The heuristic-Systematic model in its broader context. In S. Chaiken, & Y. Trope (Eds.), *Dual-process theories in social psychology*. New York, NY: Guilford Press.

Christiansen, N., Villanova, P., & Mikulay, S. (1997). Political influence compatibility: Fitting the person to the climate. *Journal of Organizational Behavior*, *18*(6), 709–730. doi:10.1002/(SICI)1099-1379(199711)18:6<709::AID-JOB811>3.0.CO;2-4

Clegg, S. (1989). *Frameworks of power*. London: Sage. doi:10.4135/9781446279267

Colin McRae Rally 2.0. (2000). Southam: Codemasters Software Company Limited.

CTRL+Z. (n.d.). Retrieved April 28, 2014, from http://sf0.org/tasks/CTRL--Z

Colvert, A. (2009). Alternate reality gaming in primary school settings. In *Breaking New Ground: Innovation in Games, Play, Practice and Theory: Proceedings of DiGRA 2009*. London. DIGRA

Connery, B. A. (1997). IMHO: Authority and egalitarian rhetoric in the virtual coffeehouse. In D. Porter (Ed.), *Internet Culture* (pp. 161–180). New York: Routledge.

Coopey, J., & Burgoyne, J. (2000). Politics and organizational learning. *Journal of Management Studies*, *37*(6), 869–885. doi:10.1111/1467-6486.00208

Covaleski, M. A., Dirsmith, M. W., Heian, J. B., & Sajay, S. (1998). The calculated and the avowed: Techniques of discipline and struggles over identity in the big six public accounting firms. *Administrative Science Quarterly*, *43*(2), 293–327. doi:10.2307/2393854

Crawford, C. (1984). *The art of computer game design*. New York: Osborne/McGraw-Hill.

Crawford, G., & Gosling, V. K. (2009). More than a game: Sports-themed video games and player narratives. *Sociology of Sport Journal*, *26*, 50–66.

CRIST. (n.d.). *Manifeste*. Retrieved October 20, 2011, from http://www.site.sociocritique-crist.org/p/manifeste.html

Cropanzano, R., Howes, J. C., Grandey, A. A., & Toth, P. (1997). The relationship of organizational politics and support to work behaviors, attitudes, and stress. *Journal of Organizational Behavior*, *18*(2), 159–180. doi:10.1002/(SICI)1099-1379(199703)18:2<159::AID-JOB795>3.0.CO;2-D

Cros, E. (2003). *La sociocritique*. Paris: L'Harmattan.

Cros, E. (2006). Spécificités de la sociocritique. *La sociocritique d'Edmond Cros*. Retrieved April 25, 2012, from http://sociocritique.fr/spip.php?article6

Crossan, M., & Guatto, T. (1996). Organizational learning research profile. *Journal of Organizational Change Management*, *9*(1), 107–112. doi:10.1108/09534819610107358

Crossan, M., Lane, H., & White, R. (1999). An organizational learning framework: From intuition to institution. *Academy of Management Review*, *24*, 522–537.

Csikszentmihalyi, M. (2008). *Flow: The psychology of optimal experience*. New York: HarperCollins.

Cummings, H. M., & Vandewater, E. A. (2007). Relation of adolescent video game play to time spent in other activities. *Archives of Pediatrics & Adolescent Medicine*, *161*(7), 684–689. doi:10.1001/archpedi.161.7.684 PMID:17606832

Dansey, N. (2013). *A grounded theory of emergent benefit in pervasive game experiences*. Retrieved from http://www.neildansey.co.uk

Davies, H. (2007). Place as media in pervasive games. In *Proceedings of the 4th Australasian Conference on Interactive Entertainment*. Melbourne, Australia: Creativity ad Cognition Studios Press. Retrieved from http://dl.acm.org/ft_gateway.cfm?id=1367963&type=pdf&CFID=82449371&CFTOKEN=37911479

DDB Amsterdam, & Flavour (Producer). (2010). *Tem de tank*. Retrieved from http://www.flavour.nl/volkswagen/

de Jong, A., Balksjö, T., & Katzeff, C. (2013). *Challenges in energy awareness: A Swedish case of heating consumption in households*. Paper presented at the ERSCP-EMSU Conference Bridges for a Sustainable Future. Istanbul, Turkey.

De la Hera Conde-Pumpido, T. (In Press). *Persuasive structures in advergames*. Utrecht, The Netherlands: Utrecht University.

De Luca, V., & Bertolo, M. (2012). Urban games to design the augmented city. *Eludamos: Journal for Computer Game Culture*, *6*(1), 71–83.

Dear Esther. (2012). Brighton, UK: thechineseroom.

Deetz, S. A. (1991). *Democracy in an age of corporate colonization: Developments in communication and the politics of everyday life*. New York: State University of New York Press.

Department of Health. (2011). *Ten things you need to know about long term conditions*. Retrieved from http://webarchive.nationalarchives.gov.uk/+/www.dh.gov.uk/en/Healthcare/Longtermconditions/tenthingsyouneedtoknow/index.htm

Diehl, M., Coyle, N., & Labouvie-Vief, G. (1996). Age and sex differences in strategies of coping and defense across the life span. *Psychology and Aging*, *11*(1), 127–139. doi:10.1037/0882-7974.11.1.127 PMID:8726378

Dixit, A. K., Skeath, S., & Reiley, D. (2009). *Games of strategy* (3rd ed.). New York: W. W. Norton & Company.

Dolak, D. (2012). *How to brand and market a commodity*. European Union Amazon Media EU.

Droomers, M. et al. (2001). Educational level and decreases in leisure time physical activity: Predictors from the longitudinal GLOBE study. *Journal of Epidemiology and Community Health*, *55*(8), 562–568. doi:10.1136/jech.55.8.562 PMID:11449013

Drory, A. (1993). Perceived political climate and job attitudes. *Organization Studies*, *14*(1), 59–71. doi:10.1177/017084069301400105

Drory, A., & Romm, T. (1990). The definition of organizational politics: A review. *Human Relations*, *43*(11), 1133–1154. doi:10.1177/001872679004301106

Dubrin, A. J. (1988). *Human relations: A job oriented approach*. Englewood Cliffs, NJ: Prentice-Hall.

Duchet, C. (1971). Pour une socio-critique ou variations sur un incipit. *Littérature*, (1), 5-14.

Duchet, C., Herschberg Pierrot, A., & Neefs, J. (1994). Sociocritique et génétique, entretien avec Anne Herschberg Pierrot et Jacques Neefs. *Genesis, Manuscrits, Recherche, Invention*, (6), 117-127.

Duchet, C. (1979). Positions et perspective. In C. Duchet (Ed.), *Sociocritique* (pp. 3–8). Paris: Nathan.

Duchet, C. (1986). La manœuvre du bélier: Texte, intertexte et idéologies dans L'Espoir. *Revue des Sciences Humaines*, (204): 107–131.

Duchet, C., & Maurus, P. (2011). *Un cheminement vagabond: Nouveaux entretiens sur la sociocritique*. Paris: Honoré Champion.

Dunne, J. (1995). What's the good of education? In W. Carr (Ed.), *The RoutledgeFalmer reader in philosophy of education* (pp. 145–160). Abingdon, UK: Routledge.

Duret, C. (2014). *Les jeux de rôle participatifs en environnement virtuel: Définition et enjeux théoriques*. (Master's thesis). Université de Sherbrooke, Sherbrooke, Canada.

Duret, C. (2013). Jeux de rôle participatifs en environnement virtuel et herméneutique conflictuelle: L'expérience ludique des joueurs dans les jeux de rôle goréens. *Communication. Lettres et Sciences du Langage*, *7*(1), 4–20.

Durkin, K., & Barber, B. (2002). Not so doomed: Computer game play and positive adolescent development. *Applied Developmental Psychology, 23*(4), 373–392. doi:10.1016/S0193-3973(02)00124-7

Düüs-Henriksen, T. (2008). Extending experiences of learning games: Or why learning games should be neither fun nor educational or realistic. In O. Leino, H. Wirman, & A. Fernandez (Eds.), *Extending experiences: Structure, analysis and design of computer game player experience*. Rovaniemi, Finland: Lapland University Press.

Egenfeldt-Nielsen, S. (2007). *Beyond edutainment exploring the educational potential of videogames*. London, UK: Continuum International Publishing Group Ltd.

Ellemers, N., Haslam, S. A., Platow, M. J., & Knippenberg, D. (2003). Social identity at work: developments, debates and directions. In S. A. Haslam, D. V. Knippenberg, M. J. Platow, & N. Ellemers (Eds.), *Social identity at work: Developing theory for organisational practice* (pp. 3–26). Hove, UK: Taylor and Francis Group.

El-Nasr, M., & Smith, B. (2006). Learning through game modding. *Computers in Entertainment, 4*(1).

Encyclopaedia Universalis. (2013). *Acculturation*. Retrieved July 25, 2013, from http://www.universalis.fr/encyclopedie/acculturation/

Enterbrain. (n.d.). *RPG Maker*. Retrieved January 26, 2014, from http://www.rpgmakerweb.com/products/programs/rpg-maker-xp

Entertainment Software Association. (2012). *Sales, demographics and usage data: Essential facts about the computer and video game industry*. Retrieved from http://www.theesa.com/facts/pdfs/ESA_EF_2013.pdf

Ernest, P. (n.d.). Varieties of constructivism: their metaphors, epistemologies and pedagogical implications. In R. Fox (Ed.), *Perspectives on constructivism* (pp. 73–92). Exeter, UK: University of Exeter School of Education.

Etzioni, A. (1998). How not to discuss character education. *Phi Delta Kappa International, 79*(6), 446–448.

Eurostat. (2011). *Energy, transport and environment indicators*. Luxemburg: Publications Office of the European Union.

Evans, E. (2008). Character, audience agency and transmedia drama. *Media, Culture & Society March, 30*(2), 197-213. doi:10.1177/0163443707086861

Ferguson, C. J., Miguel, C. S., Garza, A., & Jerabeck, J. M. (2012). A longitudinal test of video game violence influences on dating and aggression: A 3-year longitudinal study of adolescents. *Journal of Psychiatric Research, 46*(2), 141–146. doi:10.1016/j.jpsychires.2011.10.014 PMID:22099867

Ferris, G. R., Adams, G., Kolodinsky, R. W., Hochwarter, W. A., & Ammeter, A. P. (2002). Perceptions of organizational politics: Theory and research directions. In F. Dansereau, & F. J. Yammarino (Eds.), *Research in multi-level issues* (Vol. 1). Oxford, UK: Elsevier Science/JAI Press. doi:10.1016/S1475-9144(02)01034-2

Ferris, G. R., Frink, D. D., Bhawuk, D. P. S., & Zhou, J. (1996). Reactions of diverse groups to politics in the workplace. *Journal of Management*, *22*(1), 23–44. doi:10.1177/014920639602200102

Ferris, G. R., Frink, D. D., Galang, M. C., Zhou, J., Kacmar, K. M., & Howard, J. L. (1996). Perceptions of organizational politics: Prediction, stress-related implications, and outcome. *Human Relations*, *49*(2), 233–266. doi:10.1177/001872679604900206

Ferris, G. R., & Kacmar, K. M. (1992). Perceptions of organizational politics. *Journal of Management*, *18*(1), 93–116. doi:10.1177/014920639201800107

Ferris, G. R., Russ, G. S., & Fandt, P. M. (1989). Politics in organizations. In R. A. Giacalone, & P. Rosenfield (Eds.), *Impression management in the organization* (pp. 143–170). Hillsdale, NJ: Erlbaum.

Ferris, M. E., Harward, D. H., Bickford, K., Layton, J. B., Ferris, M. T., & Hogan, S. L. et al. (2012). A clinical tool to measure the components of health-care transition from pediatric care to adult care: The UNC TR(x) ANSITION scale. *Renal Failure*, *34*(6), 744–753. doi:10.3109/0886022X.2012.678171 PMID:22583152

Fine, G. A. (1983). *Shared fantasy: Role-playing games as social worlds*. Chicago: University of Chicago Press.

Fink, E., Saine, U., & Saine, T. (1968). The oasis of happiness: Toward an ontology of play. *Yale French Studies*, *41*(41), 19–30. doi:10.2307/2929663

Fiol, C. M. (2001). All for one and one for all? The development and transfer of power across organizational levels. *Academy of Management Review*, *26*, 224–242.

Fiske, J. (1987). *Television culture*. London: Routledge.

Flanagan, M. (2009). *Critical play: Radical game design*. Cambridge, MA: MIT Press.

Flusser, V. (2004). *La cultura dei media*. Milano: Bruno Mondadori.

Fogg, B. J. (2003). *Persuasive technology: Using computer to change what we think and do*. San Francisco: Morgan Kaufmann Publishers.

Foucault, M. (1977). *Discipline and punish: The birth of the prison*. New York: Vintage Books.

Fox, R. (n.d.). Constructivist views of learning. In R. Fox (Ed.), *Perspectives on constructivism* (pp. 3–16). Exeter, UK: University of Exeter School of Education.

Fox, S. (2000). Communities of practice: Foucault and actor-network theory. *Journal of Management Studies*, *37*(6), 853–867. doi:10.1111/1467-6486.00207

Frasca, G. (2003). *September 12th*. Retrieved January 26, 2014, from www.newsgaming.com/games/index12.htm

Frasca, G. (2003). Simulation versus narrative: Introduction to ludology. *Ludology.org*. Retrieved September 20, 2011, from http://www.ludology.org/articles/VGT_final.pdf

Fullerton, T. (2008). Playcentric design. *Interaction*, *15*(2), 42–45. doi:10.1145/1340961.1340971

Fullerton, T., Swain, C., & Hoffman, S. (2004). *Game design workshop: Designing, prototyping, & playtesting games*. San Francisco, CA: CMP Books.

Galbiati, M., & Piredda, F. (2012). *Visioni urbane: Narrazioni per il design della città sostenibile*. Milano: Franco Angeli.

Gamberini, L., Corradi, N., Zamboni, L., Perotti, M., Gadenazzi, C., & Mandressi, S. et al. (2011). Saving is fun: Designing a persuasive game for power conservation. In *Proceedings of the 8th International Conference of Advances in Computer Entertainment Technology, ACE'11*. ACM. doi:10.1145/2071423.2071443

GameSalad. (n.d.). *GameSalad*. Retrieved January 26, 2014, from http://gamesalad.com/

Gardner, P. (2001). *Games with a day job: Putting the power of games to work*. Retrieved from http://www.gamasutra.com/view/feature/3071/games_with_a_day_job_putting_the_.php

Gazzard, A. (2012). Re-coding the algorithm: Purposeful and appropriated play. In A. Lugmayr et al. (Eds.), *Media in the ubiquitous era: Ambient, social and gaming media*. Hershey, PA: Information Science Reference.

Gee, J. P. (2003). What video games have to teach us about learning and literacy. *Computers in Entertainment, 1*(1), 20-20. http://doi.acm.org/10.1145/950566.950595

Gee, J. P. (2007). *What video games have to teach us about learning and literacy*. Basingstoke, UK: Palgrave MacMillan.

Genette, G. (1972). *Figures III*. Paris: Éditions du Seuil.

Gentile, D. A., Choo, H., Liau, A., Sim, T., Li, D. D., Fung, D., & Khoo, A. (2011). Pathological video game use among youths: A two-year longitudinal stud. *Pediatrics, 127*(2), E319–E329. doi:10.1542/peds.2010-1353 PMID:21242221

Gentry, C. G. (1993). *Introduction to instructional development: Process and technique*. Belmont, CA: Wadsworth Publishing Company.

Giddens, A. (1984). *The constitution of society: Outline of a theory of structuration*. Cambridge, MA: Polity Press.

Giles-Corti, B., & Donovan, R. J. (2002). Socioeconomic status differences in recreational physical activity levels and real and perceived access to a supportive physical environment. *Preventive Medicine, 35*(6), 601–611. doi:10.1006/pmed.2002.1115 PMID:12460528

Gilligan, C. (1977). In a different voice: Women's conceptions of self and of morality. *Harvard Educational Review, 47*(4), 481–517.

Glanzer, P. L. (1998). The character to seek justice: Showing fairness to diverse visions of character education. *Phi Delta Kappan, 79*(6), 434–436.

Glaser, B. (1978). *Theoretical sensitivity: Advances in the methodology of grounded theory*. Mill Valley, CA: Sociology Press.

Glaser, B. (1998). *Doing grounded theory: Issues and discussions*. Mill Valley, CA: Sociology Press.

Godin, G., & Kok, G. (1996). The theory of planned behavior: A review of its applications to health-related behaviors. *American Journal of Health Promotion, 11*(2), 87–98. doi:10.4278/0890-1171-11.2.87 PMID:10163601

Goffman, E. (1959). *The presentation of self in everyday life*. New York: Garden City.

Goffman, E. (1974). *Frame analysis: An essay on the organization of experience*. New York: Harper & Row.

González Sánchez, J. L., Gutiérrez Vela, F. L., Montero Simarro, F., & Padilla-Zea, N. (2012). Playability: Analysing user experience in video games. *Behaviour & Information Technology*, *31*(10), 1033–1054. doi:10.1080/0144929X.2012.710648

Goodman, J. M., Evans, W. R., & Carson, C. M. (2011). Organizational politics and stress: Perceived accountability as a mechanism. *The Journal of Business Inquiry*, *10*(1), 66–80.

Gorean Forums. (2012). *Bringing Gor back to Gor*. Retrieved July 25, 2012, from http://www.goreanforums.net/viewtopic.php?f=230&t=6840&start=20#p162775

Gor-SL. (2011). *Gorean women: Haven't we seen them before?* Retrieved August 31, 2012, from http://www.gor-sl.com/index.php/topic,9435.msg79290.html#msg79290

Gor-SL. (2011). *Dear Dove*. Retrieved August 31, 2012, from http://www.gor-sl.com/index.php/topic,9896.msg83954.html#msg83954

Green, M., Brock, T., & Kaufman, G. (2004). Understanding media enjoyment: The role of transportation into narrative worlds. *Communication Theory*, *4*(4), 311–327. doi:10.1111/j.1468-2885.2004.tb00317.x

Greer, M. (1992). *ID project management: Tools and techniques for instructional designers and developers*. Englewood Cliffs, NJ: Educational Technology.

Gregory, P. (2003). *New scapes: Territories of complexity*. London: Springer.

Grundy, S. M., Benjamin, I. J., Burke, G. L., Chait, A., Eckel, R. H., & Howard, B. V. et al. (1999). Diabetes and cardiovascular disease: A statement for healthcare professionals from the American Heart Association. *Circulation*, *100*(10), 1134–1146. doi:10.1161/01.CIR.100.10.1134 PMID:10477542

Gustafsson, A. (2010). *Positive persuasion - Designing enjoyable energy feedback experiences in the home.* (Doctoral dissertation). Retrieved from GUPEA 21-maj-2010

Gustafsson, A., & Bang, M. (2008). Evaluation of a pervasive game for domestic energy engagement among teenagers. In *Proceedings of the ACM SIGCHI International Conference on Advances in Computer Entertainment Technology* (ACE 2008). ACM. doi:doi:10.1145/1501750.1501804 doi:10.1145/1501750.1501804

Gustafsson, A., Bang, M., & Svahn, M. (2009). Power explorer – A casual game style for encouraging long term behaviour change among teenagers. In *Proceedings of the ACM SIGCHI International Conference on Advances in Computer Entertainment Technology* (ACE 2009). Athens, Greece. ACM. doi:doi:10.1145/1690388.1690419 doi:10.1145/1690388.1690419

Gustafsson, A., Katzeff, C., & Bång, M. (2009). Evaluation of a pervasive game for domestic energy engagement among teenagers. *Computers in Entertainment*, *7*(4), 54. doi:10.1145/1658866.1658873

Habgood, J., Overmars, M., & Wilson, P. (2006). *The game maker's apprentice: Game development for beginners*. Berkeley, CA: Apress, Springer.

Hagell, A. (2012). *Health implications of new technology*. Association for Young People's Health (AYPH). Retrieved from http://www.ayph.org.uk/publications/296_RU11%20New%20technology%20summary.pdf

Hagell, A., Coleman, J., & Brooks, F. (2013). *Key data on adolescence 2013: The latest information and statistics about young people today*. Association for Young People's Health.

Hagger, M., Chatzisarantis, N., & Biddle, S. (2002). A meta-analytic review of the theories of reasoned action and planned behavior in physical activity: Predictive validity and the contribution of additional variables. *Journal of Sport & Exercise Psychology*, *24*, 3–32.

Hall, S. (1980). Encoding/decoding. In S. Hall, D. Hobson, A. Love, & P. Willis (Eds.), *Culture, media, language* (pp. 128–138). London: Hutchinson.

Hall, S. (1989). Ideology and communication theory. In B. Dervin, L. Grossberg, B. O'Keefe, & E. Wartella (Eds.), *Rethinking communication I: Paradigm dialogues*. Newbury Park, CA: Sage.

Hardy, C., & Clegg, S. R. (1996). Some dare call it power. In S. R. Clegg, C. Hardy, & W. R. Nord (Eds.), *Handbook of organization studies* (pp. 622–641). London: Sage.

Harman, G. (2012). Moral relativism explained. Unpublished, written for a volume edited by Bastian Reichard.

Harrison, G. (1976). Relativism and tolerance. *Ethics*, *86*(2), 122–135. doi:10.1086/291986

Harviainen, J. T. (2009). A hermeneutical approach to role-playing analysis. *International Journal of Role-Playing*, *1*(1).

Harwood, J. (1999). Age identification, social identity gratifications and television viewing. *Journal of Broadcasting & Electronic Media*, *43*(1), 123–136. doi:10.1080/08838159909364479

Hatch, E. (1997). The good side of relativism. *Journal of Anthropological Research*, *53*(3), 371–381.

Highet, G. (2003). Cannabis and smoking research: Interviewing young people in self-selected friendship pairs. *Health Education Research: Theory and Practice*, *18*(1), 108–118. doi:10.1093/her/18.1.108 PMID:12608688

Hochwarter, W. A., Witt, L. A., & Kacmar, K. M. (1997). *Perceptions of organizational politics as a moderator of the relationship between conscientiousness and job performance*. Paper presented at the Southern Management Association Meeting. Atlanta, GA.

Hogan, P. (1997). The politics of identity and the epiphanies of learning. In W. Carr (Ed.), *The RoutledgeFalmer reader in philosophy of education* (pp. 83–96). Abingdon, UK: Routledge.

Hogg, M. A., & Abrams, D. (1990). Social motivation, self-esteem and social identity. In D. Abrams, & M. A. Hogg (Eds.), *Social identity theory: Constructive and critical advances* (pp. 28–47). London: Harvester-Wheatsheaf.

Hrehovcsik, M. (2010). *An applied game design framework: Prioritizing game design in serious game development* [Web log message]. Retrieved September 09, 2013, from http://gamedesigntools.blogspot.kr/2013/05/2cat-framework-for-applied-game-design.html

Huber, G. (1991). Organizational learning: The contributing processes and the literatures. *Organization Science*, *2*(1), 88–115. doi:10.1287/orsc.2.1.88

Huizinga, J. (1949). *Homo ludens*. London: Routledge & Kegan Paul.

Huss, H. F., & Patterson, D. M. (1993). Ethics in accounting: Values education without indoctrination. *Journal of Business Ethics*, *12*(3), 235–243. doi:10.1007/BF01686451

Iacovoni, A. (2004). *Game zone: Playgrounds between virtual scenarios and reality*. Basel: Birkhäuser.

Interactive Software Federation of Europe. (2012). *Industry facts*. Retrieved from http://www.isfe.eu/industry-facts

ISFE. (2012). *Videogames in Europe: Consumer study*. The Media, Condent and Technology Research Specialists.

Isiklar, A. (2012). Examining psychological well being and self esteem levels of Turkish students in gaining identity against role during conflict periods. *Journal of Instructional Psychology*, *39*(1), 41–50.

IUCN/UNEP/WWF. (1991). *Caring for the earth: A strategy for sustainable living*. London: Earthscan.

Janssen, I., Boyce, W. F., & Pickett, W. (2012). Screen time and physical violence in 10 to 16-year-old Canadian youth. *International Journal of Public Health*, *57*(2), 325–331. doi:10.1007/s00038-010-0221-9 PMID:21110059

Janssen, N. M., & Genta, M. S. (2000). The effects of immunosuppressive and anti-inflammatory medications on fertility, pregnancy, and lactation. *Archives of Internal Medicine*, *160*(5), 610–619. doi:10.1001/archinte.160.5.610 PMID:10724046

Jenkins, H. (2006). *Convergence culture: Where old and new media collide*. New York: New York University Press.

Jenkins, R. (2008). *Social identity*. London: Routledge Taylor and Francis Group.

Johnson, R., & Meadows, R. (2010). Dog-walking: Motivation for adherence to a walking program. *Clinical Nursing Research*, *19*(4), 387–402. doi:10.1177/1054773810373122 PMID:20651066

Johnson, S. (2001). *Emergence: The connected lives of ants, brains, cities and software*. London: Penguin Books.

Jones, S. G. (1998). *Cybersociety 2.0: Revisting computer-mediated communication and community*. London: Sage Publications Inc.

Journey to the End of the Night 2 . (n.d.). Retrieved April 28, 2014, from http://sf0.org/tasks/Journey-To-The-End-Of-The-Night-2

Journey to the End of the Night . (n.d.). Retrieved April 28, 2014, from http://sf0.org/tasks/Journey-To-The-End-Of-The-Night

Juul, J. (2008). The magic circle and the puzzle piece. In *Proceedings of the Philosophy of Computer Games*. Academic Press.

Juul, J. (2010). *A casual revolution: Reinventing video games and their players*. Cambridge, MA: MIT Press.

Kacmar, K. M., & Baron, R. A. (1999). The state of the field, links to related processes, and an agenda for future research. In G. R. Ferris (Ed.), *Research in personnel and human resources management* (Vol. 17, pp. 1–39). Stamford, CT: JAI Press.

Kacmar, K. M., Bozeman, D. P., Carlson, D. S., & Anthony, W. P. (1999). An examination of the perceptions of organizational politics model: Replication and extension. *Human Relations, 52*(3), 383–415. doi:10.1177/001872679905200305

Kato, P. M., Cole, S. W., Bradlyn, A. S., & Pollock, B. H. (2008). A video game improves behavioral outcomes in adolescents and young adults with cancer: A randomized trial. *Pediatrics, 122*(2), E305–E317. doi:10.1542/peds.2007-3134 PMID:18676516

Katzeff, C. (2010, April 26). Engaging design for energy conservation in households. *Metering International Magazine,* 62-63.

Kaye, L. K., & Bryce, J. (2012). Putting the fun factor into gaming: The influence of social contexts on experiences of playing videogames. *International Journal of Internet Science, 7*(1), 23–37.

Kaye, L. K., & Bryce, J. (in press). Go with the flow: The experiences and affective outcomes of solo versus social gameplay. *Journal of Gaming and Virtual Worlds.*

Khaled, R., Barr, P., Biddle, R., Fischer, R., & Noble, J. (2009). Game design strategies for collectivist persuasion.*Proceedings of the 2009 ACM SIGGRAPH Symposium on Video Games* (pp. 31-38). ACM. doi:10.1145/1581073.1581078

Kilpatrick, W. H. (1972). Indoctrination and respect for persons. In *Concepts of indoctrination: Philosophical essays* (pp. 37–42). Routledge.

Kirschenbaum, H. (1995). *One hundred ways to enhance values and morality in school and youth settings.* Needham Heights, MA: Allyn and Bacon.

Kirschenbaum, H., Harmin, M., Howe, L., & Simon, S. (1977). In defense of values clarification. *Phi Delta Kappa International, 58*(10), 743–746.

Klein, M. (1946). Notes on some schizoid mechanisms. *The International Journal of Psycho-Analysis, 27*, 99–110.

Koehn, D. (1997). Business and game-playing: The false analogy. *Journal of Business Ethics, 16*(12/13), 1447–1452. doi:10.1023/A:1005724317399

Kohlberg, L. (1973). Stages and aging in moral development-some speculations. *The Gerontologist, 13*(4), 497–502. doi:10.1093/geront/13.4.497 PMID:4789527

Kohn, A. (1997). How not to teach values: A critical look at character education. *Phi Delta Kappan, 78*, 428–439.

Kong, F., Zhao, J., & You, X. (2013). Self-esteem as mediator and moderator of relationship between social support and subjective well-being among Chinese university students. *Social Indicators Research, 112*(1), 151–161. doi:10.1007/s11205-012-0044-6

Krathwohl, D. R. (2002). A revision of Bloom's taxonomy: An overview. *Theory into Practice, 41*(4), 212–218. doi:10.1207/s15430421tip4104_2

Kruglanski, A., & Stroebe, W. (2005). The influence of beliefs and goals on attitudes: Issues of structure, function, and dynamics. In *The handbook of attitudes*. Mahwah, NJ: Erlbaum.

Kumar, P., & Ghadially, R. (1989). Organizational politics and its effects on members of organizations. *Human Relations*, *42*(4), 305–314. doi:10.1177/001872678904200402

Lange, F. (2003). *Brand choice in goal-derived categories: What are the determinants?* (Doctoral Dissertation, EFI). Retrieved from http://hhs.diva-portal.org/smash/get/diva2:221410/FULLTEXT01

Lapsley, D. K., & Narvaez, D. (2006). Handbook of child psychology: Vol. 4. *Character education* (6th ed., pp. 248–296). Wiley.

Larsson, C. (2011). *Slutrapport projekt – Unga utforskar energi. (Official No. Prj.nr: 32255-1)*. Energikontoret i Mälardalen.

Lawrence, T. B., Mauws, M. K., Dyck, B., & Kleysen, R. F. (2005). The politics of organizational learning: Integrating power into the 4i framework. *Academy of Management Review*, *30*(1), 180–191. doi:10.5465/AMR.2005.15281451

Lawrence, T. B., Winn, M., & Jennings, P. D. (2001). The temporal dynamics of institutionalization. *Academy of Management Review*, *26*, 624–644.

Lea, M., Spears, R., & Rogers, P. (2003). Social processes in electronic teamwork: The central issue of identity. In S. A. Haslam, D. V. Knippenberg, M. J. Platow, & N. Ellemers (Eds.), *Social identity at Work: Developing Theory for Organisational Practice* (pp. 99–115). Hove, UK: Taylor and Francis Group.

Lee, I., Shiroma, E., Lobelo, F., Puska, P., Blair, S., & Katzmarzy, P. (2012). Effect of physical inactivity on major non-communicable diseases worldwide: An analysis of burden of disease and life expectancy. *Lancet*, *380*(9838), 219–229. doi:10.1016/S0140-6736(12)61031-9 PMID:22818936

Leonard, D. (2004). High tech Blackface-race, sports video games and becoming the other. *Intelligent Agent, 4* (4).

Licklider, J. C. R., & Taylor, R. W. (1968). The computer as a communication device. *Science and Technology*, *76*, 21–31.

Lickona, T. (1993). The return of character education. *Educational Leadership*, *51*(3), 6–11.

Lickona, T. (1996). Eleven principles of effective character education. *Journal of Moral Education*, *25*(1), 93–100. doi:10.1080/0305724960250110

Lieberman, D. (1997). Interactive video games for health promotion: Effects on knowledge, self-efficacy, social support and health. In R. L. Street, W. R. Gold, & T. R. Manning (Eds.), *Health promotion and interactive technology: Theoretical applications and future directions* (pp. 103–120). New York, NY: Routledge.

Lieberman, D. A., & Brown, S. J. (1995). Designing interactive video games for children's health education. In *Interactive technology and the new paradigm for healthcare* (pp. 201–210). Amsterdam: IOS Press.

Linderoth, J. (2009). Its not hard, just requires that you have no life: Computer games and the illusion of learning. *Digital Kompetanse*, *4*(1), 4–19.

Lin, J., Mamykina, L., Lindtner, S., Delajoux, G., & Strub, H. (2006). Fish'n'Stepts: Encouraging physical activity with an interactive computer game. *Lecture Notes in Computer Science*, *4206*, 261–278. doi:10.1007/11853565_16

Lockwood, A. (1975). A critical view of values clarification. *Teachers College Record*, *77*(1), 35–50.

Lockwood, A. (1997). *Character education: Controversy and consensus*. Thousand Oaks, CA: Corwin Press.

Longenecker, C. O., Sims, H. P., & Gioia, D. A. (1987). Behind the mask: The politics of employee appraisal. *The Academy of Management Executive*, *1*(3), 183–193. doi:10.5465/AME.1987.4275731

Luther's Gorean Essays. (n.d.). *Female Warriors #15, 5.0 Version*. Retrieved July 25, 2013, from http://www.gor-now.net/delphius2002/id31.htm

Macherey, P. (2005). *Michel de Certeau et la mystique du quotidien*. Retrieved July 25, 2013, from http://stl.recherche.univ-lille3.fr/seminaires/philosophie/macherey/macherey20042005/macherey06042005.html

MacKenzie, S. B., & Lutz, R. J. (1989). An empirical examination of the structural antecedents of attitude toward the ad in an advertising pretesting context. *Journal of Marketing*, *53*(2), 48–65. doi:10.2307/1251413

Madeira, R. N., Silva, A., Santos, C., Teixeira, B., Romão, T., Dias, E., & Correia, N. (2011). LEY! Persuasive pervasive gaming on domestic energy consumption-awareness. In *Proceedings of the 8th International Conference on Advances in Computer Entertainment Technology*. ACM. doi:10.1145/2071423.2071512

Mahan, S. (2010). *Make it soft*. Retrieved April 28, 2014, from http://sf0.org/swm/Make-it-Soft

Maingueneau, D. (1976). *Initiation aux méthodes de l'analyse du discours*. Paris: Hachette.

Make it Soft. (n.d.). Retrieved April 28, 2014, from http://sf0.org/tasks/Make-it-Soft

Marcell, K., Agyeman, J., & Rappaport, A. (2004). Cooling the campus: Experiences from a pilot study to reduce electricity use at tufts university, USA, using social marketing methods. *International Journal of Sustainability in Higher Education*, *5*(2), 169–189. doi:10.1108/14676370410526251

Marczewski, A. (2013). *Gamification a little on leaderboards*. Retrieved from http://marczewski.me.uk/2013/01/21/gamification-a-little-on-leaderboards/

Mariani, I., & Ciancia, M. (2013). The urban space as a narrative and ludic playground. In *Proceedings of CIDAG International Conference in Design and Graphic Arts*. ISEC – Instituto Superior de Educaçäo e Ciências, IPT – Instituto Politécnico de Tomar.

Maslyn, J., Fedor, D., Farmer, S., & Bettenhausen, K. (2005). *Perceptions of positive and negative organizational politics: Roles of the frequency and distance of political behavior*. Paper presented at the 2005 Annual Meeting of the Southern Management Association. Charlotte, NC.

Masterman, L. (1993). The media education revolution. *Canadian Journal of Educational Communication*, *22*(1), 5–14.

Mauthner, M. (1997). Methodological aspects of collecting data from children: Lessons from three research projects. *Children & Society*, *11*(1), 16–28. doi:10.1111/j.1099-0860.1997.tb00003.x

Mayall, B. (2000). Conversations with children: Working with generational issues. In P. Christensen, & A. James (Eds.), *Research with Children: Perspectives and Practices* (pp. 120–135). London: Falmer Press.

Mayes, B. T., & Allen, R. W. (1977). Toward a definition of organizational politics. *Academy of Management Review*, *2*(4), 672–678.

McDonagh, J. E. (2007). Transition of care: How should we do it? *Paediatrics and Child Health (Oxford)*, *17*(12), 480–484. doi:10.1016/j.paed.2007.09.007 PMID:17715444

McDonagh, J. E., Southwood, T. R., & Shaw, K. L. (2006). Growing up and moving on in rheumatology: Development and preliminary evaluation of a transitional care programme for a multicentre cohort of adolescents with juvenile idiopathic arthritis. *Journal of Child Health Care*, *10*(1), 22–42. doi:10.1177/1367493506060203 PMID:16464931

McGonigal, J. (2006). *This might be a game: Ubiquitous play and performance at the turn of the twenty-first century* [PhD Thesis]. Retrieved from http://www.avantgame.com

McGonigal, J. (2011). *Reality is broken: Why games make us better and how they can change the world.* New York: Penguin Press.

McGrath, J. E. (1976). Stress and behavior in organizations. In M. D. Dunnette (Ed.), *Handbook of industrial and organizational psychology* (pp. 1351–1395). Chicago, IL: Rand McNally.

Meroni, A., & Sangiorgi, D. (2011). *Design for services.* Farnham, MA: Gower Publishing, Ltd.

Merry, S. N., Stasiak, K., Shepherd, M., Frampton, C., Fleming, T., & Lucassen, M. F. (2012). The effectiveness of SPARX, a computerised self help intervention for adolescents seeking help for depression: Randomised controlled non-inferiority trial. *British Medical Journal*, *18*(344), e2598. doi:10.1136/bmj.e2598 PMID:22517917

Messaris, P. (1997). *Visual persuasion.* London: SAGE Publications.

Michell, L. (1997). Pressure groups: Young people's accounts of peer pressure to smoke. *Social Sciences in Health*, *3*, 3–16.

Millerand, F., Proulx, S., & Rueff, J. (2010). Introduction. In F. Millerand, S. Proulx, & J. Rueff (Eds.), *Web social: Mutation de la communication* (pp. 15–32). Québec, Canada: Presses de l'Université du Québec.

Miller, E. (1976). *Task and organization.* London: Wiley.

Miller, E. (1989). *The Leicester model: Experimential study of group and organizational processes.* London: Tavistock.

Miller, E. J., & Rice, A. K. (1967). *Systems of organization.* London: Tavistock Publications.

Mintzberg, H. (1983). *Power in and around organizations.* Englewood Cliffs, NJ: Prentice Hall.

Montola, M. (2003). Role-playing as interactive construction of subjective diegeses. In M. Gaden, L. Thorup, & M. Sander (Eds.), As LARP grows up: The book from knudepunkt (pp. 82-89). Frederiksberg: Projektgruppen KP03.

Montola, M. (2011). A ludological view on the pervasive mixed-reality game research paradigm. *Personal and Ubiquitous Computing*, *15*(1), 3–12. doi:10.1007/s00779-010-0307-7

Montola, M., Stenros, J., & Waern, A. (Eds.). (2009). *Pervasive games: Theory and design*. Morgan Kaufman.

Mulder, I., & Kort, J. (2008). Mixed emotions, mixed methods: the role of emergent technologies to study user experience in context. In S. Hesse-Biber, & P. Leavy (Eds.), *Handbook of Emergent Methods in Social Research* (pp. 601–612). New York: Guilford Publications.

Murphy, K. R., & Cleveland, J. N. (1995). *Understanding performance appraisal: Social, organizational, and goal-based perspectives*. Thousand Oaks, CA: Sage.

Murray, C. J., Richards, M. A., Newton, J. N., Fenton, K. A., Anderson, H. R., & Atkinson, C. et al. (2013). UK health performance: Findings of the global burden of disease study 2010. *Lancet*, *381*(9871), 997–1020. doi:10.1016/S0140-6736(13)60355-4 PMID:23668584

Murray, J. (1998). *Hamlet on the holodeck*. Cambridge, MA: The MIT Press.

Nakhla, M., Daneman, D., To, T., Paradis, G., & Guttmann, A. (2009). Transition to adult care for youths with diabetes mellitus: Findings from a Universal Health Care System. *Pediatrics*, *124*(6), 1134–1141. doi:10.1542/peds.2009-0041 PMID:19933731

Nash, A. A., Britto, M. T., Lovell, D. J., Passo, M. H., & Rosenthal, S. L. (1998). Substance use among adolescents with juvenile rheumatoid arthritis. *Arthritis Care and Research*, *11*(5), 391–396. doi:10.1002/art.1790110510 PMID:9830883

Oinas-Kukkonen, H. (2013). A foundation for the study of behavior change support systems. *Personal and Ubiquitous Computing*, *17*(6), 1223–1235. doi:10.1007/s00779-012-0591-5

O'Keefe, D. J. (1990). *Persuasion: Theory and research*. Newbury Park, CA: Sage.

Overmars, M. (n.d.). *Game Maker*. Retrieved January 26, 2014, from www.yoyogames.com/studio

Paavilainen, J., Korhonen, H., Saarenpää, H., & Holopainen, J. (2009). Player perception of context information utilization in pervasive mobile games. In *Breaking New Ground: Innovation in Games, Play, Practice and Theory: Proceedings of DiGRA 2009*. London, UK: DIGRA.

Packham, C. (2008). *Active citizenship and community learning*. Exeter, UK: Learning Matters Ltd.

Padol, L. (1996). *Playing stories, telling games: Collaborative storytelling in role-playing games*. Retrieved April 25, 2013, from http://www.recappub.com/games.html

Palazzi, C. E., Roccetti, M., & Marfia, G. (2010). Realizing the unexploited potential of games on serious challenges. *Computers in Entertainment*, *8*(4), 23. doi:10.1145/1921141.1921143

Palmer, B. (1979). Learning and the group experience. In W. G. Lawrence (Ed.), *Exploring individual and organizational boundaries* (pp. 169–192). London: John Wiley and Son.

Parish, M. (2007). Reflections on the administrator's role. *Organisational and Social Dynamics*, *7*(1), 61–72.

Parker, C. P., Dipboye, R. L., & Jackson, S. L. (1995). Perceptions of organizational politics: An investigation of antecedents and consequences. *Journal of Management*, *21*(5), 891–912. doi:10.1177/014920639502100505

Paske, G. H. (1986). The failure of indoctrination: A response to Wynne. *Educational Leadership*, *43*(4), 11–12.

Pendlebury, S. (1998). Feminism, epistemology and education. In W. Carr (Ed.), *The RoutledgeFalmer reader in philosophy of education* (pp. 50–62). Abingdon, UK: Routledge.

Percival, A. (1996). Invited reaction: An adult educator responds. *Human Resource Development Quarterly*, *7*(2), 131–139. doi:10.1002/hrdq.3920070204

Perloff, M. R. (2010). *The dynamics of persuasion: Communication and attitudes in the 21st century* (4th ed.). Taylor & Francis.

Pfeffer, J. (1981). *Power in organizations.* Marshfield, MA: Pitman.

Pfeffer, J. (1992). Understanding power in organizations. *California Management Review*, *34*(2), 29–50. doi:10.2307/41166692

Philpott, J. R. (2011). Transitional care in inflammatory bowel disease. *Gastroenterologia y Hepatologia*, *7*(1), 26–32. PMID:21346849

Piette, J. (1996). *Éducation aux médias et fonction critique*. Montreal, Canada: L'Harmattan.

Piette, J. (2006). La démarche d'enseignement en éducation aux médias. *Revue Vie pédagogique*, (140), 1-5.

Postmes, T. (2003). A Social identity approach to communication in organisations. In S. A. Haslam, D. V. Knippenberg, M. J. Platow, & N. Ellemers (Eds.), *Social identity at Work: Developing Theory for Organisational Practice* (pp. 81–97). Hove, UK: Taylor and Francis Group.

Primack, R. (1986). No substitute for critical thinking: A response to Wynne. *Educational Leadership*, *43*(4), 12–13.

Pritchard, I. (1988). Character education: Research prospects and problems. *American Journal of Education*, *96*(4), 469–495. doi:10.1086/443904

Puerto Rico. (2002). Bernau am Chiemsee: Alea.

Raessens, J. (2007). Playing history reflections on mobile and location based learning. In T. Hug (Ed.), *Didactics of microlearning: Concepts, discourses and examples*. Münster, Germany: Waxmann Verlag.

Raths, L. E., Harmin, M., & Simon, S. B. (1966). *Values and teaching: Working with values in the classroom*. Columbus, OH: Charles E. Merrill.

Raywid, M. A. (1980). The discovery and rejection of indoctrination. *Educational Theory*, *30*(1), 1–10. doi:10.1111/j.1741-5446.1980.tb00902.x

Reid, S. A., Giles, H., & Abrams, J. R. (2004). A social identity model of media usage and effects. *Zeitschrift für Medienpsychologie*, *16*(1), 17–25. doi:10.1026/1617-6383.16.1.17

Reis, S., & Correia, N. (2012). Playing with the weather. In *Entertainment Computing - ICEC 2012 (LNCS)* (Vol. 7522, pp. 172–184). Berlin: Springer. doi:10.1007/978-3-642-33542-6_15

Renteln, A. D. (1988). Relativism and the search for human rights. *American Anthropologist*, *90*(1), 56–72. doi:10.1525/aa.1988.90.1.02a00040

Reputation Institute. (2012). *2011 global rep trak: Results and report*. Reputation Institute.

Rice, A. K. (1966). *Learning for leadership: Interpersonal and intergroup relations*. London: Tavistock Publications.

Ricoeur, P. (1983). Temps et récit: Vol. 1. *L'intrigue et le récit historique*. Paris: Éditions du Seuil.

Ricoeur, P. (1986). *Du texte à l'action: Essais d'herméneutique II*. Paris: Éditions du Seuil.

Riediger, M., Schmiedek, F., Wagner, G. G., & Lindenberger, U. (2009). Seeking pleasure and seeking pain. *Psychological Science*, *20*(12), 1529–1535. doi:10.1111/j.1467-9280.2009.02473.x PMID:19891749

Robert, H. S. (1978). The twin rivers program on energy conservation in housing: Highlights and conclusions. *Energy and Building*, *1*(3), 207–242. doi:10.1016/0378-7788(78)90003-8

Robin, R. (1993). Pour une socio-poétique de l'imaginaire social. *Discours social/Social Discourse, 5* (1-2), 7-32.

Rogers, J., Hart, L., & Boltz, R. (1993). The role of pet dogs in casual conversations of elderly adults. *The Journal of Social Psychology*, *133*(3), 265–277. doi:10.1080/00224545.1993.9712145 PMID:8412041

Rogers, Y., Sharp, H., & Preece, J. (2011). *Interaction design: Beyond human-computer interaction*. Chichester, UK: John Wiley & Sons.

Rokeach, M. (1973). *The nature of human values*. New York: Free Press.

Royle, T. (2012). *Gamification: Unlocking hidden collaboration potential*. Retrieved from https://www-304.ibm.com/connections/blogs/socialbusiness

Ryan, K. (1993). Mining the values in the curriculum. *Education, 51*(3), 16–18.

Ryan, K. (2002). Six E's of character education. *Issues in Ethics, 13*(1).

Ryan, R. M., & Deci, E. L. (2000). Self-determination theory and the facilitation intrinsic motivation, social development, and well-being. *The American Psychologist*, *55*(1), 68–78. doi:10.1037/0003-066X.55.1.68 PMID:11392867

Saint George . (n.d.). Retrieved April 28, 2014, from http://sf0.org/tasks/Saint-George

Salen, K., & Zimmerman, E. (2004). *Rules of play: Game design fundamentals*. Cambridge, MA: MIT Press.

Sawicki, G. S., Lukens-Bull, K., Yin, X., Demars, N., Huang, I. C., & Livingood, W. et al. (2011). Measuring the transition readiness of youth with special healthcare needs: Validation of the TRAQ—Transition readiness assessment questionnaire. *Journal of Pediatric Psychology*, *36*(2), 160–171. doi:10.1093/jpepsy/jsp128 PMID:20040605

Sawyer, S. M., Afifi, R. A., Bearinger, L. H., Blakemore, S. J., Dick, B., Ezeh, A. C., & Patton, G. C. (2012). Adolescence: A foundation for future health. *Lancet*, *379*(9826), 1630–1640. doi:10.1016/S0140-6736(12)60072-5 PMID:22538178

Sawyer, S. M., Drew, S., Yeo, M. S., & Britto, M. T. (2007). Adolescents with a chronic condition: Challenges living, challenges treating. *Lancet*, *369*(9571), 1481–1489. doi:10.1016/S0140-6736(07)60370-5 PMID:17467519

Scal, P., Garwick, A. W., & Horvath, K. J. (2010). Making rheumtogrow: The rationale and framework for an internet-based health care transition intervention. *International Journal of Child and Adolescent Health*, *3*(4), 451–461.

Schell, J. (2008). *The art of game design: A book of lenses*. Boca Raton, FL: CRC Press.

Sclavi, M. (2003). *Arte di ascoltare e mondi possibili: Come si esce dalle cornici di cui siamo parte*. Milano: Mondadori Bruno.

Second Life Wiki. (2009). *Category: LSL Script/fr*. Retrieved July 30, 2013, from http://wiki.secondlife.com/wiki/

Seeman, T. E. (1996). Social ties and health: The benefits of social integration. *Annals of Epidemiology*, *6*(5), 442–451. doi:10.1016/S1047-2797(96)00095-6 PMID:8915476

Shaffer, D. W. (2004). Pedagogical praxis: The professions as models for post-industrial education. *Teachers College Record*, *10*(7), 1401–1421. doi:10.1111/j.1467-9620.2004.00383.x

Shaiken, H. (1984). *Automation and labor in the computer age*. New York: Holt, Rinehart & Winston.

Shapiro, E. R., & Carr, A. W. (1991). *Lost in familiar places: Creating new connections between the individual and society*. New Haven, CT: Yale University Press.

Shapiro, E. R., & Carr, A. W. (2006). These people were some kind of solution: Can society in any sense be understood? *Organisational and Social Dynamics*, *6*(2), 241–257.

Shapiro, E. R., & Carr, A. W. (2012). An introduction to Tavistock-style group relations conference learning. *Organisational & Social Dynamics*, *12*(1), 70–80.

Shaw, K. L., Southwood, T. R., & McDonagh, J. E. (2004). User perspectives of transitional care for adolescents with juvenile idiopathic arthritis. *Rheumatology*, *43*(6), 770–778. doi:10.1093/rheumatology/keh175 PMID:15039498

Shen, W., & Cannella, A. A. Jr. (2002). Power dynamics within top management and their impacts on CEO dismissal followed by inside succession. *Academy of Management Journal*, *45*(6), 1195–1206. doi:10.2307/3069434

Shepard, M. (2011). *Sentient city: Ubiquitous computing, architecture,and the future of urban space*. New York: Architectural League of New York.

Sheraka. (2009). Who is Tarna? *The Tahari Desert of Gor.* Retrieved July 25, 2013, from http://tahari.wordpress.com/2009/09/30/who-is-tarna

Siang Ang, C., Zaphiris, P., & Wilson, S. (2010). Computer games and sociocultural play: An activity theoretical perspective. *Games and Culture*, *5*(4), 354–380. doi:10.1177/1555412009360411

Simon, S. B., Howe, L., & Kirschenbaum, H. (1972). *Values clarification: A handbook of practical strategies for teachers and students*. New York: Hart Publishing.

Sixma, T. (2009). The Gorean community in second life: Rules of sexual inspired role-play. *Journal of Virtual Worlds Research*, *1*(3).

Solomon, S. D., & Qin, M. M. Z. Chen, M. M., & Avery, K. B. (2007). Contribution of working group I to the fourth assessment report of the intergovernmental panel on climate change. Cambridge, UK: Cambridge University Press.

Staiano, A., & Calvert, S. (2011). Exergames for physical education courses: Physical, social, and cognitive benefits. *Child Development Perspectives*, *5*(2), 93–98. doi:10.1111/j.1750-8606.2011.00162.x PMID:22563349

Stayin' Alive . (n.d.). Retrieved April 28, 2014, from http://sf0.org/tasks/Stayin-Alive

Stinson, J. N., Toomey, P. C., Stevens, B. J., Kagan, S., Duffy, C. M., & Huber, A. et al. (2008). Asking the experts: Exploring the self-management needs of adolescents with arthritis. *Arthritis and Rheumatism*, *59*(1), 65–72. doi:10.1002/art.23244 PMID:18163408

Stivale, C. J. (1997). Spam: Heteroglossia and harassment in cyberspace. In D. Porter (Ed.), *Internet culture* (pp. 133–144). New York: Routledge.

Straffin, P. D. (1996). *Game theory and strategy*. Washington, DC: The Mathematical Association of America.

Stralen, M., deVries, H., Mudde, A., Bolman, C., & Lechner, L. (2009). Determinants of initiation and maintenance of physical activity among older adults: A literature review. *Health Psychology Review*, *3*(2), 147–207. doi:10.1080/17437190903229462

Stubbé, H. E., & Theunissen, N. C. M. (2008, June). Self-directed adult learning in a ubiquitous learning environment: A meta-review. In *Proceedings - Special Track on Technology Support for Self-Organised Learners during 4th EduMedia Conference 2008 Self-Organised Learning in the Interactive Web - A Change in Learning Culture?* Salzburg, Austria: EduMedia.

Suleiman, S. R. (1983). *Authoritarian fictions: The ideological novel as a literary genre*. New York: Columbia University Press.

Sullivan, H. S. (1953). *The interpersonal theory of psychiatry*. New York: Norton.

Suris, J. C., Akré, C., Berchtold, A., Bélanger, R. E., & Michaud, P. A. (2010). Chronically connected? Internet use among adolescents with chronic conditions. *The Journal of Adolescent Health*, *46*(2), 200–202. doi:10.1016/j.jadohealth.2009.07.008 PMID:20113927

Suris, J. C., Michaud, P. A., Akre, C., & Sawyer, S. M. (2008). Health risk behaviors in adolescents with chronic conditions. *Paediatrics*, *122*(5), e1113–e1118. doi:10.1542/peds.2008-1479 PMID:18977960

Sutton, R., & Kahn, R. L. (1986). Prediction, understanding, and control as antidotes to organizational stress. In J. Lorsch (Ed.), *Handbook of organizational behavior*. Englewood Cliffs, NJ: Prentice-Hall.

Svahn, M. (2014). *Social expansion in marketing, focusing on pervasive-persuasive games for impacting energy consumption. (Unpublished PhD)*. Stockholm, Sweden: Stockholm School of Economics.

Svahn, M., & Lange, F. (2009). Marketing the category of pervasive games. In M. Montola, J. Stenros, & A. Waern (Eds.), *Pervasive games, theory and design*. Morgan Kaufman. doi:10.1016/B978-0-12-374853-9.00011-8

Swedish Energy Agency. (2009). *Save energy and make your housekeeping money last longer*. Retrieved from https://www.energimyndigheten.se/Global/Offentlig%20sektor/Energiarbete%20i%20kommun,%20l%C3%A4n%20och%20region/Kommunala%20energi-%20och%20klimatr%C3%A5dgivare/Broschyr%20andra%20spr%C3%A5k/Energispartips_en_lu.pdf

Swedish Energy Agency. (2012). *Energiläget 2012*. Energimyndigheten.

Szulborski, D. (2005). *This is not a game: A guide to alternate reality gaming. Macungie*. New Fiction Publishing.

Tackling the burden of chronic diseases in the USA. (2009). *Lancet, 373*(9659), 185. doi:10.1016/S0140-6736(09)60048-9

Tajfel, H. (1974). Social identity and intergroup behaviour. *Social Sciences Information. Information Sur les Sciences Sociales, 14*, 101–118. doi:10.1177/053901847501400204

Tajfel, H. (1978). *Differentiation between social groups*. London: Academic Press.

Tajfel, H. (1979). Individuals and groups in social psychology. *The British Journal of Social and Clinical Psychology, 18*(2), 183–190. doi:10.1111/j.2044-8260.1979.tb00324.x

Tajfel, H., & Turner, J. (1979). An integrative theory of inter-group conflict. In J. A. Williams, & S. Worchel (Eds.), *The social psychology of inter-group relations* (pp. 33–47). Belmont, CA: Wadsworth.

Tan, C. (1990). *Indoctrination, imagination and moral education*. Paper presented at the 2nd International Conference on Imagination and Education. Vancouver, Canada.

Terms. (n.d.). Retrieved April 28, 2014, from http://sf0.org/terms

Tilbury, D. (1995). Environmental education for sustainability: Defining the new focus of environmental education in the 1990s. *Environmental Education Research, 1*(2), 195–212. doi:10.1080/1350462950010206

Tomaka, J., Thompson, S., & Palacios, R. (2006). The relation of social isolation, loneliness, and social support to disease outcomes among the elderly. *Journal of Aging and Health, 18*(3), 359–384. doi:10.1177/0898264305280993 PMID:16648391

Torstensson, C. (2005). *En förstudie för att utreda förutsättningarna för forskningsprojektet young energy*. Eskilstuna, Sweden: Interactive Institute.

Tran, B. (2014). Rhetoric of play: Utilizing the gamer factor in selecting and training employees. In T. M. Connolly, T. H. Hainey, E. Boyle, G. Baxter, & P. Moreno-Ger (Eds.), *Psychology, pedagogy, and assessment in serious games* (pp. 175–203). Hershey, PA: Premier Reference Source/IGI Global.

Tran, B. (2014). Game theory vs. Business ethics: The game of ethics. In B. Christiansen, & M. Basilgan (Eds.), *Economic behavior, game theory, and technology in emerging markets* (pp. 213–236). Hershey, PA: Premier Reference Source/IGI Global.

Trist, E., & Sofer, C. (1959). *Exploration in group relations*. Leicester, UK: Leicester University Press.

Turquet, P. M. (1974). Leadership: The individual and the group. In G. S. Gibbard, J. J. Hartman, & R. D. Mann (Eds.), *Analysis of groups*. San Francisco, CA: Jossey-Bass.

Ullah, S., Jafri, A. R., & Dost, M. K. B. (2011). A synthesis of literature on organizational politics. *Far East Journal of Psychology and Business*, *3*(3), 36–49.

Unity Technologies. (n.d.). *Unity*. Retrieved January 26, 2014, from http://unity3d.com/

van Houwelingen, J. H., & van Raaij, F. W. (1989). The effect of goal-setting and daily electronic feedback on in-home energy use. *The Journal of Consumer Research*, *16*(1), 98–105. doi:10.1086/209197

VanderWijst, H. (2011). Al wandelend praat je anders. *Lopende Zaken, 57*.

Varoufakis, Y. (2001). General introduction: Game theory's quest for a single, unifying framework for the social sceinces. In Y. Varoufakis (Ed.), *Game theory: Critical concepts in the social sciences* (Vol. 1). London: Routledge.

Verstegen, D.M.L., & Hulst van der, A.H. (2000, November). *Standardized development of a needs statement for advanced training means*. Paper presented at the Interservice/Industry Training, Simulation and Education Conference (I/ITSEC). Orlando, FL.

Victor, C., Scambler, S., Bond, J., & Bowling, A. (2000). Being alone in later life: loneliness, social isolation and living alone. *Reviews in Clinical Gerontology*, *10*(4), 407–417. doi:10.1017/S0959259800104101

Vigoda, E. (2000). Organizational politics, job attitudes, and work outcomes: Exploration and implications for the public sector. *Journal of Vocational Behavior*, *57*(3), 326–347. doi:10.1006/jvbe.1999.1742

Vigoda-Gadot, E. (2001). Stress-related aftermaths to workplace politics: The relationships among politics, job distress, and aggressive behavior in organizations. *Journal of Organizational Behavior*, *23*(5), 571–591. doi:10.1002/job.160

Vigoda-Gadot, E., & Drory, A. (2006). *Handbook of organizational politics*. Cheltenham, UK: Edward Elgar. doi:10.4337/9781847201874

Viner, R. M., Ozer, E. M., Denny, S., Marmot, M., Resnick, M., Fatusi, A., & Currie, C. (2012). Adolescence and the social determinants of health. *Lancet*, *379*(9826), 1641–1652. doi:10.1016/S0140-6736(12)60149-4 PMID:22538179

Visch, V., Vegt, N., & Anderiesen, H. & vanderKooij, K. (2013). Persuasive Game Design: A model and its definitions. In *Proceedings of CHI*. Paris: ACM.

Volkswagen Group. (2011). *Green machine*. Author.

Wacker, A., Kaemmerer, H., Hollweck, R., Hauser, M., Deutsch, M. A., & Brodherr-Heberlein, S. et al. (2005). Outcome of operated and unoperated adults with congenital cardiac disease lost to follow-up for more than five years. *The American Journal of Cardiology*, *95*(6), 776–779. doi:10.1016/j.amjcard.2004.11.036 PMID:15757611

Waern, A., & Denward, M. (2009). On the edge of reality: Reality fiction in sanningen om marika. In *Breaking New Ground: Innovation in Games, Play, Practice and Theory: Proceedings of DiGRA 2009*. London, UK. DIGRA.

Walker, J. (1999). Self-determination as an educational aim. In W. Carr (Ed.), *The RoutledgeFalmer reader in philosophy of education* (pp. 74–82). Abingdon, UK: Routledge.

Walther, B. K. (2005). Reflections on the methodology of pervasive gaming. In *Proceedings of the 2005 ACM SIGCHI International Conference on Advances in Computer Entertainment Technology,* (pp. 176-179). ACM. doi:10.1145/1178477.1178501

Walz, S. P., & Ballagas, R. (2007). Pervasive persuasive: A rhetorical design approach to a location-based spell-casting game for tourists. In *DiGRA 2007 - Situated Play: Proceedings of the 3rd International Conference of the Digital Games Research Association.* Tokyo: DIGRA.

Wayne, S. J., Liden, R. C., Graf, I. K., & Ferris, G. R. (1997). The role of upward influence tactics in human resource decision. *Personnel Psychology*, *50*(4), 979–1006. doi:10.1111/j.1744-6570.1997.tb01491.x

Welsh, M. A., & Slusher, E. A. (1986). Organizational design as a context of political activity. *Administrative Science Quarterly*, *31*(3), 389–402. doi:10.2307/2392829

Wendel-Vos, W., Droomers, M., Kremers, S., Brug, J., & vanLenthe, F. (2007). Potential environmental determinants of physical activity in adults: A systematic review. *Obesity Reviews*, *8*(5), 425–440. doi:10.1111/j.1467-789X.2007.00370.x PMID:17716300

Winn, B. M. (2009). The design, play, and experience framework. In R. E. Ferdig (Ed.), *Handbook of research on effective electronic gaming in education* (pp. 1010–1024). Hershey, PA: IGI Global.

Winnicott, D. W. (1960). *The maturational processes and the facilitating environment.* New York: International Universities Press.

World Energy Outlook 2011. (2011). Paris: International Energy Agency.

World Health Organisation. (2001). *The second decade: improving adolescent health and development.* Geneva: World Health Organization.

World Health Organisation. (2013). *10 facts on the state of global health.* Retrieved from http://www.who.int/features/factfiles/global_burden/facts/en/index.html

Yeung, E., Kay, J., Roosevelt, G. E., Brandon, M., & Yetman, A. T. (2008). Lapse of care as a predictor for morbidity in adults with congenital heart disease. *International Journal of Cardiology*, *125*(1), 62–65. doi:10.1016/j.ijcard.2007.02.023 PMID:17442438

Zimmerman, E. (2012). Jerked around by the magic circle: Clearing the air ten years later. *Gamasutra Blog*. Retrieved from http://www.gamasutra.com/view/feature/6696/

Zinner, J., & Shapiro, E. R. (1972). Projective identification as a mode of perception and behavior in families of adolescents. *The International Journal of Psycho-Analysis*, *53*, 523–530.

About the Contributors

Dana Ruggiero is a senior lecturer in Learning Technology and award leader for the MA in Learning Technology at Bath Spa University, UK. She received her PhD in Learning Design and Technology at Purdue University and her MA in Education at Augsburg College. Her research focuses on persuasive game design, social action games, and learning design. She is currently developing two games using Bayesian models and open data from UNESCO, WHO, and UNICEF. Dr. Ruggiero is currently collaborating on several EU projects involving the creation of games based learning in higher education.

* * *

Maresa Bertolo is assistant professor at Design Department of Politecnico di Milano, teacher of Computer Graphics Lab and Game Design Course at the School of Design. Her research deals with Communication Design, focusing on Games Studies and Game Design, with a particular interest on ludic activity as vehicle for communication, learning, and best practices.

Wessel Bos, by the design of, and studies for, the *Travelling Rose*, graduated at the faculty of Industrial Design Engineering at the Technical University Delft in 2013. Currently, Wessel is acquiring new visions and skills in several jobs and enterprises. He is working in multiple disciplines – including sailing, custom-made design, event planning, fashion, media, and consumer products. His ambition is to set up a socially responsible trading enterprise with a focus on durable goods and durable business relations.

Peter Christiansen is a PhD Student at the University of Utah where he teaches courses in Videogame Studies and New Media. He is an avid vidogame player, prolific doodler, and occasional blogger whose research interests lie in videogame rhetoric, game design, and the Independent Games Movement. He has been making games professionally since 2005 and is currently the lead game developer for ASPIRE, the outreach program for the Utah High Energy Astrophysics Institute, where he creates educational Flash games to teach children about math and physics. In addition to his professional work, he has also worked on a number of independent projects, including a number of solo and collaborative entries for both local and global game jams.

Neil Dansey is a Senior Lecturer in Computer Games Technology at the University of Portsmouth, UK, where he recently received his PhD following a study on the emergent benefits of playing cross-media games. Since 2009, he has organised and coordinated a yearly international Game Jam event, bringing together over 300 students and staff from universities in the UK, France, USA, and Denmark. His recent research interests have included the use of ambiguity to promote creative interpretation in games, in particular to facilitate the "spookiness" that people feel when perceiving order in chaos. Going forward, he intends to investigate the development of simulated gaming companions for tabletop Role-Playing Games.

Teresa de la Hera is postdoctoral researcher and lecturer at Utrecht University, where she is member of the Center for the Study of Digital Games and Play. She started her academic career at the University of Santiago de Compostela in Spain in 2006, where she conducted research in the fields of new media and persuasive games. She was the first scholar at the USC to conduct research in the field of Game Studies and one of the first Spanish academics to study the field of Persuasive Games. In 2010, she was a visiting PhD Candidate at the University of Amsterdam, where she spent four months under the supervision of Professor Jan Simons. Later on, in 2011, she moved to The Netherlands, where she obtained an International PhD Fellowship to finish her PhD at Utrecht University. She has also taught courses and offered guest lectures on new media, persuasive games, and serious games.

Christophe Duret has a Master of Arts in Communication at Sherbrooke University (Canada), where he is currently a graduate student in French Studies. His research focuses on online role-playing games in both sociocritical and hermeneutical ways. In particular, he has interest for the processes of videogame adaptation from novels and in the strategies and tactics used by players in order to understand how they challenge or defend a specific doctrine in their role-play.

Micah Hrehovcsikis a Senior Lecturer/Game Design Researcher at the HKU, University of Arts Utrecht. Previously, he worked in the wireless entertainment industry, where he worked on trademark titles (i.e. Teenage Mutant Ninja Turtles). Currently, he functions as the Creative Director of the HKU's Games & Interaction studiolab, where he has designed serious games for the domains of health, safety, and education. His current game is being designed and developed for use in psychiatric healthcare environments.

Linda Kaye is a Senior Lecturer at Edge Hill University, UK. Her research focuses on the role of social contexts on digital gaming experiences and outcomes. In particular, she has an interest in the way in which digital games can promote positive psychosocial outcomes. Within her current research projects, she aims to study the way in which social identity operates for different gaming groups and the impact on self-esteem and psychological well-being. Additionally, she has interests in users' self-presentation in different online contexts and the implications on personality judgements.

Ilaria Mariani, PhD Student at Politecnico di Milano, Department of Design, teacher fellow in Computer Graphics Lab, Game Design Course and Augmented Reality & Mobile Exprience Course. Her research deals with the relation between Game Studies and socio-cultural aspects. The focus is on analyzing the game as a contemporary communication system for social innovation and best practices' transmission, investigating the meaningful play experiences the game can stimulate and elicit.

Victoria McArthur is a PhD candidate in the Communication & Culture programme at York University, Canada. Her research interests include Human-Computer Interaction (HCI), Actor-Networks, Feminist Technoscience, and Identity Studies. Her current research investigates the role of interface affordances on self-representational practices in social virtual worlds and Massively-Multiplayer Online Games (MMOGs).

Ingrid Mulder, PhD, is an Associate Professor of Design Techniques, Faculty of Industrial Design Engineering of Delft University of Technology, and a Research Professor in Social Innovation at Creating 010, a transdisciplinary design-inclusive research center at Rotterdam University of Applied Sciences. Before receiving her Master's degree from the Faculty of Social Sciences, University of Tilburg, she joined KPMG Strategic Vision for studying group dynamics in strategic decision-making. This experience has evoked her research interest in linking people through design. She holds a PhD in human-to-human interaction design from the University of Twente, Faculty of Behavioural Sciences. While bridging art, design, and interactive media, her current research emphasizes transformative and social design; using making and co-creation to empower people; developing methods stressing human values and social impact; and crossing boundaries of urban design and open government.

Rick Prins, PhD, worked on this project as a postdoctoral researcher at the Department of Public Health of the Erasmus Medical Center, Rotterdam, The Netherlands. His research interests include social and physical environmental influences on physical activity behaviour, the interplay between individual and environmental factors in shaping physical activity behaviour and understanding the effects of environmental interventions on physical activity behaviour. Currently, Rick is appointed as a Career Development Fellow at the UKCRC Centre for Diet and Activity Research (CEDAR) and the MRC Epidemiology Unit of the University of Cambridge.

Paschalina Skamnioti is a PhD Candidate at the Hochschule of Bremen in Germany, studying Ethics and Values Education in the design of Games for Change. Previously, she studied Cultural Informatics (Msc) and Applied Informatics (Bsc), and she also worked for 8 years as a high-school teacher of Digital Media in Greece. Believing in the educational power of games, what she calls "serious fun," she has designed games herself for children and teens. Being a member of the Gangs of Bremen, and the Mobile Game Lab, she is creating local game events, while elaborating new playful projects and research ideas for social, political, and environmental awareness (ludalina.wordpress.com).

Hester Stubbe joined TNO as a researcher and technical consultant in the department of Training & Performance Innovations. Her research focuses on self-directed learning, work-based learning, experiential learning, and the development of innovative learning solutions needed to achieve that. Hester works with the creative industry for optimal learning solutions. Currently, she is working on projects with primary schools, secondary schools, out-of-school children in Sudan, small- and medium-sized enterprises, and in the safety and security domain. Hester studied Educational Sciences at the University of Utrecht. Hester has a wide experience in the field of training and education.

Mattias Svahn is a PhD from the Stockholm School of Economics Institute of Research. He has, in collaboration with The Interactive Institute Swedish ICT, researched the media psychology of pervasive games, in particular consumers´/players´ experiences and reactions, when exposed to persuasive pervasive games for learning and advertising, and consults to large Scandinavian media houses. He has published and presented at many international conferences on the consumer psychology of ambient media and pervasive advertising. His work bridges consumer behaviour sciences, game design sciences, and play theory. He was also senior coordinator and research leader for business development of pervasive games at the EU IST Framework Programme 6 project IPerG that developed prototype pervasive games and ambient media (see www.svahn.se).

Ben Tran received his Doctor of Psychology (PsyD) in Organizational Consulting/Organizational Psychology from California School of Professional Psychology at Alliant International University in San Francisco, California, United States of America. Dr. Tran's research interests include domestic and expatriate recruitment, selection, retention, evaluation and training, CSR, business and organizational ethics, organizational/international organizational behavior, knowledge management, and minorities in multinational corporations. Dr. Tran has presented articles on topics of business and management ethics and expatriate and gender and minorities in multinational corporations at the Academy of Management, Society for the Advancement of Management, and International Standing Conference on Organizational Symbolism. Dr. Tran has also published articles and book chapters with the *Social Responsibility Journal, Journal of International Trade Law and Policy, Journal of Economics, Finance and Administrative Science*, Financial Management Institute of Canada, and IGI Global.

Josine van de Ven has worked at TNO since 2004. Dr van de Ven is a program manager at the department of Training & Performance Innovations. She earned her PhD in Cognitive Science at the University of Nijmegen, The Netherlands. Her expertise is professionals in complex information environments. This includes topics like information management, complex decision making, and training support within the crisis management domain. Serious or applied games are an important enabler for new training programs on these topics. Experience is an important factor in dealing with incidents, but experience is hard to get in this domain. Applied games can support first responders and other professionals in this domain, like mayors, to learn effective and efficient crisis management. Together with a team of specialists, Dr van de Ven explores new possibilities for applied games in crisis management and cyber security.

Valentijn Visch works as assistant professor at the faculty of Industrial Design at the Technical University Delft. He conducts and coordinates persuasive game design research, and is project leader of the Economic Affairs granted CRISP G-Motiv project (2011-2015) and the NWO granted NextLevel project (2013-2017). Both research projects contain research – as well as industry- and user-partners. Valentijn has a background in Literature (MA), Art theory (MA – Postgraduate Jan van Eijckacademy), Animation (Postgraduate NIAF Tilburg), Cultural Sciences and film studies (PhD – VU), and experimental emotion research (Geneva).

Annika Waern is professor in HCI at the University of Uppsala dept. of Informatics and Media. Annika has done research in the design and use of mobile technology with a strong focus on physical and pervasive games. Annika was formerly founder and research leader at the Mobile Life Institute in Stockholm. See more at http://annikawaern.wordpress.com/.

Andrew Wilson is a senior lecturer and programme leader for the BSc (Hons) Computer Games Technology at Birmingham City University. His interests revolve around the use of technology to support issues relating to long-term medical conditions, for example musculoskeletal disease. He is currently researching how computer games may be used to support healthcare issues in this field. He is also interested in how human and psychological factors affect the use and adoption of technology in healthcare settings.

Index

A

B

C

D

E

F

G

H

I

J

L

M

O

P

R

S

T

U

V